OPTICAL
SWITCHING

OPTICAL
SWITCHING

TAREK S. EL-BAWAB
Editor and Principal Author

 Springer

Tarek S. El-Bawab
telbawab@ieee.org

Optical Switching

Library of Congress Control Number: 2005935342

ISBN 0-387-26141-9   e-ISBN 0-387-29159-8
ISBN 978-0387-26141-6

Printed on acid-free paper.

Printed in the United States of America.

9 8 7 6 5 4 3 2 1   SPIN 11050476

springeronline.com

To my wife and my daughter

Tarek

Contents

PREFACE

The idea for this book was born in early 2001. I was preparing for an IEEE workshop on the subject of optical switching. It was the second in a series of annual optical networking workshops I have developed for the two flagship conferences of the IEEE Communications Society (ComSoc), namely the International Conference on Communications (ICC) and the Global Communications Conference (GLOBECOM). The call for participation was circulating when Alex Greene, then Senior Publisher with Kluwer Academic Publishers, saw it and called me to ask if I would be interested in editing or authoring a book on optical switching. I met Alex afterwards in March 2001 at the Optical Fiber Communications Conference (OFC) in Anaheim, California, where we discussed this idea further. The task was big and challenging. Nevertheless, the motivation to pursue this project, for a scientist interested in optical switching, was also strong.

In that time frame, the telecommunications industry in general, and the optical networking segment of it in particular, was still in a historical boom, but was about to experience unprecedented gloom. It is unbelievable how different the world is today compared with what it was before the spring of 2001 when I accepted Kluwer's invitation to undertake the task of developing this book. We telecommunications professionals were unaware of the difficult time ahead. Personally, I also did not imagine that what was supposed to be a 12-18 month assignment would take more than four years of work to complete.

It had been ten years since a book was written fully from scratch about optical/photonic switching. A lot of progress occurred during those years, both in switching-device technologies and in optical switching systems and networks. The very role envisioned for optical switching in communications networks has undergone changes. Optical switching made impressive inroads towards maturity, and had evolved from research to product development. The optical networking community came to realize that certain technologies and methods have higher practical potential than others in the short and medium terms. Some approaches once considered for imminent deployment were repositioned as long-term research topics. Some technologies turned out to be more suitable than others for certain applications. As it stands in the beginning of the 21st century, compared with the early 1990s, the subject of optical switching requires different treatment.

Thus, a lot of work was necessary to put together a new book and the job was not easy. The sharp downturn of the telecommunications industry by mid 2001, coupled with the difficulty of finding authors in certain areas and a lack of sufficient support in many situations, made the job more difficult. Obstacles became higher and delays cumulated. There were moments when it appeared to some that this book might not see the light. The fact that it is now published attests to the determination of a few who believed in the feasibility and importance of this project and granted it their time, support, and patience, especially in hard times.

The goal of this book is to provide up-to-date comprehensive coverage of the subject of optical switching. In the first part, a thorough introduction paves the way for the reader to grasp the history and foundations of this subject and better understand its terminologies. The second part of the book discusses the principles and theories of numerous optical switching technologies, along with the switching devices based upon them. Device fabrication, features, and applications are described. The third part provides a study of the principles and theories of optical switching fabrics, and how they are built and assembled into working systems. Applications of optical switching in communications networks are also discussed in this part. Optical circuit, packet, and burst switching are all covered.

Today, the field of optical switching embraces a spectrum of technologies and techniques and involves several disciplines of science and engineering. Like many contemporary fields of knowledge, it is not easy for one to master several areas in this field, much less all. Therefore, the reader is better served by an edited book, with multiple authors, as opposed to a monograph. Each of the two approaches has its merits. For instance, monographs can possess distinct harmony in their structure and have a better chance of facilitating smooth information flow. This is because they are authored by a single, or few, individuals. Owing to the multiplicity of authors, edited books can feature the richness and diversity of more contributors and may offer the reader broader coverage and more in-depth treatment of a given topic.

This book has been put together to have as many as possible of the qualities of both monographs and edited books. It has been developed with special attention to readability, logical information flow, structural organization, and pedagogy of presentation. Thorough readers will appreciate a blend of favorable features in this respect. As an edited book, it is composed of distinct collection of self-contained chapters with minimum overlaps and independent references. These chapters bring together academic and industrial contributions, analytical and descriptive treatments, and cover theories, experiments, and practical deployment. The diversity of this mix is also enriched by the fact that authors come from all over the world. On the other hand, chapters are tied together carefully to form a homogeneous fabric of knowledge. Structuring the book in parts and chapters is done in such a way as to facilitate smooth transitions from one topic to another and easy flow of information. Attention is also given to investigation and redefinition of old and new terminologies, and use of terms and notations that are as consistent as possible throughout the entire book.

Although the reader will encounter a great deal of tutorial style, some prior knowledge of electrical engineering, telecommunications, physics, and mathematics is expected. In most chapters, a senior level electrical engineering background is prerequisite. The book is intended for broad readership and, as such, can serve multiple purposes. It can be treated as a textbook for senior-undergraduate and graduate students in the fields of electrical, optical, and computer engineering. For this purpose, it is a main reference for courses in optical switching, from device technology to network applications, or an essential supplement to broader courses in communication systems and/or optical technologies, devices, and networks. Numerous combinations of chapters, in full or in parts, can constitute several courses where the primary focus may be on device technologies, systems concepts, or network architectures. The book is also a valuable reference for researchers and scientists in the fields of optical switching and networking. It is useful as well for engineers and professionals in the arenas of telecommunications, data, and video networks, and can be used there for self study.

I would like to thank several colleagues for their advice, comments, and constructive criticism of various parts and chapters of this book. In this category, I want to mention, in alphabetical order, Govind Agrawal (University of Rochester), John Bowers (University of California Santa Barbara), Sally Day (University College London), Mehran Esfandiari (SBC, Inc.), Paul Green (Tellabs, Inc., retired), Mark Karol (Avaya, Inc.), Kne-ichi Kitayama (Osaka University), Tim Murphy (Lucent Technologies), Satoru Okamoto (NTT Laboratories), Philippe Perrier (Xtera), Larry Shacklette (Harris, Inc.), Vishal Sharma (Metanoia, Inc), Olav Solgaard (Stanford University), and Munefumi Tsurusawa (KDDI Laboratories).

I would like to express my appreciation to John Midwinter (University College London), and Scott Hinton (Utah State University) for the useful discussions we had and the opportunity to benefit from their knowledge and experience in the field of optical switching.

I am deeply grateful to all chapter authors for their dedication, commitment, and quality of work. Special thanks are due to many of them who not only contributed their chapters, but also were generous enough to give me valuable advice and suggestions concerning other material. I am indebted in this regard to David Hunter, who gave me a lot of support and help. I appreciate the useful remarks of Roland Ryf, Hideaki Okayama, and Alan McGuire about a number of issues. I also want to thank Antonio d'Alessandro and Mike Harris for taking the time to draw my attention to potential improvements of some parts of the manuscript.

I thank Wanda Fox, who helped me obtain many research papers and resources when she was in charge of the old Alcatel USA library. I also thank Niel Ransom, Chief Technology Officer (CTO) of Alcatel, for personal support and approval to develop this book in 2001. My friend Ed Esposito (University of Texas at Dallas) gave me useful tips concerning this preface. I acknowledge with gratitude the influence of several discussions I had with colleagues in numerous scientific and professional forums. I am thankful to the editorial and production staff of Springer (who acquired Kluwer in 2004). In particular, Alex Greene has been a strong supporter of this project since the beginning. Melissa Guasch, Alex's assistant, has been helpful throughout the process of developing this book.

Last, but definitely not least, words may not express my appreciation to my wife Shahira and my little daughter Nadine. There is no way this book could have been completed without their unconditional love, kind understanding and continuous sacrifices. The huge proportion of my time that I put into working in this book was mostly taken off our family time and was at the expense of precious hours I could have spent with my sweet little angel and my giving wife.

Finally, this book was developed with utmost care and consideration of what is important for the readership. All the material has gone through revisions and thorough editing. Nevertheless, perfection is beyond the reach of mankind! I encourage readers to point out residual errors, and they are welcome to provide constructive criticism, if any. I am hopeful that the hard work put into this book by my colleagues, and by myself, will be rewarded by its scientific and intellectual benefit to all readers.

Tarek S. El-Bawab
telbawab@ieee.org
tarek.97@engalumni.colostate.edu
Richardson, Texas
July 2005

CHAPTER CONTRIBUTORS/AUTHORS

CHAPTER 2

Hideaki Okayama
R&D Group,
Oki Electric Industry Co. Ltd.,
550-5 Higashiasakawa, Hachioji-shi, Tokyo 193-8550,
JAPAN

CHAPTER 3

Antonio d'Alessandro
Department of Electronic Engineering,
University of Rome "La Sapienza",
Via Eudossiana, 18 - 00184 Roma,
ITALY

CHAPTER 4

R. Hauffe
K. Petermann
Technical University of Berlin,
Fachgebiet Hochfrequenztechnik,
Sekt. HFT 4-1, Einsteinufer 25,
10587 Berlin,
GERMANY

CHAPTER 5

J. Michael Harris
Corning Incorporated,
SP-AR02-2,
Corning, NY 14831,
USA

Robert Lindquist
Electrical and Computer Engineering Department,
The University of Alabama in Huntsville,
Huntsville, AL 35899,
USA

JuneKoo Rhee
School of Engineering,
Information and Communications University (ICU),
YusongGu, MunJiRo 119, DaeJeon,
SOUTH KOREA

James E. Webb
Corning Incorporated,
SP-FR-04,
Corning, NY 14831,
USA

CHAPTER 6

Roland Ryf
David T. Neilson
Vladimir A. Aksyuk
Bell Laboratories, Lucent Technologies,
Crawford Hill Laboratory,
791 Holmdel-Keyport Road,
Holmdel, NJ 07733,
USA

CHAPTER 7

Paul R. Prucnal
Ivan Glesk
Department of Electrical Engineering,
Princeton University,
Engineering Quadrangle, Olden Street,
Princeton, NJ 08544,
USA

Paul Toliver
Telcordia Technologies,
Red Bank, NJ 07701,
USA

Lei Xu[1]
NEC Labs America,
4 Independence Way,
Princeton, NJ 08540,
USA

CHAPTER 9

David K. Hunter
Department of Electronic Systems Engineering,
University of Essex,
Wivenhoe Park, Colchester CO4 3SQ,
UK

CHAPTER 10

Wojciech Kabacinski
Grzegorz Danilewicz
Mariusz Glabowski
Poznan University of Technology,
Institute of Electronics and Telecommunications,
ul. Piotrowo 3a,
60-965 Poznan,
POLAND

CHAPTER 12

Alan McGuire
British Telecom,
OP6 Polaris House, Adastral Park, Martlesham Heath,
Ipswich, Suffolk IP5 3RE,
UK

[1] Was with Department of Electrical Engineering, Princeton University, USA

CHAPTER 13

Alan E. Willner
Department of Electrical Engineering,
University of Southern California,
Room EEB 538,
Los Angeles, CA 90089-2565,
USA

Reza Khosravani
Engineering Science Department,
Sonoma State University,
1801 East Cotati Avenue,
Rohnert park, CA 94928,
USA

Saurabh Kumar
Department of Electrical Engineering,
University of Southern California,
Los Angeles, CA 90089-2565,
USA

PART I
INTRODUCTION

Chapter 1

PRELIMINARIES AND TERMINOLOGIES

Tarek S. El-Bawab

The tremendous growth of the Internet and the introduction of new communication services and applications have changed the landscape of the telecommunications industry. Today, telecommunications play a vital role in our world and this role will continue for as long as we need to communicate and deal with each other. Innovations in the field of communication technologies and networking continue to unfold and will keep changing the way we live our lives, perform our work, and interact with each other.

The principal building blocks of communications networks are communication terminals, transmission links, and switching centers. This entire book is about switching. Indeed, it is about a group of switching technologies and systems that have the potential to change the future of networking. Switching is a cornerstone of communication systems and networks and has evolved, since the inception of telegraphy and telephony in the nineteenth century until today, from one form to another while encompassing numerous enabling technologies and embracing several communication methods.

This chapter opens up our discussion of optical switching by providing the essential background and an introduction to this subject. We start by identifying the role of switching in communication systems and networks while reviewing its journey from primitiveness to maturity. We look back at the evolution of switching, both in voice networks and in data networks, and the evolution of the public telecommunications network architecture. Optical switching is positioned within this big picture. The chapter discusses the past and present status of optical switching and the drivers to introduce it in the network. The potential role of optical switching in communications networks is discussed. Optical switching methods, domains and systems are briefly examined. Numerous optical switching technologies are then presented and classified. Key terminology issues are addressed and clarified. We outline our strategy in tackling and presenting the subject of the book, describe its main parts and chapters, and provide a short overview of their contents.

1.1   EVOLUTION OF SWITCHING IN TELEPHONE NETWORKS

1.1.1   The Original Telephone Network

The evolution of switching throughout its early history was associated with developments in telephony. In the beginning, the telephone switching problem was a simple one by the standards of today. Users in a confined geographical region wanted to talk to each other using telephone sets (communication terminals) and copper wires (transmission links). Each user needed some sets and a number of copper-pair terminations. Ideally, for a total of N users, $N-1$ telephone sets and the same number of copper-wire pairs were required in every user's premises to be able to communicate with all the others. Then, the multiple user telephone sets were replaced by a single one that is augmented with some *switching* capability, namely a mechanism to connect the telephone manually to one of the copper pairs available. Yet, $N-1$ pairs of wires had to terminate at every user location and a total of $N(N-1)/2$ links were needed. As the number of users increased, along with the distances separating them, this networking approach became complex, costly, and unpractical. The need to build *switching centers*, *telephone exchanges*, *central offices*, or *switching nodes*[1] was then realized.

Switching centers were usually centered within the geographical areas they served and local telephone networks were built around them (still the general case today). Each *user* (or *subscriber*) needed only one copper pair to connect to the network via the switching center. Thus, only N links were needed in the network, which became more flexible, more scalable, and easier to maintain [1]-[2]. This development is depicted in Figure 1-1. The *analog* electrical voice signals were carried over copper pairs from subscriber to another through the switching center, which is in charge of steering these signals to their correct destinations. Every pair of wires

[1] All these terms are more or less equivalent, especially in the Public Switched Telephone Network (PSTN). The term "Central Office" (CO) is common in North America whereas the term "Telephone Exchange" is common in Europe and other parts of the world. The term "switching node" is often used in networking research literature. In this discussion, we mainly use the term "switching center" in order to maintain generality.

connecting a subscriber to the switching center represents a telephone *line*, *channel*, or *circuit*[2]. The job of the switching center is to hook up any two subscriber lines to each other, as desired, forming an end-to-end telephone circuit. This basic form of switching is known as *circuit switching*.

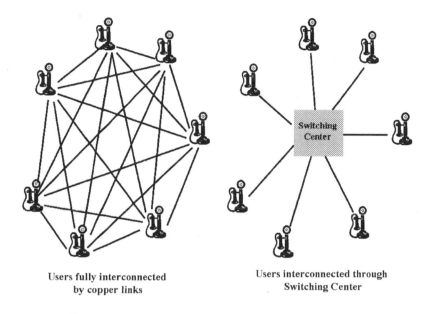

Users fully interconnected by copper links

Users interconnected through Switching Center

Figure 1-1. Introduction of switching in telephone networks

Many telephone networks started to emerge in this manner and the need for subscribers on one network to communicate with those on other networks also emerged. Therefore, switching centers were interconnected, whether via direct links, where appropriate, or by using another level of switching centers, each of which interconnects a cluster of local switching centers. The centers of the new level, in turn, had to be interconnected via trunk (long distance) national centers, and so on. The hierarchical *Public Switched Telephone Network* (PSTN) appeared everywhere. Now, a typical telephone call could undergo several switching steps through more than one switching center. At that time, a single telephone call, or circuit, reserved the entire end-to-end bandwidth of the transmission link, i.e. a pair of copper conductors. The two-wire path through a switching center was dedicated

---

[2] These three terms can be used interchangeably. However, we use "line" here to denote the connection between the subscriber and the switching center and use either of the other two terms to refer to the end-to-end, subscriber-to-subscriber, connection.

full-time to carrying both directions of the telephone conversation[3]. Figure 1-2 depicts an arbitrary PSTN architecture.

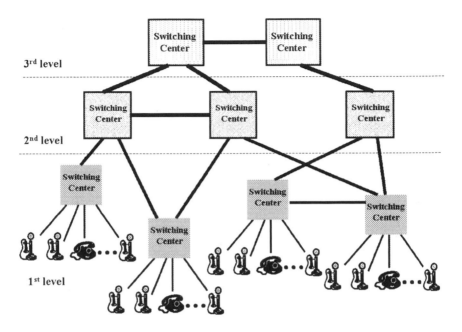

Figure 1-2. A telephone network hierarchy with arbitrary interconnection. Typically, the number of levels varies among networks, and from one country to another. It does not have to be three levels. First level switching centers and networks are usually called local or access centers/networks. Higher level centers/networks may be called junction, inter-exchange, or trunk, although other terminologies are used. This also varies among countries.

Meanwhile, efficiency in designing and maintaining the transmission plant was ensured by grouping the wires extended between two points, say two switching centers, into a single cable. The telephone channels sharing a cable represent a multiplex of channels. This form of multiplexing is carried out in the space domain. Indeed, *Space Division Multiplexing* (SDM) is the simplest form of multiplexing and is sometimes overlooked when the techniques and domains of multiplexing are listed. Switching, as explained so far, like multiplexing and transmission, was carried out entirely in the space domain.

---

[3] In most trunks, the two directions were provided on two separate paths to control voice echoes and this was known as four wire switching [1]-[3]. Digital systems, which are discussed later, also use four-wire transmission and switching, but for other reasons. Our discussion here is focusing on the evolution of circuit switching concepts and is not concerned with the detailed implementation issue of two-wire versus four-wire switching.

Switching centers and telephone networks are designed based on the assumption that not all subscribers would be engaged in telephone calls simultaneously. Studies and analyses of traffic patterns, using statistical models, and economic considerations lead to suitable switching-center designs with permissible blocking probabilities. *Blocking* is the state that occurs when the switching center can not accommodate incoming call(s) because all available circuits are already engaged in other calls. Large-scale strictly non-blocking switching matrices can be expensive to build. Therefore, switching fabrics are usually designed in the form of multi-stage architectures where smaller switching matrices are organized and interconnected according to certain rules. These designs are cost effective and scalable and can have low, or negligible, blocking probability.

The analog telephone hierarchy evolved to encompass another form of multiplexing in addition to SDM, namely *Frequency Division Multiplexing* (FDM). Two-dimensional networks, where telephone circuits were established in the space and frequency domains, offered better communication capabilities and higher utilization of resources. FDM was introduced first in the trunk network, before it penetrated lower levels of the hierarchy. Groups of telephone channels were usually combined over a single transmission link by assigning a distinct frequency band for each. Coaxial copper cables and microwave links were used as trunk transmission links [3]. At the switching center, each telephone circuit became uniquely identifiable by a space port and a frequency band. Switching centers combined the processes of frequency multiplexing/demultiplexing and space switching.

1.1.2   From Analog to Digital

With the advents of digital technologies and voice digitization, multiplexing, transmission, and switching migrated from *analog* to *digital*. Digital communications was more economical and provided improved voice quality. In principle, both analog and digital signals can be multiplexed in either the frequency or the time domains. However, it is easier and cheaper to multiplex analog signals in the frequency domain and multiplex digital signals in the time domain [1]. Therefore, *Time Division Multiplexing* (TDM) gradually replaced FDM in trunk networks and elsewhere. In TDM, signals from multiple sources are periodically sampled and samples of a certain number of sources are time multiplexed into a frame, which is transmitted all together. At the other end, samples are extracted and used to regenerate original signals. The development and wide-scale deployment of *Pulse Code Modulation* (PCM) was the cornerstone of digital telephony. PCM supplements TDM with digital quantization and coding (Figure 1-3).

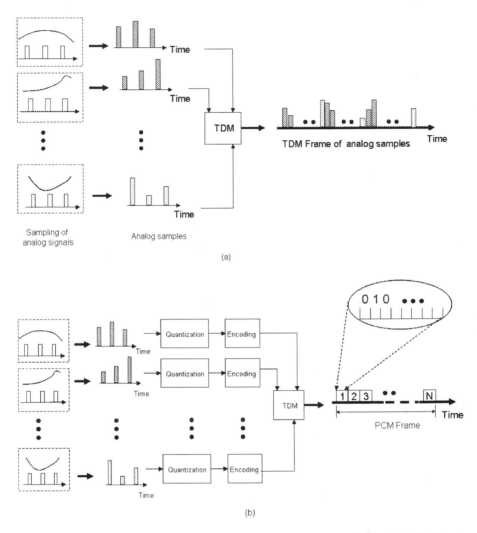

Figure 1-3. Time Division Multiplexing (TDM) of analog signals and Pulse Code Modulation (PCM): (a) TDM principle, (b) PCM principle

In PCM, transmission time is also divided into frames and each frame is divided into numbered *time slots*. Each slot represents a telephone channel. Now, instead of multiplexing analog voice samples directly in the time domain, as with straight TDM, these samples are encoded into 8-bit digital words. Each word fits into a time slot. Hence, words from different sources are multiplexed onto the frame. Two main frame types have been deployed, namely the DS-1 and J1 24-slot systems in North America and Japan, respectively, and the E-1 32-slot system elsewhere. Both frame types are similar in that the TDM sampling rate of speech signals is 8 KHz and each time slot carries 8-bit digital representations of the analog samples. This led to the ubiquitous 125 μsec frame time and the 64 Kbit/sec channel rate in telephony.

Hence, each input port to the switching center delivers a digital multiplex of telephone circuits in the form of PCM frames (and higher orders/multiplexes of these frames), and the identifiers of any telephone circuit became a space port and a time slot. For the entire duration of a telephone call, the voice samples of a subscriber are transmitted periodically over certain time slot. At the other end, samples are extracted and used to generate the original speech signals. A similar process is carried out in the reverse direction of the call using another time slot.

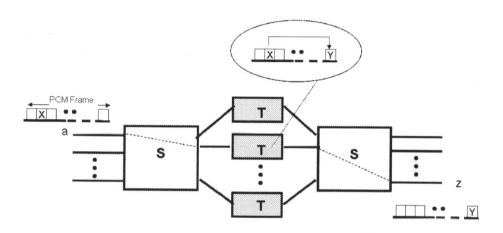

Figure 1-4. A simplified multi-Stage multi-dimensional switching fabric. This example is a Space-Time-Space (STS) architecture. A connection is shown to be established between input time slot "X" of digital highway "a" and output slot "Y" of highway "z". Connection setup depends on finding a time switch (TSI) where both slots X and Y are free. Other architectures, such as TST, are possible. A larger number of stages is possible too.

The switching process became two dimensional, requiring the capability to interchange digital words among any time slot on any input space port with those of another slot, having different number, on any output space port. In order to carry out this operation, the switching center comprises a combination of space switching stage(s) and time switching stage(s). In space switching, the switch fabric is a conducting matrix, where input and output ports are spatially interconnected at *crosspoints*. Unit switching devices are placed at these crosspoints to perform the switching process. Thus, the switching process facilitates an electrical path within the space fabric to connect the incoming and outgoing pieces of the end-to-end telephone circuit. The space switch can reconfigure on each time slot to meet the changing demand for telephone circuits. Switching in the time domain is carried out by a *Time Slot Interchanger* (TSI) which interconnects its inputs/outputs using speech and connection memories [1]-[2]. The crosspoints in this case are memory locations where digital words from the input are written cyclically and in order. Then, they are read out in a different order under control of connection memories and sent to the output. Multi-dimensional digital switching offered a new level of flexibility, scalability, and performance. It has enabled building of elegant and compact switching systems at reduced cost. Figure 1-4 depicts a simplified example.

The switching operation is composed of three main processes, namely, signaling, control, and switching itself. Signaling facilitates initiation of the call among subscribers and prompts the switch control process. The latter instructs the switch fabric to be configured in such a way so as to perform the switching process and is also in charge of holding the call until it is finished. The switching process sets up a path for the call across the switching center.

The first generation of digital-communication hierarchies is the *Plesiochronous Digital Hierarchy* (PDH). PDH was based on time multiplexing of PCM bit streams into higher order multiplexes for transmission purposes. The individual PCM streams were plesiochronous (almost synchronous) because their bit rates were close to each other, but not always identical. This hierarchy suffered from operation, scalability, maintainability, and reliability problems. Absence of worldwide PDH standards constituted a big problem since different PCM frames, and various multiplexing rules, were used in different parts of the world. Therefore, many efforts were focused on developing and standardizing synchronous digital communication systems. In the US, this led to introduction of the *Synchronous Optical Network* (SONET) standards [4]. The *Synchronous Digital Hierarchy* (SDH), i.e. a related and interoperable standard with SONET, was established soon afterwards worldwide. SDH/SONET is a set of coordinated ITU (International Telecommunication Union) and Telcordia (Bellcore) standards and specifications that defines a hierarchical set of digital-transmission rates and formats. Table 1 defines the main SDH/SONET rates. These rates embrace the basic PCM transmission rates. SDH provides global compatibility among synchronous systems, including SONET.

SDH/SONET overcame the drawbacks of PDH. The latter is not totally extinct everywhere yet, but on the verge of being so. SDH/SONET constitutes the most important pillar of today's worldwide telecommunications infrastructure [5]-[6]. SDH and SONET are optical transmission standards supported by electronic processing, signaling, and control. They are not switching technologies, but they encompass provision for cross-connection and add/drop of densely multiplexed digital signal streams and they inter-work with electronic switching centers.

SDH	SONET (Electrical/optical)	Transmission Line Rate
STM-0	STS-1/OC-1	51.840 Mb/s
STM-1	STS-3/OC-3	150.336 Mb/s
STM-4	STS-12/OC-12	622.080 Mb/s
STM-16	STS-48/OC-48	2.488 Gb/s
STM-64	STS-192/OC-192	9.953 Gb/s
STM-256	STS-768/OC-768	39.813 Gb/s

Table 1-1. SDH/SONET rates and digital hierarchy. SONET uses an acronym for the electrical signal and a corresponding acronym for its optical equivalent. STS stands for Synchronous Transport Signal and OC is for Optical carrier. SDH does not use this dual terminology. STM stands for Synchronous Transport Module.

A *cross-connect* resembles a switching center in many aspects, but possesses many differences as well. Switching centers utilize user signaling to establish fast and instantaneous non-permanent interconnections (circuits) over individual telephone lines. The lifetime of a circuit across a switching center is usually measured in minutes. Control is integrated into the switching process in order to provision the circuit in real time. A cross-connect establishes semi-permanent interconnections between its input and output ports. These interconnections are high bandwidth pipes, representing large (coarse) switching granularity. The set up of a cross-connect is done by the carrier, service provider, or operator[4]. Setup time is much longer than its switching center counterpart and the lifetime of the cross-connection can be measured in months or even years.

The technologies used to build switching devices (crosspoints) and to control them also evolved over time, along with the evolution of telephone networks. In the beginning, a person had to undertake the switching process manually. First, this was the user himself. Then, an operator, with a switchboard attached to the switching center, took over this responsibility. Control in this case was manual and the switching device was an electrical plug or key to connect the right terminals to each other. Later, automated electromechanical switching devices were introduced. Two main types of these devices were used in early switching centers, namely, the step by step rotating switch (also known as the Strowger switch, after its inventor) and the crossbar switch [2]. The term *crossbar* arises because the crosspoints

---

[4] Carriers and service providers refer to two types of telephone companies in the US. Operators is a generic term used in other countries to refer to both. For our purpose, the three terms can be used interchangeably.

were created by crossing horizontal and vertical bars[5]. Both types were based on arrays of open electromechanical contacts and control was therefore electrical (utilizing electromagnetic effects). Then, reed relays, which are sealed electromechanical contacts, were deployed in the following generation of switching devices. Electronic switching devices were introduced next. Electronics penetrated control first as some electromechanical switches used electronic wired logic for control. By the end of the 1970s, digital electronics became widely deployed in switching devices [1]-[2]. Electronic switching technologies and systems have enjoyed tremendous success. Today, electronics continue to dominate switching centers and digital cross-connect systems (DCS).

1.2 EVOLUTION OF SWITCHING IN DATA NETWORKS

1.2.1 Other Switching Methods

In ancient history, primitive forms of data communication were used to convey data *messages* across distances that could not be traveled by human voice. Messages were relayed and switched from one point to another, using horns, flags, smoke signals and other methods. Telegraphy appeared around the middle of the nineteenth century and was the first communication technology to be based on electricity. The telegraph, which led the analog telephone by few decades, carried messages in the form of dots and dashes and was therefore digital in nature. Telegraphy evolved to be based on a switching method known as *message switching*. At first, this form of digital switching was more of a manual relaying process of messages arrived at some telegraph switching center and going to other centers and destinations over different telegraph lines. The process became semi-automatic when teletypes (Telex) with paper tape punches and readers were developed. Telegraphy became fully automated with computers in the 1960s leading to *store-and-forward* message switching. Each switching center within the network stores messages in their entirety and forwards them based on the address and control information contained in their headers [2].

Other data communication technologies utilized message switching as well. Facsimile (Fax) and Teletype (Telex) are perhaps the two main examples in this respect. Fax and Telex machines were the dominant commercial data terminals before computer networks become widely

---

[5] The word "crossbar" was in fact coined before the word "crosspoint"

deployed. They operated on separate links and networks, which were derived from the transmission facilities of the PSTN [3].

Unlike the case in circuit switching, users in store-and-forward switching do not interact in real time. Messages may be delivered to their destinations after some delay and there is no need to determine the status of the destination before sending the message. The message can be stored in buffers (memories) and forwarded to the recipient when it is free, or able, to receive it. Another important difference between message switching and circuit switching is that transmission links in a message switched network are never idle while traffic is waiting to use them. In circuit switching, this can happen since the circuit is held throughout the entire duration of the call for exclusive use of its parties. The circuit may not be shared by other users of the network, regardless' of the proportion of the call time that is actually used for conversation. Hence, message switching offered better bandwidth utilization than circuit switching and was more suited to switching bursty data messages. Nevertheless, message switching was not the best possible solution for modern data networks when they emerged and gained popularity. Another form of store-and-forward digital switching, i.e. *packet switching*, soon became the switching technology of choice for these networks.

The theory of packet switching was first introduced in the early 1960s [7]-[12]. In packet switching, the message is broken down into smaller packets. Each packet has a header to carry the information necessary to forward it through the network to its destination. There are a number of reasons why it is better to break the message into smaller packets of fixed, or variable, sizes. Packetization results in fair treatment of all messages and reduces the delay encountered by them in the network, especially for shorter messages, which would otherwise experience excessive queuing delays waiting for longer messages to be fully transmitted. Packetization also simplifies data storage and processing and limits retransmissions, upon discovery of errors, to smaller packets where errors have actually occurred. In message switching, entire messages would be retransmitted in such situations.

Packet networks (Figure 1-5) are networks of storage and delay instead of blocking. However, since storage buffers have finite capacity, there is always some probability of traffic loss in packet networks. Probability of traffic loss, traffic throughput, and average delay are common performance parameters in packet networks. There are two fundamental modes of packet switching. Packets can either be switched as datagrams or over virtual circuits. A *datagram* is a packet that is treated independently from its peers. Upon arrival at a switching node, the header of each datagram is read to determine the direction in which it should be forwarded. Packets are routed accordingly to subsequent node(s) until they reach their final destinations. Headers should contain originating and destination addresses, and packets are allowed to arrive at their destination out of sequence. *Virtual circuit* packet switching is connection oriented. In this mode, a control packet carries addresses information and initiates the communication session by setting up a path between the two ends. Information pertinent to this path may be stored and maintained in a switching center. Session packets are transmitted over this path.

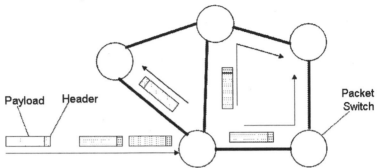

Figure 1-5. An arbitrary packet switched network

Packet switching devices have been generally based on the same technologies used to build circuit-switching devices. However, packet switching centers are not as necessary as their circuit switching counterparts since the packet switching functionality can be largely distributed over the network, as implied above. Also, the term *routing* is more familiar today than switching in packet networks, although the two are used interchangeably. Packet switches are often called *routers*. Routing in data communications is performed by setting up routing tables and packets are *forwarded* based on these tables. This process provides connectionless switching of datagrams [13]. When used with virtual circuits and in circuit switched network, the term *routing* refers to identifying the circuit path and is part of the circuit setup process.

The procedures and rules governing the movement of packets from one user to another in data networks are collectively termed *protocols* [7]. Protocols have been actually deployed in circuit switching too. Signaling and control procedures leading to establishing a telephone call constitute a protocol. The norm of a telephone conversation where it is implied, for instance, that the two parties may not talk at the same time, represents an informal protocol. The most advanced protocols in the world of circuit switching are those introduced by SONET and SDH for operation, administration, maintenance and provisioning (OAM&P)[6] purposes. Nevertheless, protocols are far more associated with packet switching than circuit switching. Since data networks interconnect computing machines, their need for sophisticated protocols to secure intelligent and reliable communications is obvious. Different types of protocols deal with various functional components of data communications. For example, there are peer-to-peer application protocols to handle end-to-end user applications. There are network level protocols, such as those for routing. There are link-level protocols along the communication path. There are protocols to secure adequate physical transmission and delivery of data, and many other protocols.

The multiplicity, diversity, and sophistication of data communication networks and their protocols made it favorable to adopt a modular design approach and lead to architectures where the communication process is broken down into delineated functional layers. This also enables the easy upgrade of network elements, equipment, and protocols. Layers are independent in that the details of operation of each of them are invisible to the others. The relationship between any two adjacent layers is governed by a client-server model. By performing its designated functions, each layer acts as a server for the client layer on top of it.

The International Standards Organization (ISO) proposed the *Open System Interconnection* (OSI) model as a framework for interconnecting heterogeneous computer systems. The model is open in the sense that it enables any two systems conforming to the model itself and its associated standards to interconnect regardless of their implementation details. While laying out a universal layered architecture for communication networks, the OSI model identified the functional areas of communications networks where protocol standards needed to be developed.

Figure 1-6 depicts the ISO/OSI model. Detailed discussion of this model and its seven layers is beyond our scope, and can be found in other references, such as [14]-[15]. However, the reader of this book needs to be aware of a terminology issue. The word *transport* is used in the telecommunications industry in a different context than in the OSI model. Historically, the two main ingredients of the PSTN were switching and transport, as in nodes and links, where transport embraced multiplexing and transmission. Transport referred to links of all types, especially between switching centers. Anything that connected switching centers was called transport, including PDH links. When SDH/SONET were introduced, and

---

[6] OAM&P is concerned with network issues such as configuration, performance management, fault detection, protection, and others. It is also referred to in several international standards as OAM (Operation, Administration and Maintenance, or, simply, Operation And Maintenance).

later when *Wavelength Division Multiplexing* (WDM) emerged, their equipment inherited this terminology because they also connected voice switching centers and many kinds of nodes. Even SDH/SONET digital cross-connects are regarded as transport systems, probably because they undertake a different type of switching which is less sophisticated and at larger granularities. As far as the OSI model is concerned, all these transport links and equipment are indeed part of the physical layer, not the transport layer. Therefore, it must be noted that terms like transport network, optical transport network (OTN), and sometimes optical transport layer which are often used in the telecommunications industry are in fact associated with the OSI layer 1 (physical layer) and should not be confused with its layer 4.

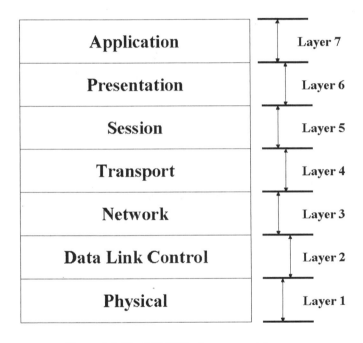

Figure 1-6. The ISO/OSI reference model

1.2.2 From ARPANET to the Internet

In 1969, ARPANET became the first packet switched data network. It connected four host computers at University of California Los Angeles, Stanford Research Institute, University of California Santa Barbara, and University of Utah. ARPANET was sponsored by the Advanced Research Projects Agency (ARPA[7]) of the US Department of Defense (DoD). The

---

[7] ARPA has changed names over time and is known today as Defense Advanced Research Projects Agency (DARPA)

network used the so called Interface Message Processors (IMPs) as packet switching centers. In two years, many other computer systems in the US were connected to ARPANET. Europe was connected to this network in 1973. At the network level, the initial ARPANET used a host-to-host protocol called the Network Control Protocol (NCP). Later, a new version of NCP that is more communication oriented was developed. The new protocol, which formally replaced NCP in 1983, evolved into the Transmission Control Protocol/Internet Protocol (TCP/IP), User Datagram Protocol (UDP), and, over time, the Internet protocol suite of today[8]. Although the Internet protocol suite is referred to as TCP/IP it actually comprises a large collection of protocols, of which IP and TCP are the most important two.

The decision made by the US National Science Foundation (NSF) to elect TCP/IP for its data networking program, i.e. NSFNET, was an important milestone in the history of data networking. NSFNET eventually became an important part of the backbone that we know today as the Internet. Another significant boost for TCP/IP came from the University of California at Berkeley when TCP/IP became part of their version of the UNIX operating system [12].

ARPANET was not the only packet-switching initiative in its time. Other initiatives also took place. By the mid 1970s there were many packet switched data networks in the world, but they were largely incompatible. Most networks at that time were serving certain research and/or professional communities with very limited access to general public. There was also little motivation to interconnect them. The introduction of the X.25 user-network interface standard by the International Telegraph and Telephone Consultative Committee (CCITT) in 1976 helped make data networking accessible to the public in many countries. The CCITT is a committee of the International Telecommunication Union (ITU), an agency affiliated with the United Nations (UN), that is known today as the ITU Telecommunication Standardization Sector (ITU-T). X.25 is a protocol that specifies an interface between host system and packet-switched network while not requiring specific packet switching mode or implementation. The X.25 standard was revised several times in the 1980s and 1990s.

Last but not least, the popularity and large-scale deployment of personal computers (PCs) and Local Area Networks (LANs) in the 1980s and 1990s pushed data communications and packet-based networks further ahead. More networks were built and interconnected. The PSTN was readily available to provide the interconnection support data networks needed to stretch their reach over the globe. As indicated earlier, the PSTN had done it before for

---

[8] IP provides Layer-3 (OSI) connectionless packet routing. TCP supports IP by a reliable assured end-to-end communication between two hosts whereas UDP provides an unreliable connectionless service between two hosts.

earlier data communication technologies. Hence, dial up into telephone lines became the primary access method to data communications. Dial up uses a data modem (modulator/demodulator) to convert digital data at one end of a connection to a form similar to speech in amplitude and bandwidth and converts it back to digital data at the other end. When the voice-like data enter the telephone network they are encoded into digital form, transmitted, and switched over telephone circuits as if they were voice signals. Conversion from digital to analog and visa versa can happen in this fashion as many times as needed. This complexity was actually one of the drivers to introduce Integrated Service Digital Networks (ISDN), which was commercially successful in parts of the world but unsuccessful in others [3].

Because of the increased interest in data communications, private data networks proliferated and competitive long-haul data-communication services soon emerged. What was born as the ARPANET grew into the Internet of today. The Internet is a huge interconnection of *Internet Protocol* (IP) routers with packet switching centers, of different types, also used as edge boxes to interconnect autonomous data networks. ARPANET was decommissioned in 1990 leaving IP as the worldwide dominant data networking protocol.

1.3 EVOLUTION OF TELECOMMUNICATIONS NETWORKS ARCHTECTURES

Throughout most of the twentieth century, communications networks architectures were based on telephony requirements. By the late 1980s, many specialized networks, dedicated to certain data services or applications, were in existence and the PSTN was used as bearer for many types of traffic. However, the circuit-switched PSTN is optimized for voice only and is not flexible enough to handle protocols and bit rates that are so diverse and inhomogeneous. Moreover, operation and maintenance of too many specialized networks, whether standing alone or sharing PSTN facilities, was costly and problematic. Therefore, there was a growing desire to introduce new broadband multi-service networking technologies that are capable of handling all traffic types efficiently [16]. The industry had to experiment and/or envision numerous switching solutions, before focusing its attention on *Asynchronous Transfer Mode* (ATM) [17].

In the beginning, the quest to merge voice, data and video[9] over unified multi-service network architectures led to considering some variations of

---

[9] We have been more focused so far on voice and data, but this discussion is equally valid for video networks, especially in the future. Video is traditionally carried over broadcast infrastructures where switching takes a different form. Today, multimedia applications are pushing for solutions where data, voice and video are communicated all together.

circuit and packet switching. Circuit switching is optimized for voice and real-time traffic and packet switching is optimized for data traffic. The goal was to modify either switching approach to make it suitable for carrying all sorts of traffic efficiently. Variations of circuit switching that were proposed include multi-rate circuit switching and fast circuit switching. Multi-rate Circuit Switching (MRCS) attempts to overcome the inflexibility of having a single fixed bit-rate in traditional circuit switching by making it possible to allocate multiple channels (slots) at a certain base rate to a connection. In Fast Circuit Switching (FSC), the circuit is only allocated to users when information is being transmitted and fast signaling is deployed to set up and release circuits in this manner. Variations of packet switching which were considered, on the other hand, include frame switching and frame relay. With relaxed error and flow control burdens, thanks to advances in transmission technology, these approaches offered advantages over traditional packet switching. Except for Frame Relay, all variations of straight circuit and packet switching enjoyed little successful deployment, if any. It is ATM, ultimately, which was positioned as the compromise between circuit and packet switching and considered capable of supporting Broadband Integrated Service Digital Networks (B-ISDN) [17].

ATM is a packet-based connection oriented technology that crosses the borders between multiplexing, transmission and switching. It uses asynchronous time division multiplexing[10] and a fixed-length packet, called a cell. Each cell consists of 53 bytes (octets), out of which five comprise a header and the rest constitute the information field. ATM does not mandate specific amount of bandwidth per channel and allows for statistical multiplexing. It has enjoyed considerable success and wide-spread deployment in carriers' networks, and has been largely relied on to carry the increasing data traffic. ATM armed carriers with advanced traffic engineering tools. It also facilitated a platform which can integrate voice and data, although it was not primarily used for this purpose. An ATM layer grew on top of the SDH/SONET cornerstone layer in the public telecommunications architecture.

Meanwhile, the tremendous growth of the Internet led to IP becoming the default data protocol in the network and an IP layer emerged on top of ATM. The growth of the Internet also created excessive demand for more bandwidth. As data transmission rates continued to rise, it became clear that electronic transmission systems would not keep up with these rates and may become too expensive as they attempt to scale beyond their current limits. Meanwhile, a lot of progress had been made in WDM technology, leading to

---

[10] The word asynchronous stems from the fact that synchronization is not required among connection ends. This leads to an ability to support any traffic irrespective of its service and signal characteristics.

cost-effective commercial Dense WDM (DWDM) transmission systems.
Thus, the demand for extra bandwidth resulted in extensive deployment of
WDM in carriers' networks. A WDM-based optical transmission layer
evolved under SDH/SONET. Figure 1-7 depicts the stack of layers forming
the network architecture in this manner. The WDM layer was sometimes
referred to in the industry as layer 0, or layer 0.5, to separate it from
SDH/SONET. However, both layers are part of the physical layer, i.e. layer
1, of the OSI model.

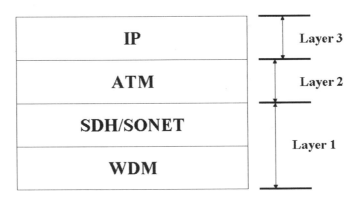

Figure 1-7. Telecommunications core networks architecture

Thus, at some terminal point in the network, IP packets are generated and
delivered to the network. As they proceed forward, these packets may be
converted to ATM cells and the latter are packed into SDH/SONET circuits
which are loaded, in turn, onto WDM transmission pipes. Of course, this is
not the only way for IP packets to be carried from one network user to
another, and it is definitely not the most efficient way to get this job done. In
the Packet over SONET (PoS) scheme, for instance, the ATM layer is
skipped and IP packets are packed directly over SDH/SONET using layer-2
framing protocols such as HDLC (High-level Data-Link Control protocol)
and PPP (Point to Point Protocol). In practice, a mix of these two approaches
is used as well as others. As for traditional voice, it is carried directly into
SDH/SONET circuits in most cases.

It must be noted that Figure 1-7 shows a fairly simplified version of
current network architectures. A more accurate picture must take into
consideration other technologies, protocols, and situations. The Ethernet
frame, for example, is the most common payload type carried by
SDH/SONET in today's networks. Hence, Ethernet is indeed part of the
detailed picture. Frame Relay is also deployed. SDH/SONET themselves are
evolving today to Next Generation (NG) SDH/SONET and may utilize the
Generic Framing Procedure (GFP), Virtual Concatenation (VC) and Link

Capacity Adjustment Scheme (LCAS) to facilitate better transport of data traffic [18]. There are also other emerging protocols and technologies that should appear in Figure 1-7 if all are to be taken into consideration. However, this is not necessary for the purpose of this discussion, especially since the four layers shown represent the major architectural ingredients in today's networks.

Hence, while it is easy to differentiate the concepts and operation principles of packet switching from their circuit-switching counterparts, it is not as easy in practice to single out, say, an end-to-end data-communication session over the Internet, and determine that it is totally, and purely, based on packet switching. Contrary to popular belief, the Internet is not entirely based on packet switching and it does use a great deal of circuit switching, not only in the core network via SDH/SONET circuits but even in access through dial up and broadband access, such as Digital Subscriber Lines (DSL). The recent advents of voice over IP (VoIP) and Video over IP (IPTV) made it also possible for voice calls and video signals to be carried over packet-switched networks. Hence, whether a communication session is a transatlantic phone call, an e-mail crossing the borders between continents, web surfing or an on-line interactive session, the chances are high that it is actually based on a mosaic of circuit switching, packet switching, and ATM.

The evolution of the network architecture to its current status was driven by a combination of factors; economic, technical, regulatory, service related, market related, and others. While the growth of the network in this manner was instrumental in the formation and success of the Internet as we know it today, and despite the fact that it led to remarkable progress in the field of telecommunications and data networking, it has left the network with major architectural problems. These problems include complexity, functional redundancy, inflexibility, and poor utilization of resources. The network architecture became difficult and costly to operate, manage, maintain, and scale. It is predominantly based on classical SDH/SONET circuits and is optimized for voice traffic at a time when data traffic is believed to dominate carriers' networks. Although ATM was successful in facilitating excellent traffic engineering tools for carriers' data networks, there are concerns about its control overhead at higher transmission rates and it became clear that it is not going to be scalable beyond a certain limit. While the nature of tomorrow's services and applications requires fast network reconfigurability to handle bursty traffic and nontraditional services, today's networks may take days, weeks or more to reconfigure their rigid topologies.

Concerns about these problems surfaced by the end of the 1990s in an atmosphere of economic prosperity and in the midst of promises of potential technological breakthroughs. Therefore, considerable efforts focused on facing these problems by innovative optical solutions. With continued demand for more bandwidth, interest in optical communications had been rising steadily throughout the 1990s. Huge resources were poured into research and development of optical devices and optical networking technologies. The goal was to extend the exploitation of optics from WDM transmission to optical networking in the general sense, which blends multiplexing, transmission, and switching. *Optical switching* became the center of utmost attention. The industry geared toward building a new bearer *optical layer* as part of the telecommunication transport infrastructure (OSI layer 1). The foundations of this layer are WDM, optical switching, and innovative network control schemes. Many research efforts even focused on the potential of a new two-layer architecture model, based on the IP and optical layers and on interworking and integration among these two layers.

The optical layer is expected to facilitate a transport infrastructure that is characterized by abundance of bandwidth, flexibility, reliability, manageability, scalability, and cost-effectiveness. It can only be fully enabled by optical switching. This layer is expected to be capable of provisioning and management of circuit switched connections, channels, or *lightpaths*, to serve the communication needs of the optical-layer clients, or optical-network users, such as IP routers, ATM switches, SDH/SONET network elements, and others. A lightpath can be an end-to-end fiber path including all wavelengths carried thereon, a subset of these wavelengths along some route which have the same end points, or a wavelength based path.

1.4   OPTICAL SWITCHING

We have seen that switching evolved over decades using manual, electromechanical, and electronic devices and systems. Optical switching presents a future alternative which is starting to be deployed. Our discussion is now turned to the subject of optical switching. In the remainder of this chapter, we examine the history of optical switching and investigate the drivers to introduce it into communications networks. We categorize optical networks, optical switching methods, optical switching domains and optical switching technologies. Finally, we try to clear up some important terminological issues pertinent to this field of science and engineering.

1.4.1   The Case for Optical Switching

Interest in optical switching appeared for the first time in the 1970s, and increased in the 1980s after the inroads made by optical fiber to the transmission arena. By the beginning of the 1990s, there was already considerable research effort on optical switching. Contributions to this effort came from researchers in areas of electronic switching, optical and opto-electronic devices, optical computing, and optical communications [19]-[21]. At that time, optical switching devices were generally classified, according to their functionality, into the so called *relational devices* and *logical devices* [19]-[20]. Relational devices establish a relation, or mapping, between the input and output ports of the device. This relation changes according to the state of the device (for example, being in cross or bar configuration) and this state, in turn, is determined by some external electrical control. Devices of this type do not use the information flowing through them to change the state of the device. Logical devices, on the other hand, are digital devices and are assumed to be capable of using the information passing through them to control the switching state of the device itself. Logical devices were mainly targeting optical logic applications and optical computing, but were also envisioned to enable building of optical digital switches for telecommunications.

The classification of optical switching devices into relational and logical types was adopted in some literature in the early 1990s, but did not live for long. Progress in what was referred to as logical devices, based on non-linear optics, has not been great enough to render them as serious deployment candidates in commercial computing and telecommunications applications. A lot of progress, however, was made in the research and development of devices that were described as relational devices. Indeed, it is this latter kind, more or less, which has the potential for large commercial deployment in the foreseeable future.

By the dawn of the 1990s, a few immature technologies were available to build optical switches for telecommunications [22]. Guided-wave electro-optic switches (mainly using Lithium Niobate with Titanium waveguides, $Ti:LiNbO_3$), macro opto-mechanical switches (such as the fiber switch), spatial light modulators (using liquid crystals), and semiconductor optical amplifier (SOA) gates were the main candidate devices. Other technologies may have existed, but were hardly noticeable.

As indicated earlier, the tremendous growth of the Internet during the 1990s led to extensive deployment of WDM [23]-[24]. WDM, in turn, provided a solution to the transmission bottleneck, but created the challenge of switching the large number of wavelength channels it enables. Since switching in carrier networks was, and still is, performed by electronics, optical to electrical to optical (OEO) conversions are required at switching-centers and cross-connect nodes. Multi-wavelength data streams have to be terminated at every node, converted to the electrical domain for switching purpose before they are converted back to the optical domain and transmitted to the next node. This approach, obviously, lacks flexibility, scalability, and suffers from serious performance bottlenecks. It is also very expensive.

Optical switching has therefore been positioned as the solution to these problems and key to switching relief [25]-[28]. As result of intense research and development effort, new optical switching technologies were introduced and old ones advanced by progress in material science and fabrication methods. Before the end of the century, many of these technologies became viable candidates for deployment in carriers' networks.

Clearly, the migration of switching to the optical domain would bring tremendous advantages to the network. Optical switching extends the reach of the virtually unlimited optical bandwidth from transmission systems to switching nodes, thereby elevating the information carrying capacity of the network to levels that are way beyond reach with electronic switching. The migration of switching to the optical domain has the potential to enhance network agility, flexibility and reliability, and can make the network more cost effective. The abundance of end-to-end bandwidth in optical-switching based networks also creates new opportunities for wavelength based services and applications. Optical switching is the cornerstone in the vision of the optical layer and is crucial for tomorrow's network architectures. As such, it is an essential part of the long-term solution to many network problems. The merits of optical switching, without doubt, constitute a strong motivation to utilize it extensively in future as soon as all technological ingredients are fully mature and cost effective, and when the telecommunications market is ready for this step.

1.4.2   Optical Networks and Optical Switching

Today, the terms optical networks and optical switching are used to denote systems and networks that are not optical in the full meaning of the word. For example, SDH/SONET systems are widely regarded in the telecommunications industry as the optical transport networks of today, although there is nothing optical in them beyond single-wavelength point-to-point transmission links. All other functional ingredients of SDH/SONET based networks, including multiplexing, cross-connection, add/drop and control, are performed by electronics. Also, some equipment vendors developed cross-connects based on electronic switching matrices with optical interfaces (OEO systems) and labeled these equipment as optical cross-connects and/or optical switching systems. By comparison, switching systems which would be based on optical switching fabrics are referred to as OOO systems. While OOO systems are called transparent, and sometimes all-optical, their OEO counterparts are often called opaque.

Transparency involves the network being able to transfer any type of information without regard to protocol and coding formats, data rates, and modulation techniques. It has many advantages. Transparency enables the optical network to support variety of higher layers concurrently. It also enables the network to be designed cost-effectively and to evolve easily in the future alongside protocols and technologies. Electrical networks are *opaque*, in contrast to *transparent*, since their performance is dependent on signal type and parameters. This is indeed associated with their ability to read and process the signals they carry, which is favorable in many applications. SDH/SONET based networks are opaque since they regenerate signals electrically between all nodes. Transparency is intrinsic to optical networks where data are transmitted and switched in the optical domain on an end-to-end basis. However, there are different levels of transparency. Full transparency implies that the network be able to transport and handle all kinds of traffic, from analog to all types of digital, concurrently without limitation and regardless of signal parameters and physical span. Since transmission media and systems are not perfect, this form of transparency is not possible in practice, at least not with today's technology. Therefore, transparency is not absolute or unlimited and may only be accomplished within certain design limits, and taking into consideration factors like fiber and components types, transmission distance, and others.

Nowadays, there are four types of networks that are all equally labeled as optical networks, but not accurately so, whether in literature or in the industrial community. In the following, we identify these types while trying to establish a terminology that makes sense for us to use:

1) SDH/SONET based networks, where only single-wavelength point-to-point transmission is carried out optically. These are actually *opto-electronic networks*.

2) Networks where WDM is used leading to multi-wavelength optical transmission, while switching and control remain entirely in the electrical domain. These are still opto-electronic networks, but they utilize advanced optical transmission technology and there is considerably more optics in them than in the first type. We consider these networks as *opaque optical networks* (OEO).

3) Networks where WDM and optical switching are used, while switch control and network control are carried out by electronics. These are *optical networks with varying degrees of transparency*, and may be classified accordingly to different types. If all nodes utilize opto-electronic conversions to regenerate signals and/or to adapt them to switching and/or transmission requirements, the network becomes *opaque* (OEO). On the other hand, if no opto-electronic conversion is utilized, the network is fully *transparent*. Between these two extremes, we have networks that are *partially transparent*. Within some design limit, data can travel through these networks transparently in the optical domain. Although control is performed by

electronics, the end-to-end data path is entirely optical. Partial transparency can take the form of islands of transparent networks which are bound by OEO conversions, or scattered OEO conversions in certain nodes in the network.

4) Networks where all operations and functions, including switch control and network control, would be performed optically. These are *all-optical* networks.

It must be noted that only the last two types of networks in this list are indeed associated with optical switching. Transparent network, partial or full, combine WDM transmission and optical switching on one hand, with electronic control on the other hand. This brings together the best of both optics and electronics. So far, it has been difficult to implement control in the optical domain. There are theories that attempt to explain the reasons behind this difficulty. Optics has a number of distinct features, the most notable of which is its tremendous bandwidth. Another important feature is that photons tend to have weak interactions with each other in transparent media (much weaker, for instance, than interactions of electrons in electrically conducting media). Indeed, this feature, coupled with bandwidth, is given credit for the qualities and benefits of optical transmission. However, strong interaction among information carriers is required for them to carry out control operations. Therefore, it is not possible, at least under the current state of the art, to control optical switches and networks by light [13], [21]. Nevertheless, photons do interact with matter and research in all-optical technologies continues to explore new possibilities. The prospect of progress and breakthroughs in this regard can not be ruled out.

1.4.3   Optical Circuit, Packet and Burst Switching

Like switching in the electrical domain, there are two main methods of optical switching, namely, *Optical Circuit Switching* (OCS) and *Optical Packet Switching* (OPS). In OCS, switching is performed at the granularity of an optical circuit or lightpath. When the lightpath is based on a wavelength, OCS is sometimes referred to as optical wavelength switching. However, this is not accurate terminology. We discuss this point in more detail in section 1.4.4. Most of Part III of this book is focused on OCS issues.

Because data traffic is dominating the network and since circuit switching is not optimized for data traffic, there has been considerable focus by the research community on OPS, where optical switching is performed at the granularity of packets. Datagram OPS must be all-optical to fully attain all of its fundamental advantages over OCS. This requires that packet header recognition, header processing, and control be fully performed in the optical domain on a packet-by-packet basis. This pure form of OPS however is beyond reach today because it requires technologies that are not readily available. Therefore, OPS is sometimes envisioned in a form that relies on electronic control. Yet, even in this form, it faces challenges, such as the need for ultra-fast scalable optical switching technologies, optical buffers/memories, optical packet delineation and synchronization.

In recent years, *Optical Burst Switching* (OBS) was introduced as a compromise between OCS and OPS, and has since then received considerable research interest. OBS is packet based, which makes it potentially more bandwidth efficient than OCS. Meanwhile, the technological requirements to implement OBS are relaxed in comparison to those of OPS. In an OBS network, packets are assembled into larger data bursts (DB). For every burst, a Burst Header Packet (BHP) is created. DB assembly and BHP generation take place in ingress OBS edge nodes. Each DB/BHP pair is routed to their destination (an egress OBS edge node) through OBS core nodes. The burst is transmitted over a data channel to the following node while its BHP is sent over a dedicated control channel to the same node. Depending on the protocol deployed BHPs may be transmitted ahead of their DBs thereby introducing an offset time to account for BHP processing delays in subsequent nodes. While data bursts are switched optically and remain in the optical domain, BHPs are converted to the electrical domain at every node for processing and recreation with new control information. Then, they are converted back to the optical domain before transmission to next node. Each BHP contains the necessary information to configure downstream optical switching fabrics for the incoming DB. At the other end, an egress OBS edge node disassembles DBs into their original packets and the latter are forwarded to their destinations outside the OBS network.

Part of the challenge facing OPS is due to the extremely small switching granularity and the desire to read/write packet headers in the optical domain. By assembling packets into larger bursts, OBS reduces the burden on control. By transmitting data and control separately and performing control electronically, the need for optical header reading/writing is avoided. However, OBS has technological issue too and architectural challenges of its own. OPS and OBS are discussed in more detail in chapter 13.

1.4.4 Optical Switching Domains

In principle, optical switching can be envisioned in the space, wavelength, and time domains. However, it has not yet attained the degree of maturity and sophistication of its electronic counterpart. Most of the progress accomplished in recent years has been in the area of *optical space switching* technologies where optical signals are switched from input space ports (fibers) to output space ports (fibers).

Reference is made in some literature to wavelength switching and/or switching in the wavelength domain. Strictly speaking, there are three main processes to handle wavelengths in WDM optical networks. The first process is wavelength separation or selection and this is usually carried out by demultiplexers and/or filters. Passive splitters and filters can also be used together to carry out this process. The second process is wavelength combining or grouping. This uses multiplexers and/or passive combiners. Most wavelength multiplexing/demultiplexing technologies that are available today provide static mapping of wavelengths to space ports, which means that they are not reconfigurable. The third wavelength handling process is *wavelength conversion*, or *wavelength translation*. This process converts (switches) the signal-carrying wavelength to another wavelength. This process is close to what wavelength switching, in the telecommunications sense, would be about.

A lot of progress has been made in recent years on wavelength conversion. In the future, development of commercial tunable/reconfigurable switching blocks which are capable of converting (switching) a number of wavelength channels simultaneously may become possible. These wavelength-domain switching fabrics could be used in conjunction with optical space switching fabrics, or reconfigurable multiplexer/demultiplexer pairs, to build multidimensional optical switching systems. For now, wavelength conversion is not easy or inexpensive enough for extensive commercial deployment. The short-term target is to build optical space switching systems that are augmented by some sparse wavelength conversion capability, say, on per-port basis. Hence, while we anticipate future progress, we suggest that the term wavelength switching be used conservatively for now. One must also note that many researchers and engineers use this very term to refer to space-switching OCS systems; simply because light in these systems is spatially switched at the wavelength granularity.

Some research efforts have been carried out in the area of optical time slot interchangers (TSIs) [19]-[21], [25]. However, optical switching in the time domain is not easy and not ready for commercial deployment. This type of switching requires viable optical memories/buffers, optical synchronization, and a capability to carry out numerous logical operations in the optical domain. Nevertheless, optical time division multiplexing and de-multiplexing of very high bit-rate streams is possible to perform today and have been demonstrated. Some of this work is indeed discussed in chapter 7.

1.4.5 Optical Switching Technologies

Optical space switching technologies can be classified in different ways. Two generic types are usually recognized, namely, the type based on guided lightwave (fiber and/or waveguides) and the type based on free-space optics. Optical switching technologies may also be classified, more specifically, based on the underlying physical effect that is responsible for the switching process. In this case, we see several technological categories. Each of these categories comprises a number of technology types depending on the manner in which the physical effect is exploited, device design, material used, and on other considerations:

1) *Electro-Optic* (EO) Switching: this category utilizes electro-optic effects. The most well-known switching technology in this category is based on optical waveguides and directional couplers which are implemented on Lithium Niobate substrates. Other technologies which can be considered electro-optic in nature include liquid-crystal optical switches utilizing polarization control and electro-holographic optical switches.

2) *Acousto-Optic* (AO) Switching: this guided-wave category exploits the acousto-optic effect and can also be implemented on Lithium Niobate.

3) *Thermo-Optic* (TO) Switching: this category utilizes the thermo-optic effect in optical waveguides. Two main types of materials are deployed to implement these switches, namely, silica and polymers.

4) *Opto-Mechanical* (OM) Switching: this category is mostly based on free-space optics. It comprises classical technologies based on moving fiber, and/or moving macroscopic optical components as well as modern technologies such as those based on optical Micro Electro Mechanical Systems (MEMS) or on planar waveguide crosspoints which are controlled by movable air bubbles.

5) *Optical-Amplifier based* Switching: both semiconductor and fiber amplifiers have been proposed as switching devices. However, Semiconductor Optical Amplifiers (SOA) are more common in optical switching applications. SOAs are also used for wavelength conversion and for multiplexing/switching in the time domain.

It must be noted that regardless of the physical effect responsible for the switching process, which is used as basis for this categorization, external control of the switching device is electrical in the vast majority of cases. Electrical control is used to trigger electro-optic effects in EO switching, to generate acoustic surface waves in AO switching, to supply heating energy in TO switching, to actuate OM switching systems, to control SOA gates, and so on.

The above five categories encompass, roughly, most optical switching technologies in existence today. In the second part of this book, we cover a large number of technologies that fall within these categories. Other technologies which do not perfectly fit within this classification may also exist in the literature, but are not as important for us.

There are a number of studies in the literature where some optical switching technologies are compared in tables, based on features like insertion loss, crosstalk, extinction ratio, switching speed, scalability, reliability, latching, and many other features. This form of comparison is usually based on performance figures reported in the literature about demonstrations that may not all share common bases, in terms of experimental conditions, fabrication accuracies, design goals and switch parameters (such as size). Also, performance figures of a technology are not independent from each other. Optimizing the switch design for one certain feature can limit the optimization of another feature. The maximum allowable insertion loss and crosstalk of a switching element, for example, vary considerably depending on fabric size [29]. Indeed, optical switching technologies can not be compared, all together, per se without regard to the specific application for which a given technology is considered. Comparisons based on best, or worst, performance figures reported in literature do not always provide the complete picture. In this book, we have adopted an approach where all technologies are covered in some depth taking into consideration the specifics of each and their suitable applications.

1.5   OPTICAL SWITCHING VERSUS PHOTONIC SWITCHING

Technologies, devices, systems, and networks that perform numerous operations and processes in the optical domain are sometimes labeled in the literature as *optical*, and at some other times as *photonic*. Today, most engineers and scientists in the field use the terms optical switching and photonic switching interchangeably and consider them equivalent. Some may elect one term over another, but it seems like a matter of preference that is not necessarily based on careful analysis of terminology. This situation intercepted me in the midst of the work on this book. The issue was to adopt one of the two terms, optical switching or photonic switching, as a title and a default term in all chapters. The intention here is not to discredit the other term, and both are indeed credible, but it seemed appropriate to initiate some genuine effort to put terminology into perspective. In the course of doing so, I had to undertake some research and analysis that is dedicated to this specific purpose. I also consulted my co-authors, since they are all experts in the field, as well as other colleagues and pioneers who lead the contributions to optical/photonic switching literature nearly fifteen years ago [19]-[21].

Unfortunately, there is no agreed upon definition of optics and photonics, as opposed to each other, in the engineering and scientific communities. For example, some scientists suggest that where light rays are simply directed from some point to another under external control, and when the process can be described, say, by geometrical optics, it is conveniently labeled as optical. On the other hand, when the process is better understood and defined by taking into consideration the quantum nature of light, it should be described as photonic. Here, the two terms are assigned different usages based on context, which is convenient.

Others believe that optics deals with the propagation of light through passive media and components where it is only affected by passive interactions with matter, such as attenuation, dispersion, etc. Photonics, according to this view, is more associated with propagation of light through active media and components where it may be amplified, filtered, wavelength converted, or subjected to other processes. However, this classification is not in harmony with the everyday use of terms like optical amplifiers, opto-electronics, and opto-electronic devices and phenomena. Indeed, many of the switching technologies discussed in the following chapters utilize physical phenomena to control the refractive index of matter, which, in turn, interacts with light and results in switching. All these technologies are named after optics, not photonics (electro-optic, acousto-optic, thermo-optic, opto-mechanical, etc.).

Other researchers tend to use optics and photonics simultaneously, based on whether the subject is device physics and engineering or systems and networks engineering. Some see that photonics is more suitable for device research whereas optics is a better term for systems research. Ironically, there are other research communities where it is the other way around.

Some lean toward one of the two terms because they view it as more general or, alternatively, more specific to telecommunications. Nevertheless, there is no agreement in this regard as well.

In the industrial community, many refer to OEO systems as optical switching systems. As a result, true optical switching systems had to search for a name. Photonic switching became this name for many in this community. Here, product differentiation and/or marketing terminologies and strategies dictate certain choice of words.

After delving into the issue for some time, it became clear that it is not possible to seek harmony by revisiting literature or by polling as many professionals in the field as possible. Hence, it appeared that help may be sought by appealing to basics!

Optics had preceded photonics by decades just like electricity preceded electronics. Indeed, these two pairs of words are somehow analogous. While optics and electricity are old veterans of physics, electronics and photonics became to be widely known only in the twentieth century. Electronics is a field of science and engineering that investigates the behavior, interaction, and motion of electrons, and deals with devices, circuits and systems based thereupon. In the context of telecommunications switching systems, electronics does it all, i.e., it is used to carry data through switched paths and to invoke control actions to configure the switching centers. Photonics, by the same token, is a field of science and engineering that investigates the behavior, interaction, and motion of photons, and deals with devices, circuits, and systems based thereupon. To be truly analogous to electronic switching, photonic switching should refer to switching systems where both data transport and control actions are carried out by photons, i.e., in the optical domain. This gets us so close to saying that photonic switching may be equivalent to all-optical switching, where control is also performed optically.

This argument comes about nicely and constitutes a reasonable basis for defining terminology. Now, as for the processes and devices that involve interactions between electrons (matter) and photons, one may choose to consider that these are opto-electronic, photo-electronic, or photonic depending on context and, if so desired, on whether taking the quantum nature of light into consideration is required, or not.

We have decided to adopt this terminological approach. Based on this decision, our choice of terms, between optical and photonic switching, is optical switching. This choice has the added advantage of fitting nicely into many themes already established in the "optical" networking community. The entire field of optical communication is predominantly based on the advents of optical fiber and fiber optics, not photonic fiber or fiber photonics. We also usually use the terms optical domain, not photonic domain; optical transmission, not photonic transmission; and optical layer, not photonic layer. Once again, the goal of this choice of terminology is to make the best possible decision for this book without meaning to discard or discredit any other terminologies.

1.6 THIS BOOK

Although other forms of optical switching will be considered throughout the following chapters, our main focus is on optical space switching for OCS networks with varying degrees of transparency. Over the past few years, optical switching has been positioned rightfully as a credible deployment candidate for cross-connection and flexible add/drop of lightpaths in carrier networks. Optical Cross-Connects (OXC) and reconfigurable Optical Add/Drop Multiplexers (OADM) represent realistic applications of optical switching in the foreseeable future. OPS and OBS remain as active areas of research.

We have differentiated between cross-connects and switching centers in the electrical domain. In principle, the same rules of differentiation apply in the optical domain. However, fundamental differences between optics/photonics and electronics render this analogy almost irrelevant. First, it is not conceivable today, and not needed, to have optical switching centers similar to the electrical ones and dealing with the same switching granularity. Indeed, cross-connection is the most meaningful form of optical switching for today. Second, given the tremendous bandwidth of optics and taking into consideration many of the anticipated next-generation services, the time scale at which an OXC will set up and tear down lightpaths must be different than its electronic counterpart. In the future, OXCs are expected to be automatically self-reconfigurable in, say, seconds. This will blur the distinction between cross-connection and switching in the optical domain, in terms of connection (lightpath) lifetimes and reconfiguration rates. Indeed, it may very well eliminate this distinction should OXCs become fast enough to reconfigure in sub-seconds. It is important to understand that optical switching will work with electronic switching, and that there is no competition among them. While optical switching will handle bulk (wholesale) traffic with large granularities, electronic switching is essential at smaller granularities and it is very likely that equipment from both worlds will have to interwork and collocate.

This book is structured into three parts. We have almost reached the end of the first part where an introduction to the subject of optical switching has been given. Part II discusses several optical space switching technologies. The theory of each is presented along with fabrication methods and principles of operation of basic switching devices. The literature is reviewed in each case and many experimental demonstrations of switching systems are described. Chapters 2, 3, and 4 are dedicated to waveguide based Electro-Optic (EO), Acousto-Optic (AO), and Thermo-Optic (TO) switching technologies, respectively. Chapter 2 discusses the principles of optical switches based on coupled waveguides in some depth. The reader will find this useful to understand many of the material in chapters 3 and 4 as well. Chapter 5 focuses on Liquid Crystal (LC) optical switching. LC switches have made considerable progress over the past few years. Chapter 6 discusses Optical MEMS switches, which is one of today's most important areas of optical switching. Chapter 7 discusses SOA based switching applications. Research in this area has been active for a long time and SOAs continue to be interesting devices. Chapter 8 concludes the second part of the book by covering several technologies, including other Opto-Mechanical types (moving fiber/component and moving air bubble), Holographic switching, and others.

Part III turns the discussion to optical switching systems and networks. Chapter 9 discusses switching fabrics. The goal is for the reader to understand how technologies and devices discussed in the second part of the book may be put together to create optical switching fabrics. Several architectures are described and light is shed on their performance characteristics. Chapter 10 addresses the issue of controlling these fabrics, how to set up paths across them, and configuration algorithms. Chapter 11 puts optical switching within the context of optical networks, and discusses its application in OXCs, Reconfigurable OADMs (R-OADM) and in other network areas. The focus in this chapter is more on the applications of optical switching in provisioning lightpaths since Chapter 12 is dedicated to its applications in protection and restoration. Chapter 13 covers OPS and OBS as important research topics which have attracted considerable attention among the research community. We conclude Part III, and the book, by chapter 14 where we look at where optical switching stands today and what lies ahead.

REFERENCES

[1] F. J. Redmill and A. R. Valdar, "SPC Digital Telephone Exchanges," Peter Peregrinus Ltd, The Institute of Electrical Engineers (IEE), Londen, UK, 1990

[2] J. Bellamy, "Digital Telephony," 3rd Edition, John Wiley & Sons, Inc, 2000.

[3] F. T. Andrews, "The telephone network of the 1960s," IEEE Commun. Mag., Vol. 40, No. 7, pp. 49-53, July 2002.

[4] R. Ballart and Y-C Ching, "SONET: Now it's the standard optical network," IEEE Commun. Mag., Vol. 27, No. 3, pp. 8-15, March 1989

[5] M. Sexton and A. Reid, "Broadband Networking: ATM, SDH and SONET," Artech House, 1997.

[6] W. Goralski, "SDH/SONET," McGraw-Hill/Osborne, 2002

[7] P. Baran, "On distributed communications networks," IEEE Trans. on Commun., Vol. 12, No. 1, pp. 1-9, March 1964.

[8] http://www.rand.org/publications/RM/baran.list.html, RAND Corporation Research Memoranda.

[9] R. D. Rosner, "Packet Switching, Tomorrow's Communications Today," Lifetime Learning Publications, 1982.

[10] D. W. Davies, "An historical study of the beginnings of packet switching," The Computer Journal, Vol. 44, No. 3, pp. 151-62, 2001.

[11] P. Baran, "The beginnings of packet switching: some underlying concepts," IEEE Commun. Mag., Vol. 40, No. 7, pp. 42-8, July 2002.

[12] B. M. Leiner, V. G. Cerf, D. D. Clark, R. E. Kahn, L. Kleinrock, D. C. Lynch, J. Postel, L. G. Roberts, and S. Wolff, "A brief history of the Internet," The Internet Society, http://www.isoc.org/internet/history/brief.shtml.

[13] L. Thylen, G. Karlsson, and O. Nilsson, "Switching technologies for future guided wave optical networks: potentials and limitations of photonics and electronics," IEEE Commun. Mag., Vol. 34, No. 2, pp. 106-13, February 1996.

[14] D. Bertsekas and R. Gallager, "Data Networks," 2nd Edition, Prentice Hall, 1992

[15] A. Tanenbaum, "Computer Networks," 4th Edition, Prentice Hall, 2002

[16] J. S. Turner, "New directions in communications (or Which way to the information age?)," IEEE Commun. Mag., Vol. 24, No. 10, pp. 8-15, October 1986.

[17] M. De Pryckcr, "Asynchronous Transfer Mode: Solution for Broadband ISDN," 3rd Edition, Prentice Hall, 1995.

[18] P. Bonenfant and A. Rodriguez-Moral, "Generic framing procedure (GFP): the catalyst for efficient data over transport," IEEE Commun. Mag., Vol. 40, No. 5, pp. 72-9, May 2002.

[19] H. S. Hinton and J. E. Midwinter (Eds), "Photonic Switching," IEEE Press, 1990

[20] H. S. Hinton (in collaboration with others), "An Introduction to Photonic Switching Fabrics," Plenum Press, 1993

[21] J. E. Midwinter (Ed), "Photonics in Switching," Vol. I and Vol. II, Academic Press, 1993.

[22] S. F. Su, L. Lou, and J. Lenart, "A review on classification of optical switching systems," IEEE Commun. Mag., Vol. 24, No. 5, pp. 50-5, May 1986[1]

[23] H. J. F. Ryan, "WDM: North American dcployment trends," IEEE Commun. Mag., Vol. 36, No. 2, pp. 40-4, Feb. 1998

[24] E. Lowe, "Current European WDM deployment trends," IEEE Commun. Mag., Vol. 36, No. 2, pp. 46-50, Feb. 1998

[25] H. T. Mouftah and J. M. H. Elmirghani (Eds), "Photonic Switching Technology-Systems and Networks," IEEE Press, 1999.

[26] T. S. El-Bawab, "Next generation switching/routing: the optical role," The 2001 IEEE International Conference on Communications (ICC 2001), Workshop # 2, Helsinki, Finland, June 11, 2001.

[27] T. S. El-Bawab, A. Agrawal, F. Poppe, L. B. Sofman, D. Papadimitriou, and B. Rousseau, "The evolution to optical-switching based core networks," Optical Networks Magazine, SPIE/Kluwer, Vol. 4, No. 2, pp. 7-19, March/April 2003.

[28] T. S. El-Bawab, "On the potential of optical switching in communication networks," Proc. SPIE, Vol. 5247, pp. 111-4, Optical Transmission Systems and Equipment for WDM Networking II, Part of SPIE International Symposium ITCom 2003 (Information Technologies and Communications), 7-11 September 2003, Orlando, Florida, USA.

[29] M. Zdeblick, "Design variables prevent a single industry standard," Laser Focus World, Vol. 37, No. 3, March, 2001

PART II
OPTICAL SWITCHING TECHNOLOGIES

Chapter 2

LITHIUM NIOBATE ELECTRO-OPTIC SWITCHING

Hideaki Okayama

This chapter reviews developments of the guided-wave *Electro-Optic* (EO) switching technology using *Lithium Niobate* (LiNbO₃ or LN) [1]. LiNbO₃ exhibits a large *electro-optic* (EO) effect, *acousto-optic* (AO) effect, *thermo-optic* (TO) effect, and nonlinear optical effects. The EO, AO and TO effects are defined by changes in the refractive index of matter due to the application of an external physical action/field. In the EO effect, the change is induced by applying an electric field. In the AO effect, the change is caused by interaction between acoustic and optical waves in the crystal. In the TO effect, the refractive-index is changed by temperature.

In the EO effect, an electric field causes electron or crystal-lattice displacements, which result in refractive index change. Very fast response is expected in devices based on this effect. LiNbO₃ is a unique EO crystal in that large wafer substrates can be grown and can exhibit small dielectric constant, fast device responses, and low power consumption. The dependence of the *refractive index n* on the electric field *E* is expressed by:

$$1/n^2 = 1/n_0^2 + rE + hE^2 + \ldots$$

(2.1)

where n_0 is the refractive index without electric field, r is the electro-optic coefficient, and h is a higher order electro-optic coefficient.

The LiNbO₃ crystal is a *ferroelectric* crystal which exhibits electric dipoles (spontaneous polarization) even without applying an electric field.

The electric dipole vanishes above a certain temperature, which is called the *Curie temperature*. The crystal shows an EO effect where the refractive index changes linearly with the applied field. This type of electro-optic effect is known as *Pockels effect*. Low loss planar waveguides can be fabricated on LiNbO$_3$ substrates by several methods. Light is confined into the core region of the waveguide where the refractive index is made larger than the surrounding region. By guiding light into planar waveguides, optical-axis alignment problems, which are associated with bulk optics, are avoided. Devices of this type are fabricated using processes similar to those used to fabricate integrated circuits (ICs). This enables mass production of relatively complex planar lightwave circuits (PLCs).

Throughout its long history, almost all guided-wave optical switching technologies have been demonstrated and verified using LiNbO$_3$. Lithium Tantalate (LiTaO$_3$ or LT) has similar crystal structure and characteristics as LN. However, its Curie temperature is lower (655 °C) making it a little difficult to fabricate good optical waveguides. Therefore, LT crystals are not used as extensively as LN in optical switching. Some LT devices are briefly described in this chapter. Other ferroelectric materials that are used in optical switches include Lead Zirconate Titanate (Pb(Zr, Ti)O$_3$, or PZT) and Lead Lanthanum Zirconate Titanate [(Pb, La)(Zr, Ti)O$_3$, or PLZT]. These materials exhibit electro-optic coefficients that are one order of magnitude larger than LN. However, difficulty of fabricating low-loss waveguides hinders their use in optical switching. Recent improvements in fabrication methods can lead to lower waveguide loss [2].

Compound semiconductors used in fabricating laser diodes and photo diodes are also used in EO switches. There are two main schemes to control the refractive index in this case. This can be done by applying an electric field, as with LN, or by injecting carriers (electrons and holes) into the waveguide layer [3]-[5]. An 8x8 switch of this type is the largest switch *matrix* fabricated so far [6]. The insertion loss tends to be higher than that of LN switches. Silicon is also used in EO switches, but it lacks Pockels effect and carrier injection is the only way to control its refractive index [7]-[8]. There are few reports on EO waveguide devices using liquid crystals [9], but these crystals are more popular in polarization controlled optical switching (chapter 5). Organic EO materials have been studied too [10]-[11], but they attract more interest for thermo-optic switching (chapter 4).

2.1 MATERIAL PROPERTIES AND DEVICE FABRICATION

2.1.1 Characteristics of Lithium Niobate

$LiNbO_3$ crystals that are large enough to obtain wafers of more than 4 inches in diameter are grown regularly. Low loss optical waveguide can be fabricated easily on $LiNbO_3$. These features have led to using $LiNbO_3$ as the substrate to fabricate EO switches from the early days of optical guided-wave device development. The Curie temperature of $LiNbO_3$ is 1120~1210 °C and its melting point is 1150~1250 °C depending on the crystal composition. The crystal structure is shown in Figure 2-1. The spontaneous polarization is along an axis about which the crystal exhibits three-fold rotation symmetry. The axis is called *c-axis*. In the Cartesian system, the z-axis is chosen to coincide with the c-axis. The y-axis lies in a plane of mirror symmetry and the x-axis is perpendicular to the mirror plane. The easiest way to fabricate a crystal, and the most popular congruent where solid composition is similar to that of the melt, has a 0.946 lithium/niobium (Li/Nb) ratio. The composition with a 1.0 Li/Nb ratio is considered ideal and is called stoichometric composition.

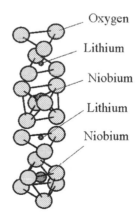

Figure 2-1. LN Crystal structure [1]

The refractive index for light polarized along the c-axis (extraordinary refractive index) is smaller than the refractive index for light with polarization perpendicular to this axis (ordinary refractive index). The index changes with material composition, temperature and wavelength. The dependency of ordinary refractive index on composition, temperature and wavelength are much smaller than the corresponding dependencies of extraordinary refractive index. The ordinary and extraordinary refractive indexes of congruent crystal at 1.55 μm wavelength are 2.21 and 2.14 respectively at room temperature.

Detailed analysis of Equation (2.1) can show that the linear refractive index change is related to electric field strength as per Equation (2.2). The first formula shows the change in optical permittivity, where E_j is the electric field vector and r_{ij} is the linear electro-optic coefficient tensor. j=1, 2, 3 correspond to the axes x, y and z respectively. By optical symmetry the number of electro-optic coefficients can be reduced to a total of eight coefficients as shown in the matrix. The elements r_{1j}, r_{2j} and r_{3j} cause refractive index change of the light polarized along the x, y and z axis, respectively. The other elements r_{42}, r_{51} and r_{61} cause the polarization rotation in the yz, xz and xy planes, respectively.

$$\Delta(1/n_i^2) = \sum_{j=1}^{3} r_{ij} E_j$$

$$r_{ij} = \begin{pmatrix} 0 & -r_{22} & r_{13} \\ 0 & r_{22} & r_{13} \\ 0 & 0 & r_{33} \\ 0 & r_{51} & 0 \\ r_{51} & 0 & 0 \\ -r_{22} & 0 & 0 \end{pmatrix}$$

(2.2)

The largest coefficient is r_{33} and the next largest is r_{51} (stoichiometric crystals have approximately 20% larger r_{33} than congruent crystals [12]). Other coefficients are less than half as big as the largest. In the congruent crystal r_{33}= 22 pico-meter/Volt (pm/V), and r_{13}= 9 pm/V at 1.52 μm wavelength. The values for the other elements are r_{51}= 28 pm/V and r_{22}= 3.4 pm/V. The magnitude of the EO effect changes with wavelength and material composition. It decreases as wavelength increases [13]. The wavelength and material composition dependencies of r_{13} are much smaller than those of r_{33}.

2.1.2   Fabrication of Lithium Niobate Electro-Optic Switching Devices

A typical device structure is shown in Figure 2-2. The fabrication of an optical waveguide on $LiNbO_3$ is performed mainly by two methods, i.e. by Titanium (Ti) diffusion or proton exchange, as depicted in Figure 2-3.

In the first method, a titanium thin film is deposited onto the LiNbO$_3$ substrate. The waveguide pattern is defined using lithographic methods. A lift-off process or etching technique can be used. The titanium thin film is diffused into the substrate at about 1000 °C. To avoid Lithium (Li) out-diffusion during the process, the chip must be placed in a platinum box. A wet atmosphere can also slow the out-diffusion of Li. The chip must be oxidized at a high temperature before ending the diffusion process or the process must be performed in atmosphere including O$_2$ to avoid crystal coloring. Colored crystals exhibit high absorption loss, high sensitivity to optical damage and large drive voltage drift.

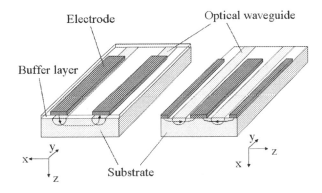

Figure 2-2. Electrode structure

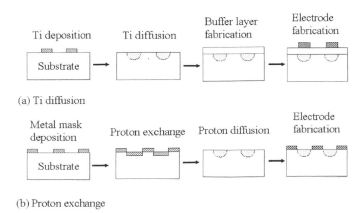

(a) Ti diffusion

(b) Proton exchange

Figure 2-3. Device Fabrication

The refractive index changes with Ti concentration (somewhat sub-linearly with extraordinary light). The Ti diffusion and distribution can be simulated using the diffusion equations. The distribution is often given as Gaussian function in depth and Error function in lateral directions. The diffusion is faster along the z-axis. Using z-cut substrate, a deeper channel waveguide is obtained than in x-cut substrate with waveguide directed along y-axis. The waveguide field distributed is deeper into the substrate with z-cut substrate resulting in smaller insertion loss. For the z-cut substrate, the –z surface is mainly used for Ti waveguide fabrication. The spontaneous polarization tends to reverse during the Ti diffusion process in the +z surface. The x-cut substrate is known to yield better channel waveguide than y-cut substrate.

The proton exchange method is performed by immersing the substrate into acids such as the benzoic acid at about 200 °C. A metal mask deposited onto the substrate is used to fabricate the waveguide. The Li in the substrate is substituted with proton. Often the sample is subsequently annealed at 300-400 °C to diffuse the proton further into the substrate to attain waveguides with sufficient depth and moderate index change. The process is simple and waveguides with high refractive index contrast against the substrate can be fabricated. Only the refractive index along the z-axis is enhanced by proton exchange and the waveguide merely supports the light polarized along this direction.

These methods lead to a high concentration of dopants in the crystal. Liquid phase epitaxial methods to fabricate waveguides with less damage to the crystal structure are developed for nonlinear optical applications. The waveguide structure obtained by Ti diffusion and proton exchange is buried into the substrate from the surface with one of the waveguide rims exposed to the air.

Electrodes are placed on the substrate surface to apply an electric field onto waveguides. When the electrode is placed above the waveguide, a low index layer called buffer layer is required to separate the metal electrode from the waveguide to prevent absorption loss of the optical signal. In most cases, silicon dioxide is used as a buffer layer. Optimized buffer layer fabrication procedures are required to minimize the drive-voltage drift phenomenon where the drive voltage may change over the electric-field application time. The permittivity and resistance differences between the buffer layer and the substrate cause piling up of space charge which screens the applied electric field. The anisotropy of the substrate permittivity generates short-term drive-voltage drift. Meanwhile, the $LiNbO_3$ crystal exhibits a pyroelectric effect which results in an electric charge on the z surface when temperature changes. Therefore, in devices using z-cut substrates, semiconductor material (typically Si) is deposited on the buffer layer to prevent non-uniform pyroelectric charge distribution and voltage shifts. In x-cut substrates the electrode gap is positioned above the waveguide. The buffer layer is not required when the electrode edge is placed far enough from the waveguide so that the light field does not overlap with the electrode.

The refractive index change induced by the EO effect can expressed as:

$$\Delta n = -\Gamma r n^3 V/(2g) \tag{2.3}$$

where V is the applied voltage, g the gap between electrodes and Γ the overlap integral of waveguide light and the applied electric field. When the electric field is induced along the z axis n= n_e and r = r_{33} are used for the extraordinary light and n= n_o and r= r_{13} for the ordinary light. When the electric field is induced along the y axis, n= n_o and r= $-r_{22}$ are used for the light polarized along the x axis and n= n_o and r= r_{22} for the light polarized along the y axis. In many cases the accumulation of phases changes so that a product of Δn and length L sets the required refractive index change Δn. The voltage-length product (proportional to ΔnL) is used as a measure of the drive efficiency of the device. The phase change is inversely proportional to wavelength and drive voltage becomes lower at shorter wavelength. Light guided with its polarization vertical to the substrate is considered *transverse-magnetic* (TM) like mode whereas polarization parallel to the substrate is considered *transverse-electric* (TE) like mode.

2.1.3 Design of Electrodes and Optical Waveguides

There are two possible electrode structures depending on the electric field direction. In the first case the electrode is placed above the waveguide. In the other structure a gap is placed in between them. The former is adopted to utilize electric fields perpendicular to the substrate and the latter utilizes the field parallel to the substrate. In the z-cut substrate the electrode is often placed above the waveguide to use the largest r_{33} electro-optic coefficient.

The induced electric field is non-uniformly distributed. It tends to be strong near the electrode and the substrate surface and weakens as it goes deeper into the substrate. The electric field is obtained using analytical formula by conformal mapping [14]-[15]. More detailed analysis which takes buffer layer and electrode width into consideration can be performed by several numerical techniques. One of the simplest methods in this regard is the relaxation method [16]. The finite element method and other techniques are also used extensively [17].

Many methods have been developed for calculating the light mode propagated in the LN based optical waveguide. Diffusion equations govern impurity (Ti or proton) profiles and the refractive index gradually changes with depth or along lateral directions. Analysis and calculation of the optical field in such a graded refractive index waveguide is much more difficult than with step index distributions. The waveguide with optical power confined only to the direction of waveguide depth (without lateral confinement) is called a *slab waveguide*. Examples of simple calculation methods for slab waveguides are the Wentzel-Kramer-Brilliouin (WKB) method and the multilayer stack theory [18]. A waveguide having light confinement in both the depth and lateral directions is called a *channel waveguide*. Optical field calculation in this case becomes more difficult. The finite element or finite difference methods are used to calculate the optical field if high accuracy is required [19]-[20]. Light propagation in optical waveguide is often analyzed using the beam propagation method (BPM) [20].

2.2 OVERVIEW OF GUIDED-WAVE ELECTRO-OPTIC SWITCHES

Chapter 1 gave an overview of optical switching technologies, including different types that fall within the electro-optic (EO) category. We now examine the guided-wave type of this category in more detail. There are two main kinds of guided-wave electro-optic switches depending on how the lightwave (lightpath) is directed. The direction of the lightwave can either be controlled by deflection, using phase control, or by diffraction.

2.2.1 Deflector-Type Implementation

Figure 2-4(a) depicts a deflector/scanner type switch where the lightpath is controlled by an analog control signal. A special case of this configuration is obtained when the device has two switching states only. In this case, control is said to be digital. In Figure 2-4(b), a multi-stage switching fabric can be built by interconnecting these elements with channel waveguides, as discussed in section 2.5. This kind of EO switching fabric has received considerable research interest.

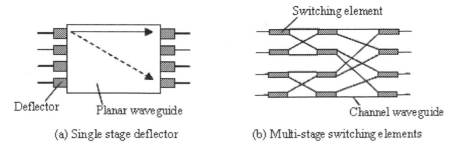

(a) Single stage deflector   (b) Multi-stage switching elements

Figure 2-4. 4x4 deflector type waveguide switches

Two configurations are used to deflect lightwaves in a guided-wave device. The first configuration uses collimators with a prism the refractive-index of which is electro-optically controlled. The second configuration uses waveguide arrays.

2.2.1.1   Electro-Optic Prism

The typical structure of this kind of switching device is shown in Figure 2-5. An optical element, such as waveguide lens, is used to collimate the light beam. The latter is launched in an EO planar waveguide and let to fall on the slanted interface of the prism. A prism shaped electrode is placed on top of the planar waveguide. By applying a voltage to the electrode the refractive index of the prism changes. The light beam is refracted according to Snell's law with an angle of refraction Θ given by

$$\Theta = \Delta n \, L/(nW) \tag{2.4}$$

where Δn is the refractive index change, L the prism length, and W the prism width. Ideally the field is applied between the electrode on top and the ground electrode under the waveguide. The spacing between electrodes must be small enough to produce sufficient field strength at moderate voltage. A device of this type has been fabricated with conducting substrate. For an LT substrate with 500-μm electrode gap and L/W= 13, a deflection angle of 4 mrad is attained at 600 V [21].

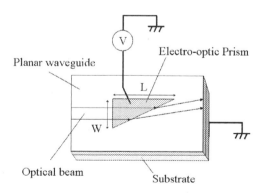

Figure 2-5. Electro-optic prism deflector

The refractive index change due to Pockels effect reverses its polarity when the direction of the electric field is reversed. A region with reversed spontaneous polarization can be fabricated. A larger refraction angle is attained by cascaded, or parallel, prisms with alternating polarizations. A deflection angle of ± 44 mrad (0.7°) is attained in LN devices with a 200-μm electrode gap and ± 90 V [22].

2.2.1.2   Waveguide Array Device

This device is shown in Figure 2-6 and is sometimes called a phaser. An electrode is placed on each waveguide. Input light is split among waveguides by an input coupler. This can be a star coupler or branching waveguides. By applying voltage to electrodes the refractive index changes and the phase shift accumulates along the electrode length as the light travels the waveguide. The amount of phase shift is set to increase from the first to the last row of the waveguide array. A planar waveguide is connected to the end of the waveguide array. Light output from the waveguide array forms a wavefront according to the phase. A plane wave is generated if the phase difference between adjacent waveguide is constant. The deflection angle Θ is given by

$$\Theta = \delta(\Delta n\, L)/(nd) \qquad (2.5)$$

where Δn is the refractive index change due to EO effect, L is the electrode length, d is the distance between adjacent waveguides at the interface and $\delta(\Delta n\, L)$ denotes the difference between adjacent waveguides in the value of $\Delta n\, L$. If the electrode length is the same for all the waveguides, we get $\Theta = \delta\Delta n\, L/(nd)$. Different voltages are applied to electrodes to attain a refractive index change difference of $\delta\Delta n$ between waveguides. If electrodes have different lengths, we get $\Theta = \Delta n\, \delta L/(nd)$ where δL denotes the electrode length difference. In this case, all electrodes are driven by the same voltage to generate a constant refractive index change Δn. The lightwave at the output waveguide is not identical to the collimated light beam of the prism electrode device, since the wavefront at the output planar waveguide is constructed from an ensemble of lightwaves from waveguides.

A strictly non-blocking NxN device is constructed by placing N deflectors (selectors) at the input side and N deflectors (selectors) at the output side [23], as shown in Figure 2-4(a). N denotes the number of input/output ports of the switching system. Each deflector has the structure shown in Figure 2-6. Any input deflector drives the light beam to a desired output deflector. The output deflector delivers it to the output waveguide. An input wave front to a deflector is aligned into a wave front that focuses the light into the waveguide by the phaser.

A 4x4 device of this kind was demonstrated in LiNbO$_3$ [23]. The device consists of 4 deflectors at the input and output ports respectively. Each deflector is composed of a 4-channel waveguide array, electrodes with different lengths and a coupler to feed the input light into the waveguide array. The length of the interconnection planar waveguide is 14 mm and the length of the electrode is 7.5 mm. The total device length is 45 mm. The required drive voltage was less than 26 V for 1.3 µm extraordinary (TM mode) light. The crosstalk was −14 to −21 dB. The deflector type switch is an analog control device in which the deflection angle is proportional to drive voltage. The deflection angle must be set with high precision for large channel numbers and the drive voltage drift should be suppressed. Three-dimensional beam steering scanning both x and y directions in the free space is also possible using an arrayed waveguide device [24].

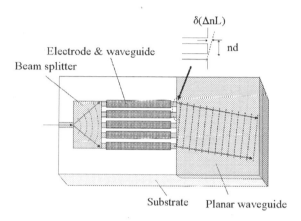

Figure 2-6. Waveguide-array deflector

2.2.2   Diffraction-Type Implementations

The EO effect can also be used for diffracting the light. An EO grating can be generated by an interdigital or comb shaped electrode. The period of the grating is defined by the period of the electrodes. The grating period cannot be changed and usually tuning of the diffraction angle is difficult. Devices of this kind are used as 1x2 or 2x2 crosspoint switching elements for multi-stage switching networks [25]. Domain reversal technology is used to generate EO grating with a strip shaped single electrode [26]. A first-order diffraction efficiency of 76% was attained at 25 V and 0.633-µm wavelength. The Bragg angle was 3.5 mrad (0.2°) and substrate thickness (electrode gap) was 300 µm. The diffraction angle is determined by the Bragg condition

$$\Theta_B = \sin^{-1}(\lambda/2\Lambda) \tag{2.6}$$

with Λ and λ being the wavelengths of the grating and the light beam respectively.

2.3   CHANNEL WAVEGUIDE DEVICES

Many types of optical switching elements based on guided-wave phenomena in channel waveguides have been studied. Most of these devices are 2x2, 1x2 or 2x1 switching elements. For the 2x2 optical switching elements two switching states exist, namely the *bar* state, where parallel paths connect the input and output ports, and the *cross* state, where crossed paths connect the inputs and outputs (Figure 2-7).

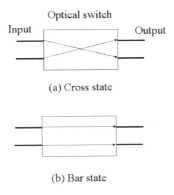

(a) Cross state

(b) Bar state

Figure 2-7. The cross and bar switching states

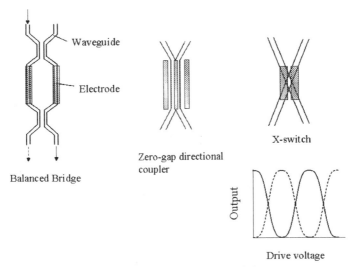

Figure 2-8. Interferometric switches

Normal modes are excited in the optical guided-wave system, similar to microwave guides. Normal modes can be distinguished by the field distribution type at the waveguide cross section. An optical field in a waveguide system is represented by the super position of normal modes.

Optical switching elements can be classified according to the type of normal mode phenomena they use. The phenomena used for switching are (a) propagation constant change, (b) field distribution change and (c) mode conversion. An example of normal modes is shown in section 2.3.1. Switching devices can be classified according to the number of modes involved. Typical examples of switching elements using single or double normal modes are shown in Figures 2-8, which depicts some *interferometric devices*, and Figure 2-9, which shows some *digital devices*. We discuss this classification below.

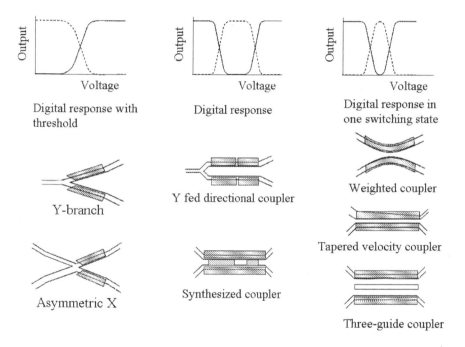

Figure 2-9. Various digital EO switch configurations

The most known switching device using single normal mode is the Y-branch optical switch. The device operates using the deformation of the optical field by changing the refractive index difference between the waveguides forming the branch. Devices using double normal modes are represented by the directional coupler, balanced bridge (Mach-Zehnder) switch and asymmetric X-branch (digital) switch. The directional coupler uses propagation constant change due to induced refractive index difference between waveguides. The balanced bridge uses normal mode conversion for switching by changing the refractive index difference between two light paths. The asymmetrical X-branch uses deformation of the optical field with changing of the refractive index difference between waveguides forming a branch similar to that of the Y-branch switch. Devices known to use more than three normal modes include the multiple waveguide directional coupler, Multi-Mode Interference (MMI) coupler, total internal reflection (TIR) and multi-leg Mach Zehnder devices.

Devices can be classified into interferometric devices and digital devices based on their switching characteristics, or switching curve. The switching curve is a plot of the optical output power (vertical axis) versus voltage, or current, (horizontal axis). The switching voltages can take numerous values in an interferometer curve (Figure 2-8). Switching states are attained at phase conditions separated by 2π. In a digital curve (Figure 2-9), the switching state may change to another state and maintain the latter above certain threshold voltage. The switching voltage range is wide for the two states (cross and bar). A switching curve that exhibits wide switching voltage ranges only for one state is also described sometimes as digital.

Digital switches are favorable in many applications due to their immunity to drive voltage variation and drift. The most basic structure for the digital optical switch is the 1x2 Y-branch optical switch. A 2x2 switching element can be constructed by connecting Y-branch switches (Asymmetric X), as shown in Figure 2-9. The Y-branch optical switch uses optical field deformation caused by refractive index changes among two waveguides. Most of the power of the fundamental mode is confined to the waveguide with higher refractive index. A digital optical switch tends to require a large refractive index change (large voltage) for switching. Design methods to reduce drive voltage have been proposed for the Y-branch switch.

In the beginning of this field of research, the directional coupler type switch was studied most because of its simple structure and low loss characteristics. In recent years, there is emphasis on digital optical switches. Each device type, with some if its varieties, is described below in greater detail. However, we start by discussing normal modes.

2.3.1 Normal Modes

Lightwave propagation in a single mode waveguide pair placed closely to each other can be described by the following coupling Equations [27]:

$$da_1/dz = j\delta a_1 - j\kappa a_2$$
$$da_2/dz = -j\delta a_2 - j\kappa a_1$$

$$(2.7)$$

Here, a_1 and a_2 are the field amplitudes in the two waveguides, and $\delta = \Delta\beta/2$ where $\Delta\beta$ is the propagation constant difference between the two waveguides. κ is the coupling coefficient between them and $-j\kappa$ is selected to satisfy the total optical power conservation condition. z is the propagation direction. When $\kappa = 0$, there is no coupling and any two lightwaves propagate independently along the two waveguides with relative propagation constants δ and $-\delta$.

Next we consider the combination of the two optical fields, which is expressed as $f_1 a_1 + f_2 a_2$. From Equation (2.7) we get:

$$d(a_1+a_2)/dz = j\delta(a_1-a_2) - j\kappa(a_1+a_2)$$
$$= j\Delta(a_1+a_2)$$

(2.8)

This is obtained assuming that

$$a_1+a_2 = (a_{10}+a_{20})\exp(j\Delta z) \qquad (2.9)$$

which is the optical field with propagation constant Δ. Constants a_1, a_2 and Δ are calculated by equating the right hand sides of Equation (2.8) and by the normalization $a_{10}^2 + a_{20}^2 = 1$, which gives:

$$a_{10} = (\kappa/|\Delta|)/[2(1-\delta/\Delta)]^{1/2}$$
$$a_{20} = -(\kappa/\Delta)/[2(1+\delta/\Delta)]^{1/2}$$
$$\Delta = \pm(\kappa^2+\delta^2)^{1/2}$$

(2.10)

There are two modes of propagation, one with the propagation constant $\Delta = -(\kappa^2+\delta^2)^{1/2}$ and the other with $\Delta = (\kappa^2+\delta^2)^{1/2}$ exhibiting symmetrical and asymmetrical fields, respectively. The optical field in the waveguide with higher refractive index is larger in the symmetric mode and smaller in the asymmetric mode. Refractive index change induced by the EO effect can control the compositions of the optical power in the waveguides. When the symmetric and asymmetric modes are both excited, the field amplitude in the first (f_1) and second (f_2) waveguide becomes

$$f_1 = f_S a_1^- + f_A a_1^+$$
$$f_2 = f_S a_2^- + f_A a_2^+$$

(2.11)

where a_1^+ and a_1^- are the asymmetric and symmetric mode amplitudes in waveguide 1 respectively. Similarly, a_2^+ and a_2^- are the asymmetric and symmetric mode amplitudes in waveguide 2 respectively. Thus, $a_i^+ = a_{i0}^+ \exp(j\Delta^+ z)$ and $a_i^- = a_{i0}^- \exp(j\Delta^- z)$, where $i = 1,2$ and Δ^+ and Δ^- denote the positive and negative values of Δ respectively. The mode excitation ratios, f_S

for symmetric and f_A for asymmetric modes, are calculated from Equation (2.11) as follows:

$$f_S = -f_1a_2^+ + f_2a_1^+$$
$$f_A = f_1a_2^- - f_2a_1^-$$

(2.12)

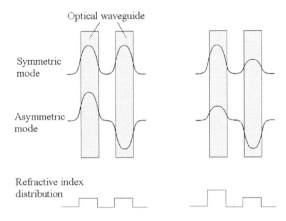

Figure 2-10. Normal modes

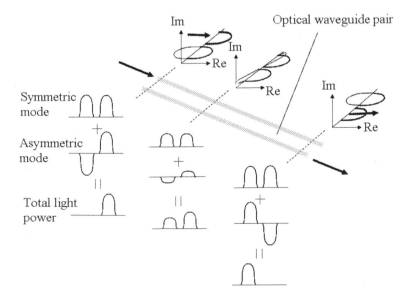

Figure 2-11. Propagation of normal modes

Both the symmetric and asymmetric modes are excited when light is launched into one waveguide ($f_S = -a_{20}{}^+$, $f_A = a_{20}{}^-$ from Equation (2-12) when $f_1 = 1$ and $f_2 = 0$). Optical power diminishes in a waveguide where the signs of field amplitudes of the two modes are apposite. The amplitude in the waveguides changes according to the phase difference between the two modes.

The interference between modes determines the output of the waveguide pair. In Figures 2-10 and 2-11, optical mode field $A_m(x)\exp(j\Phi_m)$ is depicted as a complex variable where $A_m(x)$ is the amplitude distribution along lateral direction x and $\Phi_m = \beta_m z$ is the phase of mode number m ($A_m(x)$: $a_{10}{}^\pm + a_{20}{}^\pm$, and $\Phi_m = \beta_m z = \Delta^\pm z$ in Equation (2-9)). The real part $A_m(x)\cos(\Phi_m)$ is projected on a horizontal axis and the imaginary part $jA_m(x)\sin(\Phi_m)$ on a vertical axis and the phase of the symmetric mode is used as reference. The phase difference between modes accumulates along the propagation distance z. The electric field of the lightwave is given by the real part of $A_m(x)\exp[j(\Phi_m - \omega t)]$ summed over mode number m where ω is the angular frequency of light. The optical power distribution along the x axis is given by its absolute value.

2.3.2 Directional Coupler Devices

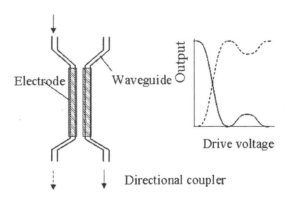

Figure 2-12. Directional coupler EO switch

This device is composed of two waveguides placed close to each other (Figure 2-12). When the distance between the two waveguides is small enough, light propagating in one of them is coupled into the other. If the two waveguides are identical, all the optical power launched into one waveguide is transferred to the other one after some propagation distance, which is known as the *coupling length* $L_c = \pi/(2\kappa)$. After the power is completely transferred, it is coupled back to the original waveguide, by the same process, at twice the coupling length. Thus, the optical power can be transferred periodically from one waveguide to the other and vice versa. If the stretch of the proximity region of the two waveguides is set to one coupling length, the optical signal fed into the input of one of the waveguides will be ejected from the output of the other waveguide. This is the cross state of the switching device. By applying voltage across one waveguide, a refractive index difference is induced by virtue of the EO effect which reduces the coupling among the two waveguides. For certain index change, coupling almost diminishes and optical power launched into a waveguide remains in the same waveguide. This is the bar state of the switch.

The operation principle of the EO directional-coupler switch can be described by the normal mode representation (Figure 2-11). Two normal modes are excited in the two-waveguide system. They are symmetrical and asymmetrical modes. The interference between these two modes as they propagate along the waveguide system results in periodical power transfer between the two waveguides. The optical field in one waveguide is either weakened or strengthened depending on mode polarities.

When light is fed into a waveguide, the two modes are excited at the input. After the light has propagated a distance equal to the coupling length, the phase difference between the two modes becomes π and the optical field in the opposite waveguide is extinguished (cross state). The coupling length is calculated from $(\Delta^+ - \Delta^-)L_c = \pi$ with $\delta = 0$. Applying voltage to the electrode changes the propagation constants as well as the optical field distributions of the normal modes. The coupling length changes (shortens) with voltage. The bar state is achieved when the device coupling region is twice the coupling length.

The bar state is attained when $(\Delta^+ - \Delta^-)z = 2v\pi$. Thus,

$$[\pi z/(2L_c)]^2 + (\delta z)^2 = (v\pi)^2 \tag{2.13}$$

where v is an integer, z the length of coupling region and 2δ the propagation constant difference between waveguides. The cross state is attained when $(\Delta^+ - \Delta^-)z = (2v+1)\pi$ and $\delta = 0$. By plotting a switching diagram with z/L_c being the vertical axis and $2\delta z/\pi$ the horizontal axis, the cross states become isolated points

$$z/L_c = -1 + 2v \tag{2.14}$$

on the vertical axis and the bar states become arcs with their radius equal to $2v$. The coupling coefficient is a function of waveguide separation g and is given approximately by:

$$\kappa = \kappa_0 \exp(-\gamma g) \tag{2.15}$$

where κ_0 and γ are constants [28] which can be obtained experimentally by measurement of coupling lengths in directional couplers with several gap values. The coupling coefficient κ can be a function of propagation distance $\kappa(z)$ and can vary along this distance z by changing g. Varying the coupling coefficient along the propagation distance (weighted coupling) can modify the switching curve of the directional coupler [29]-[30]. The sidelobe of the switching curve conspicuous in a parallel waveguide directional coupler can be decreased by introducing a structure where waveguides are brought close to each other gradually from device terminals towards its center (shown in Figure 2-9). Both the δ range and the drive-voltage range required to attain low crosstalk in the bar state become larger.

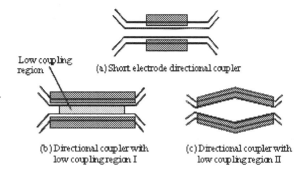

Figure 2-13. Directional couplers with variations

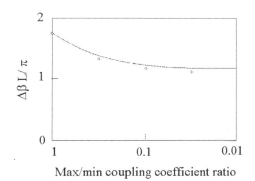

Max/min coupling coefficient ratio

Figure 2-14. Directional coupler with reduced drive voltage

The δ change required for switching in a parallel waveguide device with $z/L_c = 1$ is given by $2\delta L_c = 3^{1/2}\pi$. The electrode length L_e is the same as the coupling region length L_c. Using an electrode shorter than the coupling region, as in Figure 2-13(a), can lower the δL_e value required for switching [31]. When $L_e/L_c \ll 1$, $2\delta L_e$ is almost equal to π. A voltage-length product of 6.5 Vcm was attained at 1.56 μm wavelength TM mode (extraordinary light).

The structure requires an additional coupling region besides the electrode section. Placing low coupling strength sections in the middle of the coupling region can also lower the δL_e value required for switching, as shown in Figures 2-13(b) and 2-13(c) [32]. The structure does not require an additional coupling region. Indeed, increasing the coupling strength at this section minimizes the length of the required device coupling region. In an experiment, the low coupling region was implemented by having a large separation between the waveguides. Lowering of the switching voltage (dots in Figure 2-14) was observed in the measurement. A voltage-length product of 5 V.cm was attained at 1.3 μm wavelength extraordinary light.

2.3.3 The Reversed $\Delta\beta$ Coupler

In the directional coupler switch, the coupling region length should be exactly equal to the coupling length to attain low crosstalk in the cross state. A split electrode structure was proposed to attain low crosstalk cross-state as shown in Figure 2-15 [33]-[34]. This configuration is known as the reversed $\Delta\beta$ coupler. The adjacent electrode pairs are driven by opposite polarity voltages. The cross state is attained when optical power is excited equally in two waveguides in the middle of the reversed $\Delta\beta$ configuration. Under this condition the symmetric mode in the first half section is converted to asymmetric mode in the second half section and the asymmetric mode in the first half section is converted to the symmetric mode in the second half. By exchanging modes in the middle, the phase difference between accumulated modes in the first half is compensated in the second half section. The phase difference at the input is thereby restored. Since the waveguide systems of the first and second half sections are mirror symmetrical to each other at the center, the cross state is attained at the end of the device. The bar state is achieved when each half section is in the cross state or the bar state.

With L being the coupling region length, the cross state is achieved when

$$\sin^2\{\pi[(z/L_c)^2+(2z\delta/\pi)^2]^{1/2}/4\} = [1+(2z\delta/\pi)^2]\sin^2(\pi/4) \qquad (2.16)$$

This is obtained from $|f_2|^2=1/2$ (or $|f_1|^2=1/2$) in the middle of the device under input condition $f_1=1$ and $f_2=0$ at $z=0$. The bar state is achieved when

$$z/L_c=4v-2 \qquad (2.17)$$

or

$$(z/L_c)^2+(2z\delta/\pi)^2=(4v)^2 \qquad (2.18)$$

using the results of the uniform $\Delta\beta$ directional coupler. The cross state is attained by suitable δ value (suitable drive voltage). The voltage-length product depends on the ratio z/L_c but is typically about 10 Vcm at 1.55 μm wavelength extraordinary light. Using domain reversal, simple electrode structure can be used to attain $\Delta\beta$ reversal [35].

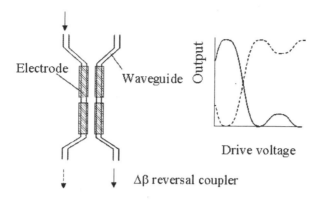

Figure 2-15. The Reversed $\Delta\beta$ coupler

The δL_e value required for switching can be lowered by the same scheme used for conventional directional coupler switches. By cascading couplers with low coupling strength at the middle of coupling region, the δL_e value required for the cross and bar states at each electrode pair can be lowered [36]. A device with digital switching curve can be attained by $\Delta\beta$ reversal [36]-[37]. The range of δ for the cross state is wide around $z/L_c=1+2v$ where $\delta= 0$. In one method, the range of δ attaining the bar state can be widened by introducing the weighted coupling for each electrode section. In another method, directional couplers with low coupling strength in the middle of the coupling region are connected in a series. This is depicted in the synthesized coupler in Figure 2-9 [36].

2.3.4 Zero-Gap Directional Coupler and Crossing-Channel Device

This configuration is depicted in Figure 2-8. As the gap between the coupler waveguides is reduced, the symmetric normal mode field between them increases while the asymmetric normal mode field remains the same. The change in propagation constant difference between normal modes due to refractive index change in the waveguide gap increases as the gap is decreased. By placing the electrode between waveguide centers, the coupling length can be changed by the applied voltage, due to the EO effect. If the coupling distance is an odd or even multiple of the coupling length, the device is in the cross or bar state.

A zero gap directional coupler that consists of a waveguide sustaining two normal modes is reported in [38]-[39]. Two branching waveguides are connected to the end of the two-mode waveguide thereby splitting the optical power into two parts. The device was originally named Birfucation Optique Active (BOA) [40].

The same switching principle is also attained using crossing channel waveguides [41]-[42]. These devices exhibit voltage-length product of about 10 Vcm at 1.3-μm wavelength extraordinary light.

2.3.5 Mach-Zehnder (Balanced Bridge) Devices

The principle of this device is depicted in Figure 2-16. By connecting directional couplers with length $L_c/2$ (3-dB coupler) via phase modulator an optical switch is implemented [43]-[44]. The phase change required for switching at the phase shifter is smaller than that for a directional coupler. The switching principle can be viewed as a process of mode exchange between symmetric and asymmetric modes at the phase shifter. In the figure, optical mode field $A_m(x)exp(j\Phi_m)$ is depicted as a complex variable where $A_m(x)$ is the amplitude distribution along lateral direction x and $\Phi_m = \beta_m z$ is the phase of mode number m. The real part $A_m(x)cos(\Phi_m)$ is projected on the horizontal axis and the imaginary part $jA_m(x)sin(\Phi_m)$ on the vertical axis. The phase of the symmetric mode is used as reference. The phase difference between modes accumulates along propagation distance z. The electric field of the lightwave is given by the real part of the sum $A_m(x)exp[j(\Phi_m - \omega t)]$ over mode number m where ω is the angular frequency of light.

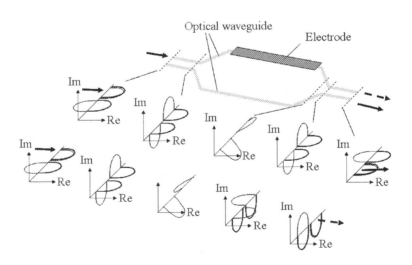

Figure 2-16. The Mach-Zehnder principle

When optical power is injected into one of its input ports, symmetric and asymmetric modes are excited in the 3-dB coupler. As light propagates through the coupler, the phase difference between modes accumulates. At the end of the input 3-dB coupler there is $\pi/2$ phase difference between modes. At the output 3-dB coupler the phase difference is further increased to π and the cross state is attained. When a π phase shift is generated by applied voltage, symmetric and asymmetric modes are exchanged before the output 3-dB coupler. The phase difference between modes vanishes at the output port resulting in the bar state.

The coupling between the waveguide pair at the phase shifter (interferometer arm with electrodes) should be sufficiently small to attain pure interferometeric response. The coupling between waveguides can be reduced with wide waveguide gap. In the basic structure, the waveguide curve structure is used to connect the narrow gap coupler and the wide gap interferometer arm sections. An etched groove between waveguides [45] or waveguides with different propagation constants [46]-[48] can be used to avoid coupling between waveguides and eliminate waveguide curve sections. The voltage-length product is typically 10-15 Vcm at 1.3-1.55 µm wavelength extraordinary light.

2.3.6 The Y-Branch Optical Switch

This device configuration is shown in Figure 2-9. At small waveguide separation, the symmetric mode optical field is almost equal in both waveguides. If voltage is applied, the symmetric mode field is concentrated in the waveguide with higher refractive index. This enables 1x2 switching functionality. The optical power of the asymmetric mode dominates the waveguide with lower refractive index.

The optical field changes its shape adiabatically as light propagates through the waveguide branch if the waveguide separation is increased gradually. The waveguide branch angle should be sufficiently small to ensure adiabatic conditions in order to suppress mode conversion between symmetric and asymmetric modes. If mode conversion occurs the extinction ratio deteriorates. The condition for adequate operation is given by [49]:

$$\Delta\beta/(\Theta\gamma) > 0.43 \tag{2.19}$$

where Θ is the branching angle and γ is a constant (recall Equation (2-15)). A small Θ (typically smaller than 1 mrad) and large $\Delta\beta$ are required. The condition shows that above some threshold of $\Delta\beta$ the switch remains in the same switching state. A digital switching curve is obtained.

When light is launched in the other direction into one of the two waveguide branches, the inverse process can take place. Symmetric normal mode is excited at the tail of the Y branch when light is launched into the waveguide with higher refractive index. Asymmetric normal mode is excited when light is launched into the waveguide with lower refractive index. The width difference between the two branches has an effect similar to refractive index difference. A 2x2 switch is implemented by connecting branching waveguide with different widths to branching waveguides with electrodes (the Asymmetric X switch shown in Figure 2-9) [50].

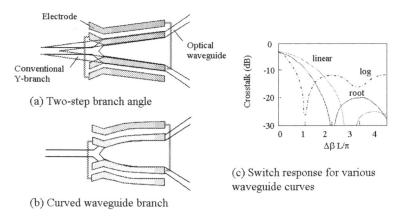

(a) Two-step branch angle

(b) Curved waveguide branch

(c) Switch response for various waveguide curves

Figure 2-17. Y-branch coupler with reduced drive voltage

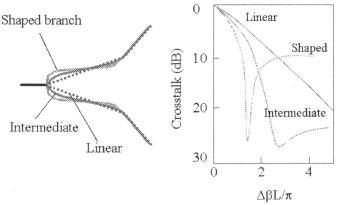

Figure 2-18. Shaped Y-branch

The $\Delta\beta$ value required for switching tends to be high in the Y-branch optical switch. There are several methods to relax this requirement however (Figures 2-17 and 2-18). In the first method [51] the small angle branching waveguide begins with a non-zero waveguide gap at the start and is connected to a branching or tapered waveguide with a higher branching angle to excite symmetric mode at the start of the branch. This structure is shown in Figure 2-17(a). The length of the device becomes shorter than the conventional structure by removing the narrow gap section. The effect of the narrow gap section on switching performance is relatively small and can be replaced by a branch with larger divergence angle. The structure is a Y-fed directional coupler with weighted coupling [52].

In the second method, a curved waveguide structure is used as depicted in Figure 2-17(b) [53]. The maximum mode conversion rate can be estimated by controlling the ratio of Δ over K, where K is the mode conversion coupling coefficient between symmetric and asymmetric modes. Δ (depends on $\Delta\beta$) is the propagation coefficient difference between these two modes. The mode conversion is large when K is large and Δ is small. The coupling coefficient is proportional to branching angle Θ and can be denoted as $K=C\Theta$, where C is a constant. The coupling coefficient K is a function of the gap between waveguides and peaks at a certain gap value. In a conventional constant branch structure, the mode conversion takes place at wide gap portion of the branch for low Δ (low applied voltage) and the narrow gap portion with large Δ (high applied voltage). The ratio of Δ over K increases as Δ is increased indicating that mode conversion becomes low at large Δ. This is the reason why the crosstalk is decreased at large Δ or high-applied voltage.

Since K is proportional to the local branching angle Θ, the crosstalk is expected to be less at low Δ when Θ is small at the wide gap region of the branch. This leads to a curved waveguide branch structure as shown in Figure 2-17(b). Calculation shows that Δ (or drive voltage) corresponding to -15 dB crosstalk can be reduced by bending the waveguide (Figure 2-17(c)). About 30 % reduction of the drive voltage is reported by using this method leading to 20-V polarization independent drive voltage for a 1-cm long electrode. Applied voltage of 5 V was required for switching only the TM mode at 1.3-μm wavelength (-15 dB crosstalk).

A waveguide curve can be generated for certain $\Delta\beta$ yielding the shortest length for a given conversion magnitude (shaped branch). The local branching angle is changed so that the coupling coefficient between normal modes is the same everywhere in the branch for a given $\Delta\beta$. The crosstalk is lowest at this value of $\Delta\beta$ and increases at larger values. By optimizing $\Delta\beta$, low switching voltage is possible but at the expense of non-digital switching response. Another structure that exhibits both a reduced switching $\Delta\beta$ and low crosstalk is depicted in Figure 2-18 (intermediate) [54].

Some designs can yield devices with crosstalk under -40 dB [55]. In the beginning of the Y-branch, where the gap is small, lateral Ti diffusion results in non-optimized refractive index distribution which generates mode conversion and results in crosstalk. The gap in this region can be extended to obtain ideal Ti distribution and improved performance. Two configurations were used to decrease crosstalk in the Y-branch EO switch. These are the redundant switch and the W-branch configurations (Figure 2-19). In the redundant switch configuration, the extra branching is utilized to route crosstalk to dummy ports. The W-branch configuration also enables the selection of an output port where crosstalk has been rejected [56]. However, better crosstalk performance is easier to attain with the redundant switch than with the W-branch. In the Y-branch EO switch, the crosstalk can be expressed as $X_{talk} = \exp(-constant.(VL))$ where V is the voltage and L is the electrode length. For a cascaded structure with the same total length, the crosstalk becomes $X_{talk2} = \exp(-2.constant.(VL/2))$. The voltage-length product to attain the same level of crosstalk is unchanged. However, the crosstalk floor in this later case is reduced by a large degree. Crosstalk less than -50 dB is reported using this scheme [55].

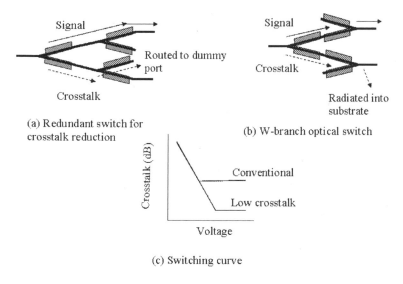

Figure 2-19. Crosstalk reduction in Y-branch switches

2.3.7   Tapered Velocity Coupler

This configuration is shown in Figure 2-9. When the width of the waveguide is gradually changed along the propagation distance the normal mode field changes its shape. At the input, the symmetrical or asymmetrical normal mode is excited when light is launched into one waveguides. At zero voltage, the light power is transferred to the other waveguide at the point where the widths of the two waveguides are equal. In this case, the output power emerge from the other waveguide. Under applied voltage, the refractive index of a waveguide can be made to be always higher or smaller than that of the other waveguide. In this case, light fed into one of the waveguides stay in the same waveguide and the bar state takes place. A wide-sense lithium-niobate digital switch was demonstrated based on this principle [57]. This device exhibits voltage-length products of 7.2 Vcm for the TM mode and 24 Vcm for the TE mode and −15-dB crosstalk at 1.32-μm wavelength.

2.3.8 Multi-Mode Electro-Optic Switching Devices

Although two normal modes are sufficient to implement 1x2 and 2x2 EO switching elements, improvements are possible if more waveguide modes are utilized. Some examples are described in the following.

2.3.8.1 Directional Coupler

Directional coupler switches can be realized with more than two waveguides. The number of normal modes in this case is equal to the number of waveguides. However, the switch operation becomes more complex than a device with two waveguides [58]-[59]. The simplest form of this category of EO switches is a device using three waveguides and having three normal modes. When light is launched into a waveguide, the three modes are excited with different propagation constants. As light propagates through the coupler, the phase relations between normal modes change due to propagation-constant differences. If the differences between the fundamental and first order mode is the same as between first-order mode and second-order modes, the field distribution can become the mirror symmetry of the input light at some propagation distance. A cross state is attained at this distance. If voltage is applied, the refractive index change alter the propagation constant difference turning the switch to the bar state [60].

A polarization independent three-waveguide LiNbO$_3$ coupler switch was demonstrated [61]. The device tolerates variations in length and voltage that are larger than those of its two-waveguide counterparts. Other studies have reported using large numbers of coupled waveguides [62].

2.3.8.2 Multi-Mode Interference Coupler

This device is shown in Figure 2-20(a). It uses the interference of normal modes in multi-mode waveguides. An interferometric optical switch similar to the zero-gap directional coupler can be implemented by placing a thin electrode in the middle of the multi-mode waveguide [63]. For a multi-mode waveguide with width W, thickness b and refractive index n, the propagation constants of the low-order normal modes are approximated by:

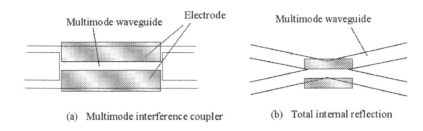

(a) Multimode interference coupler   (b) Total internal reflection

Figure 2-20. Multi-mode EO switching devices

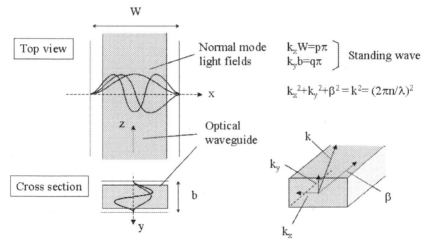

Figure 2-21. Multi-mode waveguide modeling

$$\beta = [(2\pi n/\lambda)^2 - (\pi p/W)^2 - (\pi q/b)^2]^{1/2} \qquad (2.22)$$

with p and q being the mode numbers (Figure 2-21). In the simplest case, where the waveguide is single mode in depth (q= 1), the phase difference between modes at distance z is approximated by $\Delta\beta_{ij}z = [\pi\lambda z/(4W^2 n)](p_i^2 - p_j^2)$ for modes far from cutoff. Between adjacent modes (p_i= p+1, p_j=p), we have:

$$\Delta\beta_{m,m+1} z = [\pi\lambda z/(4W^2 n)](2p+1) \qquad (2.23)$$

so that at $z=4nW^2/\lambda$ the phase difference between symmetric and asymmetric mode is $(2p+1)\pi$. At this distance the field distribution of the asymmetric mode changes its sign compared to the symmetric mode leading to the cross state. The symmetric mode propagation constant is changed by applying the voltage. A phase shift of π can reverse the phase relation between symmetric and asymmetric modes and attain the bar state. The switching function is attainable for devices with input and output waveguides that are multimodal both in width and depth.

2.3.8.3   Total Internal Reflection

This device is composed of crossing waveguides with electrodes placed at their intersection as depicted in Figure 2-20(b). When the voltage-induced refractive index difference is large enough, light is totally reflected at the "boundary" of the low refractive index region (the region under voltage). The switching curve characteristics vary depending on design [64]. When the crossing angle is small and when the waveguide supports a few modes, the switching curve has sinusoidal interferometric-type characteristics. This is attained when the crossing angle is smaller than the value corresponding to a phase shift of π between the fundamental (m= 0) and first order (m= 1) modes. When the angle is large and with highly multimodal waveguide, the curve exhibits a threshold at total internal reflection (TIR) and digital switching become attainable. A large refractive index change is required in this case.

The device has been demonstrated using LiNbO$_3$ at 0.6328 μm wavelength with 50 V drive voltage [65]. At the 1.3 μm wavelength, 60-V drive voltage is reported [66]. Similar devices can be constructed using a MMI coupler with an electrode placed along half side of the coupler.

2.3.9 Polarization Independent Structures

In optical networks, the polarization state of lightwaves changes randomly due to several perturbations. Therefore, it is important that that devices inserted into the optical-signal transmission path be as polarization insensitive as possible. Electro-optic materials have different properties in this regard. In the LiNbO$_3$ crystal, the EO coefficient r$_{33}$ for light polarized in parallel to the ferroelectric dipole axis (c-axis) is several times larger than the coefficient r$_{13}$ for light polarized perpendicular to this axis. The crystal is birefringent; meaning that the refractive index also changes with the polarization of light.

Many methods have been proposed to overcome polarization sensitivity as shown in Figure 2-22. These methods can be classified according to the EO coefficient(s) utilized and according to switch operation. An optical signal may be split into two orthogonal polarization components at the input. These components are treated separately and combined at the output of the device [67]-[68]. Polarization beam splitters (PBS) are often used in this respect. In another method cascaded devices each handling a separate polarization component are used [69]. Polarization conversion of r_{51} can be utilized as well [70]. This principle can be applied to the directional coupler, reversed $\Delta\beta$ coupler, and Mach-Zehnder type EO switches [71].

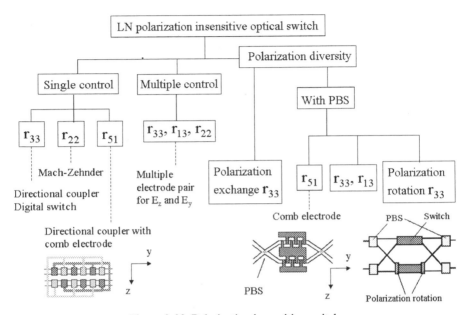

Figure 2-22. Polarization insensitive switches

Two methods are usually used overcome polarization sensitivity. The first one is common in Mach-Zehnder devices using x-cut substrate and light propagating along the z-axis [72]-[73]. The device uses the r_{22} EO coefficient that is one third or less of the largest r_{33}. The magnitude of the EO coefficients becomes equal for both orthogonal polarizations in this configuration (with different polarities). The refractive indexes are also almost made equal and a polarization-independent device can be obtained. The second method is common in digital switching devices with z-cut substrate where waveguides having good light confinement can be fabricated easily [50], [74]. The largest coefficient r_{33} and second largest r_{13} are used. Since the magnitude of r_{13} is at least three times smaller than r_{33}, digital switching (Figure 2-9) is required for both polarizations. The large voltage range of the digital switch enables the device to optically switch different polarizations with different EO coefficients simultaneously.

Insertion loss and dispersion in optical communication systems should also be polarization independent. Polarization exchange across optical devices can average out polarization dependent characteristics for two input orthogonal polarizations [69], [83]. Figure 2-23 describes this scheme in a simplified manner.

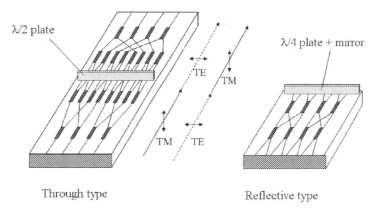

Through type   Reflective type

Figure 2-23. Polarization exchange schemes

2.4 OTHER COMPONENTS OF ELECTRO-OPTIC SWITCHING

To construct a guided-wave optical switch *matrix* (or *array*), components other than optical switching elements are required. The optical switching elements are connected by curved waveguides to minimize the interconnection length. In many architectures, waveguide intersections are necessary. A spot size converter is sometimes integrated at the interface between the fiber and the waveguide to minimize insertion loss due to mode size mismatch [75].

Curved waveguide are used to guide light to the desired directions [76]. There are two causes of optical power loss in curved waveguides. One is a radiation loss of light into the substrate. The other cause is mode mismatch between sections with different radiuses of curvature. The radius of the curvature of waveguides must be sufficiently large to prevent radiation loss. Methods have been developed to modify the equivalent index distribution across the waveguide width [77] and to lower the refractive index at the outer rim of the waveguide [78]. These methods can increase the index barrier between the waveguide outer rim and the substrate. In the low diffusion speed x-cut LiNbO$_3$ substrate, a two-step diffusion process is adopted to attain high index contrast [79]. First, a thin titanium waveguide pattern film is diffused. Then, a second (similar) film is aligned and deposited onto the first waveguide pattern and this is followed by a diffusion process. The procedure avoids creation of rough surfaces which may result from diffusion of a thick titanium film into x-cut substrate. A curved waveguide with small radius of curvature using highest index contrast waveguide in LiNbO$_3$ is also obtained using the proton exchange technique

[80]. Special curves with no discontinuity in the radius of the curvature are possible [81]-[82].

2.5   ELECTRO-OPTIC SWITCHING FABRICS

Switching fabrics are multi-port switching blocks which maybe constructed using any of the switching elements described so far. A Fabric can be made in the form of a single matrix/array or multiple matrices which are arranged into switching stages. An EO switching matrix can be fabricated into a chip by integrating composite switching elements and interconnecting them via waveguides.

Optical switching fabrics can be classified according to their blocking characteristics, or their ability to establish lightpaths between their input and output ports. The subjects of switch fabric architectures and their blocking characteristics is discussed in detail in chapter 9 and in other chapters within Part III of this book. In the following, we only discuss this chapter briefly and highlight some fabric architecture issues that are pertinent to lithium-niobate electro-optic switching and to several experiments and field trials in this regards.

Strictly non-blocking fabrics can establish a path in between any pair of input and output ports in any order. In a *wide-sense non-blocking* switching fabric, these paths can be established only by a specialized switching algorithm. In a *rearrangeably non-blocking* fabric, all new paths can be established by rearranging already existing ones. Finally, in a *blocking* fabric some connection requests can be denied because new paths are not available.

Large non-blocking LN EO switching architectures tend to be complex and difficult to integrate on one chip. Crosstalk figures below –40 dB are required in many cases, especially in Wavelength Division Multiplexing (WDM) networks. EO switching elements are often used in pairs to attain crosstalk levels that are not possible with single elements. The additional switch is used to reject or bypass crosstalk. Architectures with fewer switching stages and fewer waveguide crossings are advantageous for their low insertion loss and small size.

The largest non-blocking LN matrix fabricated so far is the 16x16 tree structure [83]. The most complex LN switching system, integrating many components other than the switching elements, was reported within the MONET project (see chapter 11). Table 2-1 summarizes a selection of EO LN switching fabrics that are reported so far in literature. Typically, the voltage-length product is 5-10 Vcm at 1.3-1.55 µm wavelength with polarization along the largest EO coefficient. The insertion loss ranges from 5 to 15 dB depending on fabric size. Typically the crosstalk is better than -15 dB, but figures that are lower than –40 dB are attainable.

Blocking	Rearrangeable Nonblocking	Strictly Nonblocking
Banyan 16x16 (Δβ) [92] banyan 32x32 (MZ) [84]	Benes 16x16 (Δβ) [85] dilated Benes 8x8 (Δβ) [98] dilated Benes 16x16 (Δβ) [86] N-stage 4x4 (Δβ) [97] **Customized tree 8x8 (Y) [99]	crossbar 4x4 (Br) [25], (Δβ) [87], (Δβ) [100],(Δβ) [88], (X) [101] crossbar 8x8 (Δβ) [89] PI-LOSS 4x4 (Δβ) [111]
1xN switch		**tree 4x4 (Y) [74]
1x4 (Δβ) [93] 1x16 (MZ) [94] *1x16 (DC) [96] **1x16 (Y) [95] **1x32 (Y) [95]		**tree 8x8 (Y) [90] **tree 16x16 (Y) [83], [102] *simplified tree 4x4 (Y, AX) [103] **simplified tree 8x8 (Y, AX) [91] **simplified tree 8x8 (MZ) [73], (DC)[104]

Table 2-1. LN switch fabric demonstrations (Br: Bragg, MZ: Mach-Zehnder, Δβ: Reversed Δβ, DC: Directional coupler, Y: Y-branch, AX: Asymmetric X-branch, X: X-switch, *Polarization independent, **Digital response and, Polarization independent)

2.5.1 Single Chip Architectures

In the following discussion, a prior knowledge of switch fabric architecture is assumed. For the reader who is not familiar with these architecture, the subject is discussed in more detail in chapter 9.

Construction of NxN fully connected matrices using 2x2 switching elements requires $\log_2 N$ stages of elements and $\log_2 N - 1$ stages of interconnections. Due to the limited size of the substrate, the matrix size is also limited and the design goal is to have as a switch that is as large as possible for a given physical size. A representative interconnection for meeting this goal is the *banyan* network, which has been used to build large non-blocking optical switching architectures. A single chip 32x32 blocking switch is reported [84], [106]. In this switch, 26 waveguide crossings generate extra loss of 10 dB. The device operates in the 1.3-μm TM mode with drive voltages of ± 12 V for the cross and the bar states. The crosstalk is -18 dB.

A rearrangeably non-blocking 16x16 LN *Benes* switch matrix (N/2 rows of switching elements arranged in $2\log_2 N - 1$ stages) has been reported for TM-mode 1.3-μm wavelength [85]. Since the switch is fabricated on a single chip resulting in short switching elements, the drive voltage was as high as 60 volts. Other rearrangeably non-blocking 16x16 and 8x8 LN *dilated-Benes* switching fabrics have been also reported for TM mode 1.3-μm wavelength [107], [86]. Two chips were connected to enable lengthier switching elements and to lower the drive voltage (9 and 12 volts for the 8x8 and the 16x16 systems respectively). The insertion loss was 8-13 dB, and was higher in a 16-port matrix. Crosstalk was better than -20 dB in most cases.

Representative architectures for strictly non-blocking switching include the *crossbar*, *tree* and *simplified tree*. Almost all LN switches are demonstrated for the TM-mode 1.3 μm wavelength light although some have been demonstrated at 1.55 μm since the early nineties. Crossbars (N parallel inputs, N parallel outputs, and N^2 crosspoints) are simple and involve no waveguide crossings other than the crosspoints where switching devices are placed. EO crossbars have been proposed (4x4 [87], [88] and 8x8 [89]), but they are seldom used now in LiNbO$_3$ due to its long device length. Insertion loss can be lowest for crossbars (around 5 dB). For relatively large switching arrays where the light signal travels through many switching elements, however, insertion loss and crosstalk can be high.

The NxN tree router/selector architecture is constructed by an interconnection of 1xN and Nx1 *binary tree* switches. The tree structure exhibits the highest insertion loss (8 to 12 dB have been reported depending on the number of ports and on polarization). The main contribution to insertion loss in a tree structure comes from waveguide crossings. Special arrangements are often pursued to reduce these crossings in LN switches. Longer switching elements and lower drive voltages are possible in tree architecture because of having fewer stages ($2\log_2 N$). 16x16 [83], 8x8 [90] and 4x4 [91] LN tree switching architectures have been reported. The 16-port device was demonstrated for 1.3~1.6 μm and the others were demonstrated for 1.3 μm wavelength. In the 16x16 switch, two chips were connected to allocate sufficient length to switching elements and lower the drive voltage (40~55 V). Crosstalk is down to -30 dB in these reports.

In the middle stage of a 4x4 or 8x8 tree structure, a unit with 4 switches and 4 interconnection waveguides exhibits a switching function similar to a 2x2 switch. This unit can be replaced by a single 2x2 switching element. The resulting architecture is called a *simplified tree* architecture. The number of switching elements and total length are reduced compared to the tree architecture. An 8x8 switch has been demonstrated [79], [91]. The drive voltage for polarization independent operation was around 40 V at 1.3 and 1.55 µm wavelength.

2.5.2 Multiple-Chip Architectures

A 128-port LN *Clos* architecture was partially demonstrated in [108]. A polarization insensitive 8-channel simplified tree structure was used in each switch module. The total loss of the 5-stage system was 50 dB and optical amplifiers were used to compensate for this loss. Recent demonstration of 3-stage 32-port Clos network yielded crosstalk figures that are less than –40 dB as well as 28-dB insertion loss [109]. Polarization insensitive 8-channel trees were also used for each switch module. The total power consumption was under 30 W.

A low-crosstalk 16x16 strictly nonblocking system based on the *Extended Generalized Shuffle* (EGS) switching architecture was demonstrated in [92]. The demonstration used eight dual binary tree 1x8 (1x7), a 16x16 banyan, and eight dual binary tree 8x1 (7x1) switch arrays. The system is designed for 1.53 µm wavelength. The insertion loss was high (22 dB) due to the fact that it is a 3-stage structure. Crosstalk was -60 dB on average. The experimental system operated continuously for 20 month.

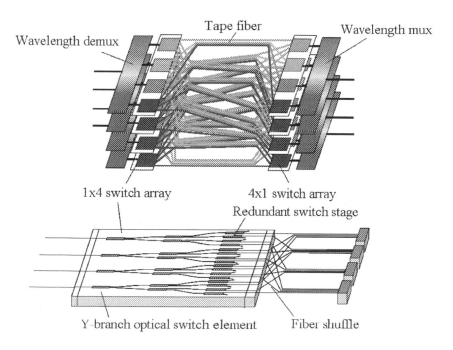

Figure 2-24. A wavelength routing architecture

Figure 2.24 depicts the switching system reported in [110]. Individual wavelengths are routed through a 1x4 switch matrix modules which are integrated onto a chip. Fiber ribbons interconnect the different stage chips in order to reduce on-chip connections. The architecture is modular and can be scaled to larger sizes. It can be enhanced with passive splitters to support multi-casting and by wavelength conversion to realize flexible cross-connect nodes. The fabricated 1x4 switch array is designed for the 1.55-µm wavelength. System crosstalk of -50 dB was attained. The drive voltage was 25 V for polarization independent operation. The 4-array 1x4 device was 1-cm wide and 5-cm long.

2.6 ADVANTAGES AND LIMITATIONS OF LITHIUM-NIOBATE ELECTRO-OPTIC SWITCHING

LN switches based on the EO effect involve no moving parts and are therefore reliable. They have low power consumption. They are capable of very fast switching speeds. Switching times in microseconds can be expected and several experiments reported switching times of tens of nanoseconds. Applications that require such high switching speed include optical packet switching, optical burst switching, fast demultiplexing/multiplexing of ultra-high-capacity optical signals, and others. However, switching of optical packets also requires optical buffer memory the development of which has been proven to be very difficult. Today, optical fiber delay lines are used for this purpose, but they are bulky, unstable and not easy and/or flexible to manage. Optical burst switching can be a suitable future application for LN EO switches. Optical packet switching and optical burst switching are discussed in chapter 13.

LN EO switches can also enable fast and flexible optical cross-connects (OXCs) of relatively small scale. They are suitable for reconfigurable optical add/drop multiplexers (R-OADMs). These network elements are discussed in chapter 11.

EO LN switching fabrics are limited in size today to a maximum of 16x16 or 32x32. There are some applications for which this port count is adequate, such as R-OADMs. New designs, including those discussed in this chapter, can make it possible to build simple optical switching systems with larger port counts in the future.

REFERENCES

[1]   R. S. Weis and T. K. Gaylord, "Lithium Niobate: Summary of physical properties and crystal structure," Appl. Physics A, Vol. 37, No. 4, pp. 191-203, 1985.

[2] K. Nashimoto, H. Moriyama, S. Nakamura, M. Watanabe, T. Morikawa, E. Osakabe, and K. Haga, "PLZT electro-optic waveguide and switches," Optical Fiber Communication Conference (OFC2001), Vol. 4, Paper PD10, 2001.

[3] M. Renaud, M. Bachmann, and M. Erman, "Semiconductor optical space switches," IEEE J. Select. Topics in Quantum Electron., Vol. 2, No. 2, pp. 277-88, 1996.

[4] G. A. Fish, L. A. Coldren, and S. P. DenBaars, "Suppressed modal interference switches with integrated curved amplifiers for scaleable photonic crossconnects," IEEE Photon. Technol. Lett., Vol. 10, No. 2, pp. 230-2, 1998.

[5] S. Yu, M. Owen, R. Varrazza, R. V. Penty, and I. H. White, "Demonstration of high-speed optical packet routing using vertical coupler crosspoint space switch array," Electron. Lett., Vol. 36, No. 6, pp. 556-8, 2000.

[6] K. Hamamoto, T. Anan, K. Komatsu, M. Sugimoto, and I. Mito, "First 8x8 semiconductor optical matrix switches using GaAs/AlGaAs electro-optic guided-wave directional couplers," Electron. Lett., Vol. 28, No. 5, pp. 441-3, 1992.

[7] R. A. Soref, "Silicon-based optoelectronics," Proceedings of IEEE, Vol. 81, No. 12, pp. 1687-706, 1993.

[8] B. Li and S-J Chua,, "2x2 optical waveguide switch with bow-tie electrode based on carrier-injection total internal reflection in SiGe alloy," IEEE Photon. Technol. Lett., Vol. 13, No. 3, pp. 206-8, 2000.

[9] H. Terui, M. Kobayashi, and T. Edahiro, "4x4 guided wave liquid crystal optical switch," National Convention of IECE, Paper 1047, 1983 (in Japanese).

[10] E. van Tomme, P. van Daele, R. Baets, G. R. Mohlmann, and M. B. J. Diemeer, "Guided wave modulators and switches fabricated in electro-optic polymers," J. Appl. Physics, Vol. 69, No. 9, pp. 6273-6, 1991.

[11] M-H. Lee, S. Park, J. J. Ju, S. K. Park, and J. Y. Do, "Multi-arrayed polymeric waveguide devices," Optoelectronics and Communications Conference (OECC2002), Paper 12A4-1, pp. 532-3, 2002.

[12] T. Fujiwara, M. Takahashi, M. Ohama, A. J. Ikushima, Y. Furukawa, and K. Kitamura, "Comparison of electro-optic effect between stoichiometric and congruent $LiNbO_3$," Electron. Lett., Vol. 35, No. 6, pp. 499-501, 1999.

[13] Y. Kondo, Y. Yamashita, T. Fukuda, T. Takano, Y. Furukawa, and K. Kitamura, "Wavelength dependence of electro-optic coefficients in Lithium Niobate crystals with different composition," European Conference on Integrated Optics (ECIO 2001), p. 185, 2001.

[14] O. G. Ramer, "Integrated optic electro-optic modulator electrode analysis," IEEE J. Quantum Electron., Vol. QE-18, No. 3, pp. 386-92, 1982.

[15] M. Goano, F. Bertazzi, P. Caravelli, G. Ghione, and T. A. Driscoll, "A general conformal-mapping approach to the optimum electrode design of coplanar waveguides with arbitrary cross section," IEEE Transactions on Microwave Theory and Techniques, Vol. 49, No. 9, pp. 1573-80, 2001.

[16] H. Green, "The numerical solution of some important transmission problems," IEEE Transactions on Microwave Theory and Techniques, Vol. 12, No.5, pp. 676-92, 1965.

[17] T. Kitazawa and T. Itoh, "Propagation characteristics of coplanar-type transmission lines with lossy media," IEEE Transactions on Microwave Theory and Techniques, Vol. 39,No. 10, pp. 1694-700, 1991.

[18] H. Kogelnik, "Theory of optical waveguides," Chapter 2 in Guided-Wave Optoelectronics, Springer-Verlag, 1987.

[19] H. Okayama, Oki Electric Ind. internal research report, 1986.

[20] R. Scarmozzino, A. Gopinath, R. Pregla, and S. Helfert, "Numerical techniques for modeling guided-wave photonic devices," IEEE J. Select. Topics in Quantum Electron., Vol. 6, No. 1, pp. 150-62, 2000.

[21] Q. Chen, Y. Chiu, D. N. Lambeth, T. E. Schlesinger, and D. D. Stancil, "Guided-wave electro-optic beam deflector using domain reversal in LiTaO₃," IEEE J. Lightwave Technol., Vol. 12, No. 8, pp. 1401-3, 1994.

[22] M. Yamada, M. Saitoh, and H. Ooki, "Electric-field induced cylindrical lens, switching and deflection devices composed of the inverted domains in LiNbO₃ crystals," Appl. Physics Letters, Vol. 69, No. 24, pp. 3659-61, 1996.

[23] H. Okayama and M. Kawahara, "Experiment on deflector-selector optical switch matrix," Electron. Lett., Vol. 28, No. 7, pp. 638-9, 1992.

[24] H. Okayama, "Ring light beam deflector," International Conference on Optics-photonics Design and Fabrication (ODF2002), Paper TP11, pp. 73-4, 2002.

[25] R. A. Becker and W. S. C. Chang, "Electro-optical switching in thin film waveguides for a computer communication bus," Appl. Optics, Vol. 18, No. 15, pp. 3296-300, 1979.

[26] H. Gnewuch, C. N. Pannell, G. W. Ross, P. G. R. Smith, and H. Geiger, "Nanosecond response of Bragg deflectors in periodically poled LiNbO₃," IEEE Photon. Technol. Lett., Vol. 10, No. 12, pp. 1730-2, 1998.

[27] A. Yariv, A., "Coupled mode theory for guided wave optics," J. Quantum Electron., Vol. 19, No. 9, pp. 919-33, 1973.

[28] C. H. Bulmer and W. K. Burns, "Polarization characteristics of LiNbO₃ channel waveguide directional couplers," J. Lightwave Technol., Vol. LT-1, No. 1, pp. 227-36, 1983.

[29] R. Alferness, "Polarization-independent optical directional coupler switch using weighted coupling," Appl. Physics Letters, Vol. 35, No. 10, pp. 748-50, 1979.

[30] O. G. Ramer, C. Mohr, and J. Pikulski, "Polarization-independent optical switch with multiple sections of Δβ reversal and Gaussian taper function," IEEE J. Quantum Electron., Vol. QE-18, No. 10, pp. 1772-9, 1982.

[31] L. McCaughan and S. K. Korotky, "Three-electrode Ti:LiNbO₃ optical switch," IEEE J. Lightwave Technol., Vol. LT-4, No. 9, pp. 1324-7, 1986.

[32] H. Okayama, T. Ushikubo, and T. Ishida, "Directional coupler with reduced voltage-length product," IEEE J. Lightwave Technol., Vol. 9, No. 11, pp. 1561-6, 1991.

[33] H. Kogelnik and R. V. Schmidt, "Switched directional coupler with alternating Δβ," IEEE J. Quantum Electron., Vol. QE-12, No. 7, pp. 396-401, 1976.

[34] R. V. Schmidt and P. S. Cross, "Efficient optical waveguide switch/amplitude modulator," Optics Lett., Vol. 2, No. 2, pp. 45-7, 1978.

[35] S. A. Samson, R. F. Tavlykaev, and V. Ramaswamy, "Two-section reversed Δβ switch with uniform electrodes and domain reversal," IEEE Photon. Technol. Lett., Vol. 9,No. 2, pp. 197-9, 1997.

[36] H. Okayama, T. Kamijoh, and T. Tsuruoka, "Reversed and uniform Δβ directional coupler with periodically changing coupling strength," Japanese J. Appl.. Physics, Vol. 39, No. 3B, pp. 1512-5, 2000.

[37] S. Thaniyavarn, "A synthesized digital switch using a 1x2 directional coupler with asymmetric Δβ phase reversal electrode," Topical Meeting on Integrated and Guided Wave Optics (IGWO'88), Paper TuC6, 1988.

[38] R. A. Forber and E. Marom, "Symmetric directional coupler switches," IEEE J. Quantum Electron., Vol. QE-22, No. 6, pp. 911-4, 1986.

[39] J. Ctyroky, "Voltage-length product of x and z-cut Ti:LiNbO₃ directional coupler and BOA switches: comparison," J. Optical Communication, Vol. 7, pp. 139-43, 1986.

[40] M. Papuchon, A. M. Roy, and D. B. Ostrowsky, "Electrically active optical bifurcation," Appl. Physics Letters, Vol. 31, No. 4, pp. 266-8, 1977.

[41] H. Nakajima, I. Sawaki, M. Seino, and K. Asama, "Bipolar-voltage-controlled optical switch using Ti:LiNbO₃ intersecting waveguides," International Conference on Integrated Optics and Optical Communication (IOOC'83), Paper 29C4-5, 1983.

[42] A. Neyer, W. Mevenkamp, M. Kretzschmann, and H. Richer, "Optical switching system using a nonblocking Ti:LiNbO₃ switch array," European Conference on Integrated Optics (ECIO'87), pp. 32-5, 1987.

[43] V. Ramaswamy and R. D. Standley, "A phased, optical coupler-pair switch," Bell System Technical Journal, Vol. 55, No. 6, pp. 767-75, 1976.

[44] V. Ramaswamy, M. D. Divino, and R. D. Standley, "Balanced bridge modulator switch using Ti-diffused LiNbO₃ strip waveguides," Appl. Physics Letters, Vol. 32, No. 10, pp. 644-6, 1978.

[45] M. Minakata, "Efficient LiNbO₃ balanced bridge modulator/switch with an ion-etched slot," Appl. Physics Letters, Vol. 35, No. 1, pp. 40-2, 1979.

[46] O. Mikami and S. Zembutsu, "Modified balanced-bridge switch with two straight waveguides," Appl. Physics Letters, Vol. 35, No. 2, pp. 145-6, 1979.

[47] J. L. Jackel and J. J. Johnson, "Nonsymmetric Mach-Zehnder interferometers used as low-drive-voltage modulators," IEEE J. Lightwave Technol., Vol. 6, No. 8, pp. 1348-51, 1988.

[48] H. Okayama, H. Yaegashi, I. Asabayashi, and M. Kawahara, "Balanced bridge optical switch composed of mode splitters," Microoptics Conference (MOC/GRIN'93), Paper G19, pp. 190-3, 1993.

[49] W. K. Burns and A. F. Milton, Waveguide transitions and junctions. In Guded-wave optoelectronics, chapter 3., Springer series in electronics and photonics 26., Springer-Verlag, 1988.

[50] Y. Silberberg, P. Perlmutter, and J. E. Baran, "Digital optical switch," Appl. Physics Letters, Vol. 51, No. 16, pp. 1230-2, 1987.

[51] H. Okayama and M. Kawahara, "Y-fed directional coupler with weighted coupling," Electron. Lett., Vol. 27, No. 21, pp. 1947-8, 1991.

[52] S. Thaniyavarn, "Modified 1x2 directional coupler waveguide modulator," Electron. Lett.., Vol. 22, No. 18, pp. 941-2, 1986.

[53] H. Okayama and M. Kawahara, "Reduction of Voltage-length product for Y-branch digital optical switch," IEEE J. Lightwave Technol., Vol. 11, No. 2, pp. 379-87, 1993.

[54] W. K. Burns, "Shaping the digital switch," IEEE Photon. Technol. Lett., Vol. 4, No. 8, pp. 861-3, 1992.

[55] R. Krahenbuhl and W. K. Burns, "Enhanced crosstalk suppression for Ti:LiNbO$_3$ digital optical switches," Topical Meeting on Photonics in Switching, Paper PFA1, pp. 110-2, 1999.

[56] R. Hauffe, F. Kerbstadt, U. Siebel, J. Bruns, and K. Petermann, "Reliable low cross talk digital optical switch based on cascaded like structure with small dimensions," Topical Meeting on Integrated Photonics Research (IPR '01), Paper ITuH2-1, 2001.

[57] S. K. Kim and V. Ramaswamy, "Tapered both in dimension and in index, velocity coupler: theory and experiment," IEEE J. Quantum Electron., Vol. 29, No. 4, pp. 1158-67, 1993.

[58] R. G. Peall and R. R. A. Syms "Further evidence of strong coupling effects in three-arm Ti:LiNbO$_3$ directional couplers," IEEE J. Quantum Electron., Vol. 25, No. 4, pp. 729-35, 1989.

[59] S. Rushin and D. Meshulach, "Voltage-controlled NxN coupled waveguide switch and power splitter," IEEE Photon. Technol. Lett., Vol. 5, No. 2, pp. 203-6, 1993.

[60] H. A. Haus and C. G. Fonstad, "Three-waveguide couplers for improved sampling and filtering," IEEE J. Quantum Electron., Vol. QE-17, No. 12, pp. 2321-5, 1981.

[61] H. Okayama, T. Ushikubo, and T. Ishida, "Three guide directional coupler as polarization independent optical switch," Electron. Lett., Vol. 27, No. 10, pp. 810-1, 1991.

[62] T. Pertsch, T. Zengtraf, U. Peschel, A. Brauer, and F. Lederer, "Beam steering arrays," Appl. Physics Letters, Vol. 80, No. 18, pp. 3247-9, 2002.

[63] J. C. Campbell and T. Li, "Electro-optic multimode waveguide modulator or switch," J. Appl. Physics, Vol. 50, No. 10, pp. 6149-54, 1979.

[64] G. E. Betts and W. S. C. Chang, "Crossing-channel waveguide electro-optic modulators," IEEE J. Quantum Electron., Vol. QE-22, No. 7, pp. 1027-38, 1986.

[65] C. S. Tsai, B. Kim, and F. R. El-Akkari, "Optical channel waveguide switch and coupler using total internal reflection," IEEE J. Quantum Electron., Vol. QE-14, No. 7, pp. 513-7, 1978.

[66] H. Okayama, A. Matoba, R. Shibuya, and T. Ishida, "Ti: LiNbO$_3$ total internal reflection switch," Spring national convention of IEICE, Paper C-488, 1989 (in Japanese).

[67] P. J. Duthie and C. Edge, "A polarization independent guided-wave LiNbO$_3$ electro-optic switch employing polarization diversity," IEEE Photon. Technol. Lett., Vol. 3, No. 2, pp. 136-7, 1991.

[68] T. Pohlman, A. Neyer, and E. Voges, "Polarization-independent switches on LiNbO$_3$," Topical Meeting on Integrated Photonics Research (IPR'90), Paper MH7, 1990.

[69] H. Okayama, A. Matoba, R. Shibuya, and T. Ishida, "Polarization independent optical switch with cascaded optical switch matrices," Electron. Lett., Vol. 24, No. 15, pp. 959-60, 1988.

[70] D. W. Yoon. O. Eknoyan, and H. F. Taylor, "Polarization-independent LiTaO$_3$ guided-wave electro-optic switches," IEEE J. Lightwave Technol., Vol. 8, No. 2, pp. 160-3, 1990.

[71] H. Okayama, "Optical switching device using coupled waveguides in phase matching structure," Microoptics Conference (MOC'01), Paper H21, pp. 210-3, 2001.

[72] J. L. Nightingale, J. S. Vrhel, and T. E. Salac, "Low-voltage, polarization -independent optical switch in Ti-indiffused Lithium Niobate," Topical Meeting on Integrated and Guided Wave Optics (IGWO '89), Paper MAA3, pp. 10-3, 1989.

[73] Y. Nakabayashi, "Polarization independent-DC drift free Ti: LiNbO$_3$ matrix optical switches," Optoelectronics and Communications Conference (OECC '98), Paper 13C2-1, pp. 60-1, 1998.

[74] P. Granestrand, B. Lagerstrom, P. Svensson, L. Thylen, B. Stolz, K. Bergvall, J.-E. Falk, and H. Olofsson, "Integrated optics 4x4 switch matrix with digital optical switches," Electron. Lett., Vol. 26, No. 1, pp. 4-5, 1990.

[75] K. Komatsu, S. Yamazaki, M. Kondo, and Y. Ohta, "Low-loss broad-band LiNbO$_3$ guided-wave phase modulator using Titanium/Magnesium double diffusion method," IEEE J. Lightwave Technol., Vol. LT-5, No. 9, pp. 1239-45, 1987.

[76] L. D. Hutcheson, I. A. White, and J. J. Burke, "Comparison of bending losses in integrated optical circuits," Optics Lett., Vol. 5, No. 6, pp. 276-8, 1980.

[77] S. K. Korotky, E. A. J. Marcatili, J. J. Veselka, and R. J. Bosworth, "Greatly reduced losses for small-radius bends in Ti: LiNbO$_3$ waveguides," Appl. Physics Letters, Vol. 48, No. 2, pp. 92-4, 1986.

[78] B. Shuppert, "Reduction of bend losses in Ti: LiNbO$_3$ waveguide through MgO double diffusion," Electron. Lett., Vol. 23, No. 15, pp. 797-8, 1987.

[79] Y. Nakabayashi, M. Kitamura, and T. Sawano, "DC-drift free-polarization independent Ti: LiNbO$_3$ 8x8 optical matrix switch," European Conference on Optical Communication (ECOC'96), Paper ThD2.4, Vol. 4, pp. 157-60, 1996.

[80] Y. Okabe, H. Okayama, and T. Kamijoh, "Fabrication of large scale LiNbO$_3$ 1xN optical switches," Pacific Rim Conference on Laser and Electro-optics (CLEO/Pacific Rim), Paper FH3, 1997.

[81] W. J. Minford, S. K. Korotky, and R. Alferness, 1982 "Low-loss Ti: LiNbO$_3$ waveguide bends at $\lambda = 1.3$ μm," IEEE J. Quantum Electron., Vol. QE-18, No. 10, pp. 1802-6, 1982.

[82] R. Ganguly, J. C. Biswas, and S. K. Lahiri, "Modelling of titanium indiffused lithium niobate channel waveguide bends: a matrix approach," Optics Communications, Vol. 155, No. 1, pp. 125-34, 1998.

[83] T. O. Murphy, S. -Y. Suh, B. Comissiong, A. Chen, R. Irvin, R. Grencavich, and G. Richards, "A strictly non-blocking 16x16 electro-optic photonic switch module," European Conference on Optical Communication (ECOC'2000), Vol. 4, pp. 93-4, 2000.

[84] H. Okayama and M. Kawahara, "Prototype 32x32 optical switch matrix," Electron. Lett., Vol. 30, No. 14, pp. 1128-9, 1994.

[85] P. J. Duthie and M. J. Wale, "16x16 single chip optical switch array in Lithium Niobate," Electron. Lett., Vol. 27, No. 14, pp. 1265-6, 1991.

[86] T. O. Murphy, C. T. Kemmerer, and D. T. Moser, "A 16x16 Ti: LiNbO$_3$ dilated Benes photonic switch module," Topical meeting on Photonic Switching, Paper PD3, 1989.

[87] G. A. Bogert, E. J. Murphy, and R. T. Ku, "Low crosstalk 4x4 Ti: LiNbO$_3$ optical switch with permanently attached polarization maintaining fiber array," IEEE J. Lightwave Technol., Vol. LT-4, No. 10, pp. 1542-5, 1986.

[88] P. P. Pedersen, J. L. Nightingale, B. E. Kincaid, J. S. Vrhel, and R. A. Becker, "A high-speed 4x4 Ti: LiNbO$_3$ integrated optic switch at 1.5 μm," IEEE J. Lightwave Technol., Vol. 8, No. 4, pp. 618-21, 1990.

[89] P. Granestrand, B. Stoltz, L. Thylen, K. Bergvall, W. Doldissen, H. Heinrich, and D. Hoffmann, "Strictly nonblocking 8x8 integrated optical switch matrix," Electron. Lett., Vol. 22, No. 15, pp. 816-8, 1986.

[90] P. Granestrand, B. Lagerstrom, P. Svensson, H. Olofsson, J. –E. Falk, and B. Stolz, "Pigtailed tree-structured 8x8 LiNbO$_3$ switch matrix with 112 digital optical switches," IEEE Photon. Technol. Lett., Vol. 6, No. 1, pp. 71-3, 1994.

[91] H. Okayama and M. Kawahara, "Ti: LiNbO$_3$ digital optical switch matrices," Electron. Lett., Vol. 29, No. 9, pp. 765-6, 1993.

[92] E. J. Murphy, T. O. Murphy, A. F. Ambrose, R. W. Irvin, B. H. Lee, P. Peng, G. W. Richards, and A. Yorinks, "16x16 strictly nonblocking guided-wave optical switching system," IEEE J. Lightwave Technol., Vol. 14, No. 3, pp. 352-8, 1996.

[93] K. Habara and K. Kikuchi, "Optical time-division space switches using tree-structured directional couplers," Electron. Lett., Vol. 21, No. 14, pp. 631-2, 1985.

[94] A. C. O'Donnell and N. J. Parsons, "1x16 Lithium Niobate optical switch matrix with integral TTL compatible drive electronics," Electron. Lett., Vol. 27, No. 25, pp. 2367-8, 1991.

[95] A. C. O'Donnell, "Polarisation independent 1x16 and 1x32 Lithium Niobate optical switch matrices," Electron. Lett., Vol. 27, No. 25, pp. 2349-50, 1991.

[96] J. E. Watson, M. A. Milbrodt, and T. C. Rice, "A polarization-independent 1x16 guided-wave optical switch integrated on Lithium Niobate," IEEE J. Lightwave Technol., Vol. LT-4, No. 11, pp. 1717-21, 1986.

[97] D. Hoffman, H. Heidrich, H. Ahlers, and M. K. Fluge, "Performance of rearrangeable nonblocking 4x4 switch matrices on LiNbO$_3$," IEEE J. Select. Areas of Commun., Vol. 6, No. 6, pp. 1232-40, 1988.

[98] J. E. Watson, M. A. Milbrodt, K. Bahadori, M. F. Dautartas, C. T. Kemmerer, D. T. Moser, A. W. Schelling, T. O. Murphy, J. J. Veselka, and D. A. Herr, "A low-voltage

8x8 Ti:LiNbO$_3$ switch with a dilated-Benes architecture," IEEE J. Lightwave Technol., Vol. 8, No. 5, pp. 794-801, 1990.

[99] T. O. Murphy, E. J. Murphy, and R. W. Irvin, "An 8x8 Ti:LiNbO$_3$ polarization-independent photonic switch," European Conference on Optical Communication (ECOC'94), Vol. 2, pp. 549-52, 1994.

[100] M. Kondo, N. Takado, K. Komatsu, and Y. Ohta, "32 switch-elements integrated low-crosstalk LiNbO$_3$ 4x4 optical matrix switch," International Conference on Integrated Optics and Optical Communication with European Conference on Optical Communication (IOOC-ECOC'85), pp. 361-4, 1985.

[101] E. Voges and A. Neyer, "Integrated-optic devices on LiNbO$_3$ for optical communications," IEEE IEEE J. Lightwave Technol., Vol. LT-5, No. 9, pp. 1229-38, 1987.

[102] S. Thaniyavarn, J. Lin, W. Dougherty, T. Traynor, K. Chiu, G. Abbas, M. LaGasse, W. Chaczenko, and M. Hamilton, "Compact, low insertion loss 16x16 optical switch array modules," Optical Fiber Communication Conference (OFC'97), Paper TuC1, Vol. 6, pp. 5-6, 1997.

[103] H. Nishimoto, S. Suzuki, and M. Kondo, "Polarisation-independent LiNbO$_3$ 4x4 matrix switch," Electron. Lett., Vol. 24, No. 8, pp. 1122-3, 1988.

[104] H. Nishimoto, M. Iwasaki, S. Suzuki, and M. Kondo, 1990 "Polarisation independent LiNbO$_3$ 8x8 matrix switch," IEEE Photon. Technol. Lett., Vol. 2, No. 9, pp. 634-6, 1990.

[105] H. Okayama, Y. Okabe, T. Kamijoh, and N. Sakamoto, "Optical switch array using banyan network," IEICE Transactions on Communications, Vol. 82-B, No. 2, pp. 365-72, 1999.

[106] E. J. Murphy, C. T. Kemmerer, D. T. Moser, M. R. Serbin, J. E. Watson, and P. L. Stoddard, "Uniform 8x8 Lithium Niobate switch arrays," IEEE J. Lightwave Technol., Vol. 13, No. 5, pp. 967-70, 1995.

[107] C. Burke, M. Fujiwara, M. Yamaguchi, H. Nishinoto, and H. Honmou, "128 line photonic switching system using LiNbO$_3$ switch matrices and semiconductor traveling wave amplifiers," IEEE J. Lightwave Technol., Vol. 10, No. 5, pp. 610-5, 1992.

[108] M. Yamashita, Y. Nkabayashi, K. Asahi, and C. Konishi, "A strictly non-blocking 32x32 optical switch built by LiNbO$_3$ optical switch matrices," IEICE communication society conference, Paper B-12-19, 1999 (in Japanese)

[109] H. Okayama, T. Arai, and T. Tsuruoka, "Arrayed 1xN switch for wavelength routing," IEICE Transactions on Electronics, Vol. E83-C, No. 6, pp. 920-6, 2000.

[110] I. Sawaki, T. Shimoe, H. Nakamoto, T. Iwama, T. Yamane, and H. Nakajima, "Rectangularly configured 4x4 Ti:LiNbO$_3$ matrix switch with low drive voltage," IEEE J. Select. Areas of Commun, Vol. 6, No. 6, pp. 1267-72, 1988.

Chapter 3

ACOUSTO-OPTIC SWITCHING

Antonio d'Alessandro

3.1   THE ACOUSTO-OPTIC EFFECT

Optical switches can be made by exploiting the *acousto-optic* (AO) effect where the refractive index of an optical medium is modulated by acoustic waves [1]. An acoustic or elastic wave traveling in a medium induces periodical deformation in the form of alternating compressions and rarefactions. This results in a periodical strain in the medium. The vibrations of the molecules due to this strain alter the optical polarizability of the material and, consequently, its refractive index. An acoustic wave can be generated for this purpose by the *piezoelectric effect*[1] either in the bulk of the material (leading to bulk acoustic waves) or on its surface (leading to *surface acoustic wave*, or SAW). This is done by applying a radio-frequency (RF) electric field. The electrodes across which the field is applied to induce the AO effect are called acoustic transducers.

Refractive index modulation due to the acousto-optic effect induces a dynamic phase grating (acousto-optic grating) which can diffract light beams. In *isotropic* materials[2], diffraction due to AO effect results in light beam deflection. In *anisotropic* materials this deflection comes along with variation in light polarization. Beam deflection and polarization conversion

---

[1] The piezoelectric effect is a phenomenon where the application of mechanical stress on substance results in electrical polarization. It is a property that can be utilized in certain crystals to convert a mechanical signal (such as a pressure wave or sound wave) into an electrical signal, and vice versa.

[2] An isotropic material have constant physical properties in all directions. An anistropic material, on the other hand, is characterized by physical properties and parameters which vary depending on direction.

by the acousto-optic effect can be exploited to make several optical devices such as deflectors, tunable filters, switches, modulators, spectrum analyzers, and frequency shifters [2]. In integrated optical devices, where light is confined in slab or channel waveguides, surface waves (SAW's) provide higher acousto-optic efficiency than bulk waves because surface-wave power is concentrated in a thin surface layer where optical waveguides are placed. A SAW can be excited by applying a low power RF signal to an interdigital transducer (IDT) electrode. The structure of IDT's, patterned on piezoelectric substrates or a thin films, is particularly suitable for integrated planar lightwave circuit (PLC) fabrication.

Lithium niobate (LN or $LiNbO_3$) has been widely used to make acousto-optic waveguide devices since it is characterized by large acousto-optic effect and because low-loss waveguides can be fabricated on it using well-established fabrication technologies [3]. Furthermore, lithium niobate is a good piezoelectric material which means that RF energy can be efficiently converted into acoustic energy. Due to their capability to perform wavelength-selective optical switching, LN AO switches were thoroughly studied in the past decade for wavelength division multiplexing (WDM) applications and networks [4].

Acousto-optic devices have been fabricated also using III-V semiconductors such as Gallium Arsenide (GaAs) and Indium Gallium Arsenide Phosphide (InGaAsP), where monolithic integration may be obtained easily. While these compound semiconductors are used to fabricate excellent lasers, photo-detectors and optical waveguides, AO devices made from them require higher driving power and yield much lower diffraction efficiencies than their $LiNbO_3$ counterparts [5]. Furthermore since III-V semiconductors are not good piezoelectric materials, a thin piezoelectric film of zinc oxide (ZnO) must be deposited between electrodes and waveguides to get efficient excitation of SAW's. Low loss and low driving power routing switches are also demonstrated by exploiting the acousto-optic effect in all-fiber directional couplers [6].

3.1.1 Modulation of Refractive Index by Acoustic Waves

The strain, represented by a tensor S_{kl}, resulting from a traveling acoustic wave in a medium, induces change in the optical impermeability tensor (inverse of the relative permittivity tensor) $\Delta \eta_{ij}$. This can be described by the following linear tensor relationship [1]:

$$\Delta \eta_{ij} = \Delta \left(\frac{1}{n_{ij}^2} \right) = p_{ijkl} S_{kl} \quad i, j, k, l = 1, 2, 3 \qquad (3.1)$$

where p_{ijkl} represents a fourth-rank tensor defined as the strain-optic tensor. In the cartesian coordinate system, the values of the indices $i, j, k, l = 1, 2, 3$ denote the coordinates x, y, z, respectively. Since both $\Delta\eta_{ij}$ and S_{kl} are symmetric tensors, a permutation of indices is possible and Equation (3.1) can be rewritten using the contracted indices $1 = (11), 2 = (22), 3 = (33), 4 = (23) = (32), 5 = (13) = (31), 6 = (12) = (21)$ as follows:

$$\Delta\left(\frac{1}{n_i^2}\right) = p_{ij}S_j \qquad i, j = 1, 2, \ldots, 6 \tag{3.2}$$

where p_{ij} is a 6x6 matrix representing the coefficients of the strain-optic tensor usually defined in the principle coordinate system, in which n_x, n_y, and n_z are the principle indices of refraction and $n_{xy} = n_{yz} = n_{xy} = 0$ [1].

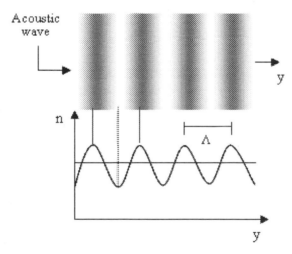

Figure 3-1. Variation of the refractive index n induced by a sound wave represented by a sequence of compressions (dark areas) and rarefactions (white areas) propagating in the medium along the y direction with period Λ

An acoustic wave of wavelength Λ which is propagating in a medium modulates its refractive indices creating a dynamic phase grating whose period is Λ, as sketched in Figure 3-1. The velocity of sound V_s (typically in the order of 10^3 m/s) is much lower than that of light ($c = 3 \times 10^8$ m/s), therefore an acoustic wave is seen as a stationary wave by the traveling *photons*.

Consider a linearly polarized shear acoustic wave, which is producing a strain in the x direction while traveling along the y direction with angular frequency Ω and period Λ. Since the only non-vanishing strain tensor component in this case is S_6, the modulation of refractive indices according to Equation (3.2) can be approximated by:

$$\Delta n_i \cong -\frac{1}{2} n_i^3 p_{i6} S_6 \sin(\Omega t - K y) \qquad (3.3)$$

where $K = 2\pi/\Lambda$ is the acoustic wave number assuming small index changes [7].

3.1.2 Acousto-Optic Bragg Diffraction

Diffraction of light by acoustic waves can give rise to either many diffracted beams or just a single diffracted beam, depending on wavelength, the refractive indices of the medium, the period of the acoustic wave and the acousto-optic interaction length [1]. The case where many beams are diffracted is often referred to as Raman-Nath diffraction. When a single beam is diffracted, this is called Bragg diffraction. Bragg diffraction is more interesting for device applications because of its higher diffraction efficiencies.

Bragg diffraction can be described by means of a simple particle picture, which clarifies the laws of acousto-optic interaction in a simplified way. Electromagnetic waves of angular frequency ω and wavelength λ propagating in a medium with refractive index n can be represented by photons of energy $\hbar\omega$ and momentum $\hbar\mathbf{k}$, where $\hbar = h/2\pi$, h is Planck's constant, and $\mathbf{k}$ (with modulus $|\mathbf{k}| = k = 2\pi n/\lambda$) is the wave vector. An acoustic wave of angular frequency Ω and wavelength Λ can be described by *phonons*[3] of energy $\hbar\Omega$ and momentum $\hbar\mathbf{K}$, where $\mathbf{K}$ (with modulus $|\mathbf{K}| = K = 2\pi/\Lambda$) is the wave vector. Light diffraction due to acoustic waves can be described as the sum of processes of collisions between photons and phonons where a photon/phonon collision results in diffracted photon. The photon/phonon collision is governed by the conditions of conservation of momentum, also called phase-matching, and the conservation of energy:

$$\mathbf{k}_d = \mathbf{k}_i \pm \mathbf{K} \qquad (3.4)$$

$$\omega_d = \omega_i \pm \Omega \qquad (3.5)$$

where d refers to the diffracted photon and i to the incident one. The sign + or − is used to describe a phonon that is absorbed with a frequency upshift or emitted with frequency downshift respectively. In an isotropic medium, $|\mathbf{k}_d|$ is always nearly the same as $|\mathbf{k}_i|$ and momentum conservation implies a beam deflection as depicted in Figure 3-2. The corresponding vector diagram is an

---

[3] A phonon is a quantized mode of vibration occurring in a rigid crystal lattice, or a quantum of vibrational energy which can be treated as a quasi-particle in quantum mechanics.

isosceles triangle where the diffracted and undiffracted wave vectors are coplanar forming an angle $2\theta_B$, where θ_B is called the Bragg angle.

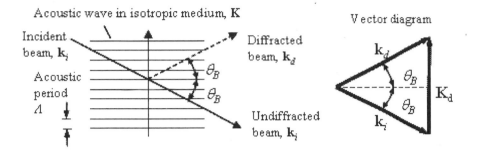

Figure 3-2. Light deflection in an isotropic medium due to an acoustic Bragg diffraction grating of period Λ in a coplanar acousto-optic interaction. The corresponding vector diagram is also shown.

From the vector diagram, the Bragg angle is given by:

$$\theta_B = \sin^{-1}\left\{\frac{1}{2n}\left(\frac{\lambda}{\Lambda}\right)\right\} \qquad (3.6)$$

which implies that a monochromatic light beam can be deflected at different angles by changing the period of the acoustic grating.

A birefringent medium, like lithium niobate, has two refractive indices, an extraordinary index n_e along the optical axis c and an ordinary index n_o in the plane orthogonal to c [8]. Since the two refractive indices are related to two linearly polarized orthogonal propagation modes, namely the extraordinary wave with phase velocity c/n_e and the ordinary wave with phase velocity c/n_o [1], it may result that beam deflection is accompanied by polarization conversion with $|\mathbf{k}_i| \neq |\mathbf{k}_d|$. In the case of collinear co-directional acousto-optic interaction, since $\mathbf{k}_d$, $\mathbf{k}_i$ and $\mathbf{K}$ are collinear with the same orientation, as depicted in Figure 3-3, the conservation of momentum given by (3.4) can be written as follows:

$$\left|k_d - k_i\right| = \frac{2\pi}{\lambda}\left|N_e - N_o\right| = K = \frac{2\pi}{\Lambda} \qquad (3.7)$$

where N_e and N_o are the refractive indices of the birefringent crystal for the two modes.

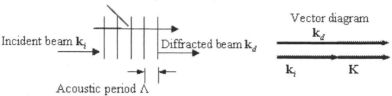

Figure 3-3. Light diffraction in an anisotropic medium through an acoustic Bragg grating of period Λ in a collinear codirectional acousto-optic interaction and the corresponding vector diagram.

According to the conservation of momentum, the acoustic wave compensates the momentum mismatch between the two orthogonal polarized modes propagating in the medium, leading to polarization conversion. If light propagates in channel waveguides, located just under the surface of lithium Niobate, a SAW gives rise to mode conversion between TE-like (Transverse-Electric like) and TM-like (Transverse-Magnetic like) guided modes [9], whose effective indices are N_e and N_o.

The following condition of resonance, or phase-matching, can be derived from (3.7):

$$\Lambda = \frac{V_s}{F_s} = \frac{\lambda}{\Delta n} \tag{3.8}$$

where $\Delta n = |N_e - N_o|$ is the optical birefringence, V_s and F_s are the velocity and the frequency of the acoustic wave respectively. According to Equation (3.8), in a birefringent material a conversion of polarization occurs when the condition $\Lambda = L_b$ is met, where $L_b = \lambda/\Delta n$ is the beat length of the two orthogonal linear polarizations at wavelength λ. For a LiNbO$_3$ acousto-optic device operating at 1.55-μm wavelength, Equation (3.8) gives $F_s = V_s/\Lambda = V_s \Delta n/\lambda = 177$ MHz, for $\Delta n = 0.074$ and $V_s = 3.7$ km/s.

3.2   OPERATION OF ACOUSTO-OPTIC SWITCHES

Acousto-optic switches can operate by exploiting beam deflection (Equation (3.6)) or collinear mode conversion (Equation (3.7)). Figure 3-4 sketches the operation principle of a basic 2x2 switch consisting of an acousto-optic beam deflector (AOBD) in the cross and bar states. In the bar-state, an input monochromatic light beam (lightpath) through port 1 (port 2) is Bragg deflected and routed to output 1` (output 2`) by applying an RF control signal which excites acoustic waves. The cross-state is obtained when the input beam through port 1 (port 2) is routed to port 2` (port 1`) without any deflection since no RF signal is applied. Guided wave acousto-optic switches based on beam deflection have been demonstrated in LiNbO$_3$, GaAs, and InP [5].

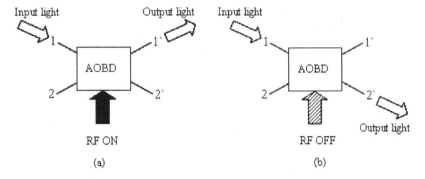

Figure 3-4. Optical switch based on an acousto-optic beam deflector (AOBD) with monochromatic input light through port 1: (a) Bar-state with beam deflection, (b) Cross-state with no deflection.

An 8 × 8 space switch module based on LN AOBD's has been demonstrated. The module is designed by placing each output port at a given Bragg angle (Equation (3.6)), so that outputs can be selected by changing the actuating RF frequency [10].

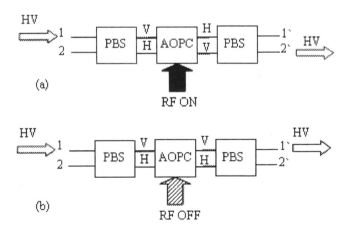

Figure 3-5. Block scheme of a polarization independent optical switch made of an AOPC and two PBS's. The arrow (HV) indicates unpolarized light: (a) Cross-state (b) Bar-state.

Figure 3-5 shows the block diagram of a polarization independent acousto-optic switch based on collinear polarization conversion. The switch is made of an *acousto-optic polarization converter* (AOPC) and two identical polarization beam splitters (PBS's). The two PBS's allow the acousto-optic switch to operate according to the polarization diversity scheme used in LN polarization-independent acousto-optic tunable filters [11]. The first PBS decomposes an arbitrarily polarized monochromatic light beam into two orthogonal linear polarization components, denoted generally as horizontal (H) and vertical (V). The PBS works in such a way that H polarization goes from the upper (lower) input port towards the lower (upper) output port while V polarization is routed from the upper (lower) input to the upper (lower) output port. Figure 3-5(a) represents the switch in the cross-state where a light signal is switched from input port 1 to output port 2`. In fact, when an RF signal is applied, the AOPC rotates the linear polarizations by 90°, then the output PBS, acting as a beam combiner, routes the combined light beam to output port 2`. Similarly, an input beam can be routed from input port 2 to output port 1`. Figure 3-5(b) shows the switch in the bar-state where no RF-control is applied and, hence, no polarization conversion takes place. In this case, an input signal may be routed from input port 1 to output port 1` or from input 2 to output 2`.

Acousto-optic switches have been also realized by exploiting mode conversion in fiber directional couplers [6]. All-fiber 2 × 2 and 3 × 3 acousto-optic switches based on null couplers are demonstrated [12]-[13]. A 2 × 2 fiber coupler is obtained by pre-tapering two identical fibers along a short length before both fibers are fused and elongated together to form the coupler. The central part of the coupler where the two fibers are fused together is sometimes referred to as the waist of the coupler. The two single-mode fibers have mismatched diameters so that the resultant coupler does not actually couple any light. Light in one input fiber excites the fundamental mode in the narrow waist. Light in the other input fiber excites just the second mode in the waist. When an acoustic wave propagates along the coupler, it modulates the refractive index in the waist. At resonance, which occurs when the beat length of the modes is matched by the acoustic period, a mode conversion takes place in the waist and light is routed to the second fiber (cross-state). If no acoustic wave is generated, light from one fiber propagates undisturbed through the waist to the output of the same fiber (bar-state). Polarization insensitive operation can be obtained by adding a twist induced birefringence in the coupler waist [14].

3.2.1 Optical Response of an AOPC

Acoustic Bragg diffraction can also be described by the theory of mode coupling in periodic media, where the refractive index varies periodically [15]. In an anisotropic medium, the change of polarization induced by the AO effect can be studied in terms of the coupling between two orthogonal linearly polarized modes. For collinear codirectional interaction along the y axis, mode coupling can be described by the following coupled mode differential Equations [1]:

$$\frac{dE_1}{dy} = -j\frac{\beta_1}{|\beta_1|}\kappa E_2 e^{j2\delta y}$$

$$\frac{dE_2}{dy} = -j\frac{\beta_2}{|\beta_2|}\kappa E_1 e^{-j2\delta y}$$

(3.9)

where E_1 and E_2 are the complex amplitudes of the electric fields of the two polarizations whose corresponding propagation constants are β_1 and β_2, κ is the coupling coefficient and 2δ is the detuning from the phase matching condition:

$$2\delta = \beta_1 - \beta_2 \pm K$$

(3.10)

where $K = 2\pi/\Lambda$ is the acoustic wave number. For the acousto-optic effect, κ is given by [1]:

$$\kappa = \frac{\omega p S(n_1 n_2)^{3/2}}{4c}\Gamma$$

(3.11)

where p is a coefficient of the strain-optic tensor which depends on the direction of the acousto-optic interaction in the medium, S is the amplitude of the linearly polarized acoustic wave, and n_1 and n_2 are the effective indices of the modes. The factor Γ, ranging from 0 to 1, is the overlap integral of the optical and acoustic fields [3]. In the case of optical guided modes, Γ takes the value of 1 when single mode waveguides are used and the acoustic wavelength Λ is larger than the optical waveguide thickness. Furthermore, from Equation (3.11) κ^2 is proportional to the acoustic wave intensity $I_S = 1/2\rho V_s S^2$, where ρ is the mass density. Hence, κ^2 is proportional to the applied drive RF power.

The solution of (3.9) is obtained by integration from 0 to L, where L is the length of the acousto-optic interaction region:

$$E_1(L) = \frac{e^{j\delta L}}{\mu}\left[(\mu\cos\mu L - j\delta\sin\mu L)E_1(0) - j\kappa E_2(0)\sin\mu L\right]$$

(3.12)

$$E_2(L) = \frac{e^{-j\delta L}}{\mu}\left[-j\kappa^* E_1(0)\sin\mu L + (\mu\cos\mu L + j\delta\sin\mu L)E_2(0)\right]$$

where $\mu^2 = \kappa^*\kappa + \delta^2$. κ^* is the complex conjugate of κ. Assuming $E_1(0) \neq 0$ and $E_2(0) = 0$, which means that certain linear polarization exists at the beginning of the interaction region, then the conversion efficiency T, defined as the fraction of polarized light power converted into the other linear polarization at the end of the interaction region, is given by:

$$T = \frac{|E_2(L)|^2}{|E_1(0)|^2} = \frac{\kappa^2}{\kappa^2 + \delta^2} \sin^2\left(L\sqrt{\kappa^2 + \delta^2}\right) \qquad (3.13)$$

Equation (3.13) implies that the maximum conversion efficiency at resonance, for which $\delta = 0$, is 100% when $\kappa L = \pi/2$. Hence L is inversely proportional to κ and, consequently, to the square root of $P_{100\%}$, which is the acoustic driving power required to reach full polarization conversion.

The sinc-squared function of Equation (3.13) shows that polarization conversion is wavelength dependent and that T represents the passband spectrum or the optical response of an AOPC. The function $(1-T)$ represents the notch spectrum. Linear plots of T and $1-T$ versus normalized detuning are shown in Figure 3-6 for $\kappa L = \pi/2$.

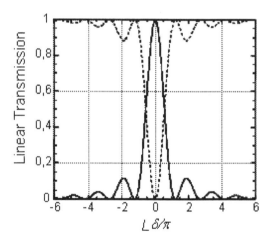

Figure 3-6. Passband (continuous line) and notch (dashed line) spectra of an AOPC versus normalized detuning for $\kappa L = \pi/2$.

The optical response of the AOPC exhibits high wavelength selectivity which can be exploited to devise tunable optical filters and wavelength-selective switches. The optical bandwidth, defined as the full width half maximum (FWHM) response, can be calculated from Equation (3.13), for $\kappa L = \pi/2$:

$$\Delta\lambda_{FWHM} = \frac{0.8\lambda^2}{|\Delta n|L} \qquad (3.14)$$

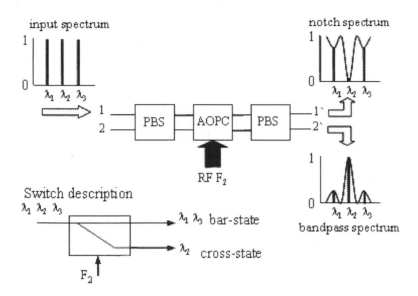

Figure 3-7. AOPC surrounded by two PBS's and acting as a polarization independent AO selective switch where wavelength λ_2 is routed to a different output port than λ_1 and λ_3 by applying resonant RF drive frequency F_2.

A typical bandwidth of a guided wave LN AOPC switch is about 1.6 nm for an interaction length of 20 mm [16]. The value of the AO interaction length (L) is the result of a compromise between device length, wavelength resolution and drive power. The passband of an AOPC can be tuned by controlling the RF frequency F_s, according to Equation (3.8). A tunable AOPC can be used as a wavelength-selective optical switch as shown in Figure 3-7. This switch can behave as a multi-state device, whose bar-state and cross-state, for different wavelengths, may coexist at the same time. When several passbands are active, the corresponding wavelengths are switched to their diverse destinations simultaneously. This unique property makes AO switches attractive for WDM optical networks. Therefore, guided wave acousto-optic switches based on polarization conversion in lithium niobate have been developed and tested for this application [4].

3.3   INTEGRATED LITHIUM-NIOBATE ACOUSTO-OPTIC SWITCHES

Guided wave acousto-optic switches have been fabricated mainly with LiNbO$_3$. A fully integrated 2×2 polarization-insensitive acousto-optic LiNbO$_3$ switch is shown in Figure 3-8. The switch operation is based on acousto-optic polarization conversion and on the polarization diversity scheme of Figure 3-5.

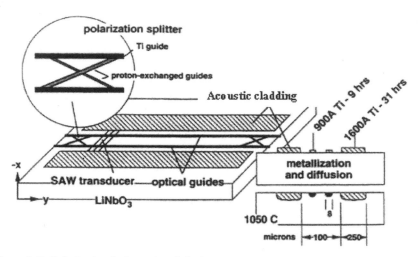

Figure 3-8. Polarization-independent fully-integrated AO switch made on x-cut y-propagating LiNbO₃ substrate. The circled inset shows the input proton-exchanged PBS. The cross section of the device shows the dimensions of both the optical and SAW waveguides before (upper section) and after (lower section) thermal diffusion [4] (© 1996 IEEE).

The device in Figure 3-8 includes two parallel titanium (Ti) waveguides for propagation of two orthogonal optical polarizations; a TE-like mode and a TM-like mode. In fact, unpolarized input optical signals are decomposed into two orthogonal polarization components by the input annealed proton-exchanged X-junction, which acts as an integrated PBS. An identical X-junction at the output then recombines the two polarization components into one signal that is routed towards certain output. Single mode Ti waveguides of the type depicted in Figure 3-8 can be obtained from titanium stripes of 8 μm width and 900 Å thickness, deposited by e-beam evaporation and then diffused for about 9 hours in air at 1050 °C. Thermal diffusion is a well established technology to obtain good quality Ti:LiNbO₃ optical waveguides. In particular, low-loss waveguides can be obtained by preventing the phenomenon of LiO_2 out-diffusion, by placing the sample in a closed platinum crucible during diffusion and by using x-cut lithium niobate substrates [17]. Most integrated acousto-optic devices in LiNbO₃ are fabricated by using x-cut y-propagation wafers where the optical axis coincides with the z axis [8].

The proton exchange technique is used to make integrated PBS's. This process is performed in a melted pure benzoic acid at 220 °C followed by annealing at 360 °C [18]. The benzoic acid is a source of protons H^+, which replace Li atoms in LiNbO₃. Proton exchange makes the extraordinary refractive index n_e, along the z axis, higher than in Ti:LiNbO₃ waveguides, while reducing the ordinary index n_o along the x axis. In an optical channel waveguide of an x-cut LiNbO$_3$ substrate, the effective refractive index of the TE-like mode, whose electric field vector lays along the z axis, depends on n_e, while the effective refractive index of the TM-like mode, whose electric field vector is oriented along the x axis, depends on n_o [3]. Hence a TE-like mode ("H" polarization in Figure 3-5) coming from an input Ti waveguide is routed towards the other waveguide through the proton exchanged X-junction, due to its higher n_e. An input TM-like mode ("V" polarization in Figure 3-5) propagates only in the Ti waveguide since its n_o is higher than that of the proton-exchanged guides.

All-titanium PBS's have also been used in monolithic acousto-optic switches, using either an optical directional coupler [11] or a bow-tie-type waveguide crossing [19]. Another all-titanium PBS design is based on two-mode interference in a zero gap optical directional coupler [20].

Figure 3-8 also shows two wider titanium in-diffused stripes acting as acoustic cladding, to confine SAW's into the area where the waveguides are located. In such a configuration the overlap integral (Γ) in Equation (3.11) is approximately 1. The IDT's of the switch in Figure 3-8 consist of pairs of 5 µm wide gold fingers. The dimensions of the fingers are designed in such a way to launch SAW's whose period Λ is about 20 µm, corresponding to F_s = 175 MHz in LiNbO$_3$. This is to match the beat length L_b of TE-like and TM-like optical modes in Ti waveguides within the 1.55-µm optical transmission window. The IDT fingers are also tilted by about 5° with respect to the z axis, to compensate for SAW walk-off from the y-direction in x-cut LiNbO$_3$ [21]. This tilted IDT fingers excite SAW's which propagate collinearly with the optical waves [22]. Such a tilt angle also improves RF to SAW power conversion efficiency by about 40%. Devices with drive power of about 10 mW were fabricated [23]. Two acoustic absorbers made of common rubber cement must be placed before the IDT's and the output PBS. The first absorber prevents the propagation of acoustic power towards the input ports, the second one terminates the interaction length L. Conversion efficiencies higher than 99% were measured in such integrated acousto-optic switches. Switching speed, depending on the time which an acoustic wave takes to travel along L, is typically less than 10 µs for L = 20 mm [24].

An alternative approach to monolithic fabrication of fully integrated devices is to use a hybrid approach based on deployment of external bulk PBS's connected to Ti:LiNbO$_3$ waveguides by polarization-preserving fiber (PPF). PPF permits the rotation of one the input polarizations in such a way that initial polarization components are preserved at the input of the Ti waveguides. In this way the polarization-diversity structure of the hybrid AO switch appears as if it has a single polarization, as far as the acousto-optic interaction is concerned. In this case, the orthogonal signal components are treated exactly the same way, even with respect to frequency shifts, due to the energy conservation law of the acousto-optic interaction (Equation (3.5)). This frequency shift is opposite for TE-like/TM-like and TM-like/TE-like conversions, which causes self-beating if both polarizations are mixed again in the output PBS of the fully integrated AO switch. This self-beating is an undesired effect in optical network applications [4].

3.4   ACOUSTO-OPTIC SWITCHES FOR WDM OPTICAL NETWORKS

Acousto-optic switches can be driven by many RF signals to switch several lightpaths simultaneously. Multiwavelength operation of acousto-optic switches within the erbium doped fiber amplifier gain spectrum has been experimentally demonstrated [25]. Therefore, acousto-optic tunable switches have been considered for optical cross-connects in WDM networks. In this application, the main challenge is crosstalk. Although the discussion below refers mainly to collinear LiNbO$_3$ AOPC based switches, it applies also to integrated AO switches made of other materials or AO switches based on single-mode fiber tapers and couplers [6].

3.4.1   Crosstalk and Optical Response in Acousto-Optic Switches

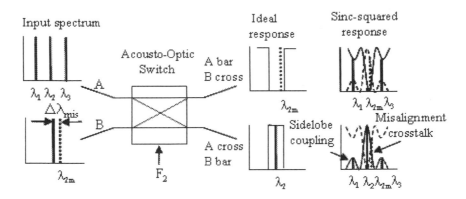

Figure 3-9. Depiction of the crosstalk problem for AO switching in a WDM systems and a comparison between an ideal square switch passband with a classical sinc-squared passband. The latter exhibits significant off-resonant leakage due to prominent sidelobes and incomplete switching due to slight misalignment $\Delta\lambda_{mis}$

Typically, the optical response of a collinear integrated acousto-optic switch takes the shape of the sinc-squared function plotted in Figure 3-6, which is quite different from an ideal rectangular passband. In fact, sidelobes of the sinc-squared function are the source of out-of-band crosstalk in WDM applications, as shown in Figure 3-9. In this figure, the RF frequency F_2

causes the switch to route only wavelength λ_2 from input port A to the lower output port, along with undesired fractions of the signals at λ_1 and λ_3 due to sidelobe coupling. The sharp peaks of the optical transmission response, coupled with possible laser-wavelength drifts, impose tight constraints in regards to wavelength misalignments. In the example of Figure 3-9, the misaligned signal at λ_{2m} is only partly switched from input B to the selected upper output, leading to coherent crosstalk, i.e. beating among optical signals at the same wavelength, at the lower output port.

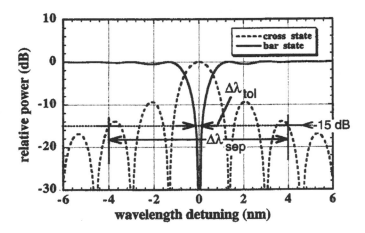

Figure 3-10. Bar-state (solid line) and cross state (dashed line) transmission characteristics for a 20 mm-long LiNbO$_3$ AO switch at 1550 nm. The wavelength-misalignment tolerance and interchannel separation corresponding to –15 dB crosstalk are identified [3] (© 1996 IEEE).

The bar-state and cross-state transmission characteristics, versus wavelength detuning, are shown in Figure 3-10. The Figure indicates that a crosstalk level of –15 dB is pertinent to a minimum wavelength separation $\Delta\lambda_{sep}$ of ±4 nm and a maximum wavelength misalignment $\Delta\lambda_{tol}$ of only ±0.14 nm. The undesirable large wavelength separation between channels to maintain low crosstalk level is a direct result of the high sidelobes in the optical transmission response. This tight wavelength tolerance can be relaxed by flattening the top of this response curve. Acousto-optic switch performance with smaller $\Delta\lambda_{sep}$ separation and larger $\Delta\lambda_{tol}$ can be obtained by non-uniform acousto-optic interaction as discussed next.

3.4.2 Model for Passband Engineering

The sinc-squared response of a standard acousto-optic switch is due to uniform interaction between an optical wave and a SAW the amplitude of which is constant throughout the interaction length. The optical response of the switch can be modified by using a SAW with amplitude that varies along the interaction region.

A numerical model has been implemented to study the optical transmission response for non-uniform acousto-optic interaction. A varying coupling coefficient $\kappa = \kappa(y)$ and a varying birefringence Δn, or equivalently a varying detuning from phase-matching $\delta = \delta(y)$, are considered, assuming a collinear acousto-optic interaction along the y axis. The model is based on Jones' matrix calculus [26] and on decomposition of the interaction length L to N subsections. Each i-th section acts as a short AOPC of length $\Delta L = L/N$, with an acousto-optic coefficient κ_i and detuning δ_i, as shown in Figure 3-11.

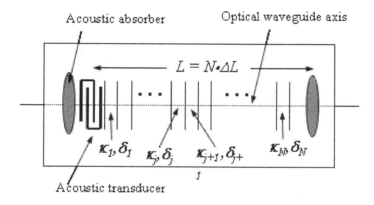

Figure 3-11. AO interaction length divided into N subsections with different acousto-optic coupling coefficients κj and detuning δj [27] (© 1996 IEEE).

For the i-th section, the polarization transformation of the electric field can be expressed by Equations (3.12) in the matrix form:

$$\mathbf{E}_i = \mathbf{M}_i(\kappa_i, \delta_i, y_i)\mathbf{E}_{i-1} \qquad i = 1, 2, ..., N \qquad (3.15)$$

where the vectors of the electric field and the matrix $\mathbf{M}_i$ are given by:

$$\mathbf{E}_{i-1} = \begin{vmatrix} E_1(y_{i-1}) \\ E_2(y_{i-1}) \end{vmatrix}, \quad \mathbf{E}_i = \begin{vmatrix} E_1(y_i) \\ E_2(y_i) \end{vmatrix} \qquad (3.16)$$

and

$$\mathbf{M}(\kappa_i, \delta_i, y_i) =$$

$$= \begin{pmatrix} e^{j\delta_i \Delta L}\left(\cos\mu_i \Delta L - j\dfrac{\delta_i}{\mu_i}\sin\mu_i \Delta L\right) & -je^{j\delta_i \Delta L}\dfrac{\kappa_i}{\mu_i}\sin\mu_i \Delta L \\ -je^{-j\delta_i \Delta L}\dfrac{\kappa_i^*}{\mu_i}\sin\mu_i \Delta L & e^{-j\delta_i \Delta L}(\mu_i \cos\mu\Delta L + j\delta_i \sin\mu_i \Delta L) \end{pmatrix}$$

$$(3.17)$$

where $\Delta L = L/N = y_i - y_{i-1}$, $\delta_i = \delta(y_i)$ and $\kappa_i = \kappa(y_i)$. This model was successfully used to explain the asymmetry of the optical response measured in standard acousto-optic tunable filters due to variation of optical birefringence in the waveguides along the interaction length [28]. A similar approach was also used to describe the passband of weighted coupling integrated acousto-optic tunable filters [29].

3.4.3 Acousto-Optic Switches with Sidelobe-Suppressed Passband

Several prototypes of acousto-optic tunable polarization converters were experimentally demonstrated with significant sidelobe suppression. All these solutions were based on apodization of acousto-optic interaction intensity [30]-[34]. In particular, sidelobe-suppressed acousto-optic tunable filters and switches were obtained by sinusoidal acousto-optic interaction strength [35]. In this case the coupling coefficient can be expressed as

$$\kappa(y) = \kappa_0 \sin (\pi y/L) \tag{3.18}$$

where κ_0 is the acousto-optic coupling coefficient peak. The coefficient in (3.18) was obtained by embedding an optical waveguide in one arm of a SAW directional coupler with coupling length of $2L$, as sketched in Figure 3-12, which also shows a regular AOPC for comparison. The acoustic intensity varies sinusoidally along the optical waveguide and so does the acousto-optic coefficient as indicated simply by "the center of mass" of the SAW front traced in Figure 3-12 [27]. The SAW is launched by applying RF field on the arm of the SAW coupler not involved in the acousto-optic interaction (this arm has a "dummy" optical waveguide to obtain structural symmetry for correct operation of the coupler). Full polarization conversion over half of the coupling length L occurs if the following condition is fulfilled

$$\int_0^L \kappa(y)dy = \frac{\pi}{2} \tag{3.19}$$

Using $\kappa(y)$ of Equation (3.18), the left side of Equation (3.19) yields

$$\int_0^L \kappa_0 \sin\left(\frac{\pi y}{L}\right)dy = 2\kappa_0 \frac{L}{\pi} \tag{3.20}$$

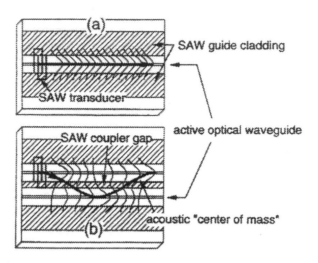

Figure 3-12. Integrated AOPC with: (a) uniform acousto-optic interaction strength, (b) SAW-coupler based apodized interaction strength [27] (© 1996 IEEE).

This apodized SAW-coupler based AOPC may be compared qualitatively with a uniform SAW intensity based AOPC, by introducing an effective coupling coefficient κ_{eff} and an effective interaction length L_{eff} for which 100% polarization conversion occurs:

$$\kappa_{eff}L_{eff} = \pi/2 \qquad\qquad (3.21)$$

From Equations (3.19), (3.20) and (3.21), the driving power requirement of an apodized AOPC is $(\pi/2)^2$ times the driving power for a uniform AOPC of the same length with coupling coefficient κ_{eff}, bearing in mind that κ^2 is proportional to the driving power. On the other hand, if the same driving power is used for both AOPC's ($\kappa_0 = \kappa_{eff}$), L_{eff} is shorter than L by a factor of about $\pi/2$, implying a larger bandwidth for an apodized AOPC, compared to the uniform type, according to Equation (3.14). Direct calculations of FWHM yield a factor of 1.4 in this regard. The optical transmission response of an apodized AOPC can be calculated by applying the model for passband engineering with a sinusoidal κ as per Equation (3.18). Optical transmission spectra of uniform and apodized AOPC for a given interaction length L are plotted in Figure 3-13. This result is experimentally confirmed by the measurements shown in Figure 3-14.

The apodized AOPC response has lower sidelobes by about 10 dB in theory and larger bandwidth in comparison with the uniform AOPC response. The asymmetry of the optical response in Figure 3-14 is due to birefringence variation along the interaction length, which can be expected according to the passband engineering model [28].

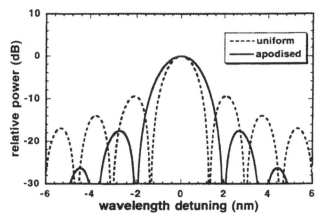

Figure 3-13. Comparison of calculated SAW-coupler-apodized AOPC transmission response to the uniform-SAW-intensity AOPC response. LN parameters are assumed at 1550 nm. The interaction length is 19 mm [27] (© 1996 IEEE).

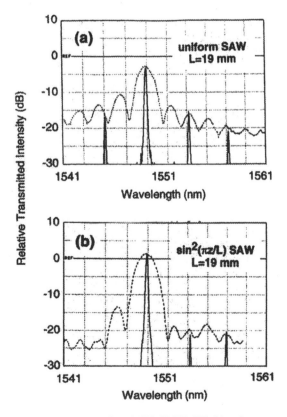

Figure 3-14. Cross-state spectra for a fixed 175.5 MHz RF drive frequency for (a) a uniform 19 mm active length AOPC and (b) an apodized AOPC of the same length. Dash line spectra refer to an input white light source and solid line spectra refer to a WDM input signal consisting of 4 nm spaced wavelengths emitted by four DFB lasers [27] (© 1996 IEEE).

A complete apodized 2 × 2 acousto-optic switch was made by using the hybrid approach, where two waveguide apodized AOPC's are combined with bulk optical PBS's and polarization preserving fibers [3]. The passband engineering model shows that a gaussian SAW profile reduces the sidelobes even further. A quasi-gaussian acousto-optic interaction was realized by using a tapered acoustical directional coupler [36] and was included in a fully packaged 2 × 2 acousto-optic switch, whose passband revealed sidelobes that are about –18.5 dB below the transmission peak [37].

3.4.4 Passband Flattening

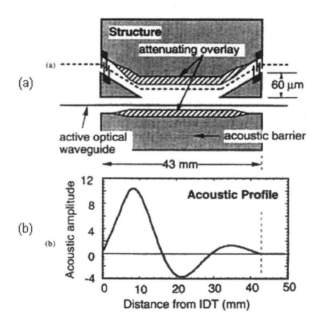

Figure 3-15. (a) Dual linear tapered zero-gap coupler with an attenuating overlay and (b) corresponding calculated SAW amplitude [3][4] (© 1996 IEEE).

Wavelength-misalignment tolerances in AO switches can also be improved by reducing the sharpness of the optical transmission response peaks. In particular, a fourth-order Butterworth profile of the acousto-optic coupling coefficient yields a rectangular profile of the response curve [38]. A quasi-rectangular response was computed by modifying the SAW coupler in such a way to have a tapered onset zero-gap coupler with 60 μm gap tapering down to zero over 10 mm and a 6 dB/cm acoustic absorber as depicted in Figure 3-15. The optical response in this case is shown in Figure 3-16 [39]. These theoretical results suggest that a flat-top shape of the optical transmission response curve may be obtained by slowly decreasing the amplitude and by periodic phase reversal.

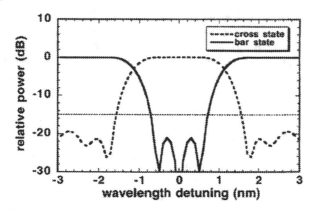

Figure 3-16. Flat optical response of the device of Figure 3-15 [4] (© 1996 IEEE).

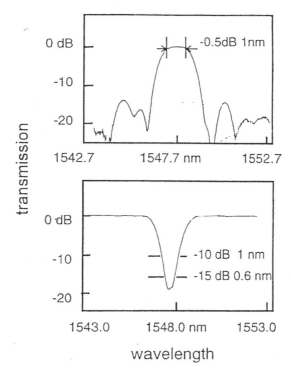

Figure 3-17. Experimental flat optical response of a double length SAW-coupler based AOPC: the upper trace is the passband, the lower trace is the notch spectrum [40] (© 1994 IEEE).

An experimental flat top passband is shown in Figure 3-17. This was demonstrated by using a double length SAW coupler-based AOPC [40]. The device length was nearly equal to the coupling length. In another experiment an even better rectangular bar-state spectrum was obtained by using a temperature compensated, tapered-onset, zero-gap apodized AOPC [41].

3.5 MULTIWAVELENGTH PERFORMANCE OF ACOUSTO-OPTIC SWITCHES

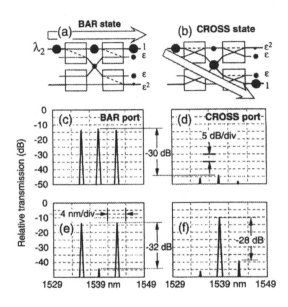

Figure 3-18. Dilated 2x2 switch made of four polarization-insensitive apodized acousto-optic switches in the bar state (a) and cross state (b), as seen from the upper input port. The dashed lines represent crosstalk path. Three wavelengths are used and none has been switched in (c) and (d), but wavelength λ_2 is selected in (e) and (f) [35] (© 1996 IEEE).

Sidelobe suppression of the optical response of acousto-optic switches for WDM applications reduces out-of-band crosstalk and passband flattening lowers coherent crosstalk, caused by wavelength misalignment. Although passband engineering techniques can greatly improve the optical response of LN AOPC switches, crosstalk levels due to sidelobes are still above the required level to prevent performance degradation in WDM optical networks (about -35 dB) [42]. Crosstalk can be further reduced from a value of ε to ε^2 by space switch dilation [43] as in electro-optical directional coupler switch matrices [44]. Space dilation permits routing of single-stage crosstalk signals to unused output ports, as shown for a 2×2 dilated acousto-optic switch in the upper side of Figure 3-18. The dilated switch consists of four single-stage hybrid apodized fiber-pigtailed acousto-optic switches connected as shown in the Figure. The output spectrum of the switch using three DFB laser channels separated by 4 nm is also shown. Interchannel crosstalk measured at cross-port was always better than −28 dB [35].

In the case of closely spaced WDM channels, many acoustic gratings can coexist in a single interaction length, but, unfortunately, this effect comes along with mutual time-dependent interaction among passbands. This phenomenon is observed experimentally [45]-[46]. Such an interaction arises from the composite acoustic grating, which is a traveling beat pattern that is related to the differences among the different driving acoustic frequencies. The beat pattern induces an interaction strength which varies along the device length and the resulting passband is strongly time dependent. This, in turn, produces a channel inter-modulation effect. This effect is reduced in sidelobe suppressed acousto-optic switches but becomes stronger with closely spaced wavelength channels which are switched by acoustic frequencies having small differences. The time dependence of the passband induced by a composite acoustic grating can be studied in detail by using the mathematical formalism developed for passband engineering, where the coupling coefficient κ is modulated in time and space. Bit-error-rate degradation due to this intermodulation effect was also calculated by the coupled mode equations [47].

Wavelength dilation can be used to overcome the intermodulation effect in WDM AO switches. Wavelength dilation implies subdividing the WDM spectrum of the system into interleaved subsets of channels and dilating the switch architecture into subswitches, each of which handles a WDM subsets [48]. This approach was applied to an acousto-optic switch [49]. In this AO switch operating with eight wavelengths spaced by 2 nm, a crosstalk ranging from −25 to −35 dB has been reported [50]. Nevertheless, the benefits gained by using both space and wavelength dilation come at the expense of switch complexity and extra power loss [51]. Dilation and passband engineering which are often demonstrated with LN AO switches can be applied to integrated AO switches based on other materials and to AO fiber-coupler switches.

Under the current state of the art, coherent crosstalk and time dependent intermodulation effects limit the use of AO switches to WDM systems where wavelength spacing is at least 2 nm. Meanwhile, AO switching matrices with only a limited number of ports are feasible by means of multistage LN integrated architectures. This is because of insertion losses; at least 3 dB per stage. Nevertheless, multiwavelength operation makes acousto-optic switching unique. In general, acousto-optic switches, unlike other optical switches, can be tuned nearly at will over their wavelength range and are not limited to specific set or number of wavelengths. Acousto-optic switching is wavelength-selective with many advantages, including sub-millisecond switching time, polarization-insensitive performance and driving power in the range of a few mW. Acousto-optic single-mode fiber tapers and couplers, still at research level, have promising potentialities with insertion loss of 0.1 dB and driving power that is less than 1 mW. This is due to excellent overlap between acoustic and optical modes in the fiber coupler waist.

REFERENCES

[1]   A. Yariv and P. Yeh, "Optical waves in crystals", John Wiley & Sons, 1984.

[2]   C. S. Tsai, "Guided wave acousto-optics", Springer-Verlag, 1990.

[3]   H. Nishihara, M. Haruna, and T. Suhara, "Optical Integrated Circuits", McGraw-Hill, 1989.

[4]   D. A. Smith, R. S. Chakravarthy, Z. Bao, J. E. Baran, J. J. Jackel, A. d'Alessandro, D. J. Fritz, S.H. Huang, X. Y. Zou, S. M. Hwang, A. Willner, and K. D. Li, "Evolution of the acousto-optic wavelength routing switch," J. Lightwave Technol., Vol. 14, No. 6, pp. 1005-19, 1996.

[5]   C. S. Tsai, "Integrated acousto-optic and magneto-optic devices for optical information processing," Proceedings of IEEE, Vol. 84, No. 6 pp. 853-69, 1996.

[6]   T. A. Birks, P. St. J. Russell, and D. O. Culverhouse, "The acousto-optic effect in single-mode fiber tapers and couplers," J. Lightwave Technol., Vol. 14, No. 11, pp. 2519-29, 1996.

[7]   B. E. A. Saleh and M. C. Teich, "Fundamentals of Photonics," John Wiley & Sons, 1991.

[8]   R. S. Weis and T. K. Gaylord, "Lithium Niobate: Summary of Physical Properties and Crystal Structure," Appl. Phys. A, Vol. 37, No. 4, pp. 191-203, 1985.

[9]   Y. Ohmachi and J. Noda, "LiNbO$_3$ TE-TM converter using collinear acousto-optic interaction," IEEE J. Quantum Electron., Vol. 13, No. 2, pp. 43-6, 1977.

[10] A. Kar-Toy and C. S. Tsai, "8 x 8 Symmetric nonblocking integrated acoustooptic space switch module on LiNbO$_3$," IEEE Photon. Technol. Lett., Vol. 4, No. 7, pp. 731-34, 1992.

[11] D. A. Smith, J. E. Baran, K. W. Cheung, and J. J. Johnson, "Polarization independent acoustically tunable optical filters," Appl. Phys. Lett., Vol. 56, No. 3, pp. 209-11, 1990.

[12] T. A. Birks, D. O. Culverhouse, S. G. Farwell, and P. St. J. Russell, "2 x 2 Single-mode fiber routing switch," Opt. Lett., Vol. 21, No. 10, pp. 722-4, 1996.

[13] D. O. Culverhouse, T. A. Birks, S. G. Farwell, and P. St. J. Russell, "3 x 3 All-fiber routing switch," IEEE Photon. Technol. Lett., Vol. 9, No. 3, pp. 333-5, 1997.

[14] D. O. Culverhouse, R. I. Laming, S. G. Farwell, T. A. Birks, and M. N. Zervas, "All fiber 2 x 2 polarization insensitive switch," IEEE Photon. Technol. Lett., Vol. 9, No. 4, pp. 455-7, 1997.

[15] H. Kogelnik, "Coupled mode theory for thick holograms," Bell System Tech. Journal, Vol. 48, No. 9, pp. 2909-47, 1969.

[16] A. d'Alessandro, D. A. Smith, and J. E. Baran, "Polarization independent low-power acousto-optic tunable filter/switch using APE/Ti polarization splitters on lithium niobate," Electron. Lett., Vol. 29, No. 20, pp. 1767-9, 1993.

[17] A. Neyer and T. Pohlmann "Fabrication of low-loss titanium-diffused LiNbO$_3$ waveguides using a closed platinum crucible," Electron. Lett., Vol. 23, No. 22, pp. 1187-8, 1987.

[18] J. E. Baran and D. A. Smith, "Adiabatic 2 x 2 polarization splitter on LiNbO$_3$," IEEE Photon. Technol. Lett., Vol. 4, No. 2, pp. 39-40, 1992.

[19] T. Pohlmann, A. Neyer, and A. Voges, "Polarization independent Ti:LiNbO$_3$ switches and filters," IEEE J. Quantum Electron., Vol. 27, No. 7, pp. 602-7, 1991.

[20] F. Tian, Ch. Harizi, H. Herrmann, V. Reimann, R. Ricken, U. Rust, W. Sohler, F. Wehrmann, and S. Westenhöfer, "Polarization-independent integrated, optical, acoustically tunable double-stage wavelength filter in LiNbO$_3$," IEEE J. Lightwave Technol., Vol. 12, No. 7, pp. 1192-7, 1994.

[21] D. P. Morgan, "Surface-Wave devices for signal processing," Elsevier, 1985.

[22] B. L. Heffner, D. A. Smith, J. E. Baran, and K. W. Cheung, "Integrated-optic acosutically-tunable infrared optical filter," Electron. Lett., Vol. 24, No. 25, pp. 1562-3, 1988.

[23] D. A. Smith and J. J. Johnson, "Low drive-power integrated acousto-optic filter on X-cut Y-propagating LiNbO$_3$," IEEE Photon. Technol. Lett., Vol. 3, No. 10, pp. 923-5, 1991.

[24] D. A. Smith and J. J. Johnson, "Switching speed of an integrated acosutically tunable optical filter," Electron. Lett., Vol. 27, No. 23, pp. 2102-3, 1991.

[25] A. d'Alessandro, D. A. Smith, and J. E. Baran, "Multichannel operation of an integrated acousto-optic wavelength routing switch for WDM systems," IEEE Photon. Technol. Lett., Vol. 6, No. 3, pp. 390-3, 1994.

[26] R. C. Jones, "New Calculus for the treatment of optical systems I. Description and Discussion of the Calculus," J. Opt. Soc. Am., No. 31, p. 488-93, 1941.

[27] D. A. Smith, A. d'Alessandro, J. E. Baran, D. Fritz, J. Jackel, and R. S. Chakravarthy "Multiwavelength performance of an apodized acosuto-optic switch," IEEE J. Lightwave Technol., Vol. 14, No. 9, pp. 2044-51, 1996.

[28] D. A. Smith, A. d'Alessandro, J. E. Baran, and H. Herrmann, "Source of sidelobe asymmetry in integrated acousto-optic tunable filters," Appl. Phys. Lett., Vol. 62, No. 8, pp. 814-6, 1993.

[29] A. Kar-Roy and C. S. Tsai, "Integrated acousto-optic tunable filters using weighted coupling," IEEE J. Quantum Electron., Vol. 30, No. 7, pp. 1574-86, 1994.

[30] D. A. Smith and J. J. Johnson, "Sidelobe suppression in an acousto-optic filter with a raised cosine interaction strength," Appl. Phys. Lett., Vol. 61, No. 9, pp. 1025-7, 1992.

[31] H. Hermann and H. Schmid, "Integrated acousto-optical mode-converters with weighted coupling using surface wave directional couplers," Electron. Lett., Vol. 28, No. 11, pp. 979-80, 1992.

[32] Y. Yamamoto, C. S. Tsai, and K. Esteghamat, "Guided-wave acousto-optic tunable filters using simple coupling weighting technique," IEEE Proc. Ultrason. Symp., Vol. 2, pp. 605-8, 1990.

[33] Y. Yamamoto, C. S. Tsai, and K. Esteghamat, and H. Nishimoto, "Suppression of sidelobe levels for guided-wave acousto-optic tunable filters using weighted coupling," IEEE Trans. Ultrason. Ferroelect. Freq. Cont., Vol. 40, No. 6, pp. 814-8, 1993.

[34] A. Kar-Roy and C. S. Tsai, "Low-sidelobe weighted-coupled integrated acoustooptic tunable filter using focused surface acoustic waves," IEEE Photon. Technol. Lett., Vol. 4, No. 10, pp. 1132-5, 1992.

[35] D. A. Smith and J. J. Johnson "Surface-Acoustic-Wave directional coupler for apodization of integrated acousto-optic filters," IEEE Trans. Ultrason., Ferroelect., Freq. Cont., Vol. 40, No. 1, pp. 22-5, 1993.

[36] H. Hermann, K. Schafer, and W. Sohler, "Polarization independent, integrated optical acoustically tunable wavelength filters/switches with tapered acoustical directional coupler," IEEE Photon. Technol. Lett., Vol. 6, No. 11, pp. 1335-8, 1994.

[37] F. Wehrmann, C. Harizi, H. Herrmann, U. Rust, W. Sohler, and S. Westenhöfer, "Integrated optical, wavelength selective, acoustically tunable 2 x 2 switches (add-drop multiplexers) in LiNbO$_3$," IEEE J. Select. Top. Quantum Electron., Vol. 2, No. 2, pp. 263-9, 1996.

[38] G. H. Song, "Toward the ideal codirectional Bragg filter with an acousto-optic filter design," IEEE J. Lightwave Technol., Vol. 13, No. 3, pp. 470-80, 1995.

[39] R. S. Chakravarthy, D. A. Smith, A. d'Alessandro, J. E. Baran, and J. L. Jackel, "Passband Engineering of acousto-optic tunable filters," Proc. 7th European Conf. on Integrated Optics, Delft, April 3-6, 1995.

[40] J. L. Jackel, J. E. Baran, A. d'Alessandro, and D. A. Smith, "A passband-flattened acoust-optic filter," ," IEEE Photon. Technol. Lett., Vol. 7, No. 3, pp. 318-20, 1995.

[41] D. A. Smith, H. Rashid, R. S. Chakravarthy, A. M. Agboatwalla, A. A. Patil, Z. Bao, N. Imam, S. W. Smith, J. E. Baran, J. L. Jackel, and J. Kallman, "Acousto-optic switch with a near rectangular passband for WDM systems," Electron. Lett. Vol. 32, No. 6, pp. 542-3, 1996.

[42] E. L. Goldstein, L. Eskjldsen, and A. F. Elrefaie, "Performance implications of component crosstalk in transparent lightwave networks," IEEE Photon. Technol. Lett., Vol. 6, No. 5, pp. 657-60, 1994.

[43] D. A. Smith, A. d'Alessandro, J. E. Baran, D. J. Fritz, and R. H. Hobbs "Reduction of crosstalk in an acousto-optic switch by means of dilation," Opt. Lett., Vol. 19, No. 2, pp. 99-101, 1994.

[44] J. E Watson, M. A. Milbrodt, K. Bahadori, M. F. Dautartas, C. T. Kemmemerer, D. T. Moser, A. W. Schelling, T. O. Murphy, J. J. Veselka, and D. A. Herr "A low-voltage 8 x 8 Ti:LiNbO3 switch with a dilated Benes architecture," IEEE J. Lightwave Technol., Vol. 8, No. 5, pp. 794-801, 1990.

[45] D. A. Smith, R. S. Chakravarthy, L. Troilo, and A. d'Alessandro, "Passband collisions and multi-channel crosstalk in acousto-optic filters and switches," ECIO '95 Proc. of the 7th European Conf. on Integrated Optics, Delft, Vol. 1, pp. 509-11, April 3-6, 1995.

[46] J. L. Jackel, J. E. Baran, D. A. Smith, R. S. Chakravarthy, and D. J. Fritz, "Observation of modulated crosstalk in multichannel acousto-optic switches," Proc. of Integrated Photonics Research Conf., postdeadline paper, PD5, Dana Pt., Ca, Feruary 23-25, 1995.

[47] F. Tian and H. Herrmann, "Interchannel interference in multiwavelength operation of integrated acousto-optical filters and switches," IEEE J. Lightwave Technol., Vol. 13, No. 6, pp. 1146-54, 1995.

[48] J. Sharony, K. W. Cheung, and T. E. Stern, "The wavelength dilation concept in lightwave networks-Implementation and system considerations," IEEE J. Lightwave Technol., Vol. 11, No. 5-6, pp. 900-7, 1993.

[49] J. L. Jackel, J. E. Baran, G. K. Chang, M. Z. Iqbal, G. H. Song, W. J. Tomlinson, and R. Ade, "Multi-channel operation of AOTF switches: reducing channel-to-channel interaction," IEEE Photon. Technol. Lett., Vol. 7, No. 4, pp. 370-2, 1995.

[50] J. L. Jackel, M. Goodman, J. Gamelin, W. J. Tomlinson, J. E. Baran, G. II. Song, C. A. Brackett, D. J. Fritz, R. Hobbs, K. Kissa, R. Ade, and D. A. Smith, "Simultaneous and independent switching of 8-wavelength channels with 2 nm spacing using a wavelength-dilated acousto-optic switch," IEEE Photon. Technol. Lett., Vol. 8, No. 11, pp. 1531-3, 1996.

[51] K. Padmanabhan and A. N. Netravali, "Dilated network for photonic switching," IEEE Trans. on Commun., Vol. 35, No. 12, pp. 1357-65, 1987.

Chapter 4

THERMO-OPTIC SWITCHING

Ralf Hauffe
Klaus Petermann

Although the principle of thermo-optic (TO) switching is simple and the technology is not difficult, research in this field has started relatively recently. This was probably due to the slow switching speed of TO switches compared to other technologies. Early work in guided-wave devices was more focused on utilizing the electro-optic (EO) effect to realize fast optical switches, modulators and other devices. Thermal effects were regarded as undesirable and usually degraded the performance of these devices.

In the 1980, early thermo-optic devices were fabricated using ion exchanged glass [1] and Titanium diffused Lithium Niobate (Ti:LiNbO$_3$) [2]. However, since the TO effect in these materials was relatively weak, devices based on them were not efficient. Polymers became candidates to fabricate thermo-optic devices due to their large thermo-optic coefficients and low thermal conductivity [3]-[4]. In the late 1990s, a lot of research was devoted to thermo-optic switches especially in polymers. Thermally-optimized silica-based interferometric TO switches became also popular. Virtually, all thermo-optic switching principles and device concepts, including directional couplers, Mach-Zehnder interferometers (MZI) and digital switches [5], had been used before in EO switching and were adapted with slight modifications for TO switching.

Thermo-optic waveguide switches have the advantages of relying on guided lightwaves (no free space optics) and not having moving parts. They are robust, enjoy long-term stability and their electrical driving circuitry can be very simple. Today, TO switching speeds ranging from 100 μs to several milliseconds, depending on switch design, are possible. Several switching devices can be interconnected into integrated multi-stage optical switching fabrics. In a single chip, 16x16 TO waveguide switching matrices have been fabricated, and these matrices can be used as building blocks of larger modular switching systems. Thermo-optic switches are particularly suitable for small-to-medium scale optical switching applications.

4.1   PRINCIPLES OF THERMO-OPTIC SWITCHING

4.1.1   Thermo-Optic Effect in Optical Waveguides

Interaction of electro-magnetic lightwaves with the core and cladding of a dielectric waveguide is governed by the geometry of the waveguide and its refractive indices (n_{cor}, n_{clad}). *Thermo-optic switching* utilizes the temperature dependence of the refractive index dn/dT to realize switching functionality. Small thermally-induced changes of the refractive index can significantly alter the light-intensity distribution within waveguides, especially in coupled systems. By using micro heaters, temperature gradients can be induced within these waveguide structures leading to changes in the refractive index profile, which, in turn, is used to switch lightpaths.

In general the refractive index is a second rank tensor which becomes complex if optical losses are taken into consideration [6]. As an approximation, it is often acceptable to ignore these losses and treat the refractive index as a real scalar value (lossless isotropic media). Stress and the associated birefringence along the principle axis of the waveguide are accounted for by a semi vectorial approximation of the waveguide equation and by using two different values of the refractive index n for the *transverse electric* (TE) and *transverse magnetic* (TM) fields.

Material	Refractive Index (at 1550nm)	$C_{to}=dn/dT$ [$10^{-4}K^{-1}$]	Thermal Expansion Coefficient γ [$10^{-6}K^{-1}$]	Thermal Conductivity σ_{th} [$WK^{-1}m^{-1}$]
Polymers	1.3 - 1.7	(-1) – (-4)	10 - 220	0.1 - 0.3
Silica	1.5	0.1	0.6	1.4
Silicon	3.5	1.8	2.5	168

Table 4- 1 Properties of polymers, silicon, and silica glass at room temperature [7]-[10]

As shown in Table 4-1, the thermo-optic coefficient $C_{to}=dn/dT$ varies considerably among thermo-optic material. In terms of thermal properties, polymers make ideal waveguides for thermo-optic applications. Due to the high magnitude of their thermo-optic coefficient and their low thermal conductivity, *polymer* devices exhibit high power-conversion efficiency. This makes them suitable for making digital optical switches (where large refractive-index change is required).

Devices based on *silica* or silicon are usually restricted to interferometric switches. In the case of silicon, this is not due to low thermo-optic coefficient, since the silicon coefficient is comparable to that of polymers, but because of high thermal conductivity. Despite its low thermo-optic coefficient, silica glass is often used for thermo-optic switches due to its very low waveguide loss and because of its well established fabrication techniques. These techniques are widely used to fabricate arrayed waveguide gratings (AWGs) and other devices, which can be integrated with TO switches.

Polymers and silica glass are the most prominent materials for fabricating thermo-optic switches. We focus our attention on them throughout this chapter.

The relation between the refractive index n and temperature T is given approximately by:

$$n(T)=n(T_0)+C_{to}\cdot(T-T_0) \tag{4.1}$$

where $n(T_0)$ is the refractive index at the reference temperature T_0 and C_{to} is given approximately by [11]-[12]:

$$C_{to} = \frac{dn}{dT} = \frac{(n^2-1)(n^2+2)}{6n}\left(\underbrace{\frac{1}{\alpha}\frac{d\alpha}{dT}}_{\substack{\text{Polarizability} \\ \text{coefficient}}} - \underbrace{3\gamma}_{\substack{\text{Thermal} \\ \text{expansion}}} \right) \tag{4.2}$$

In the analysis leading to Equation (4.2), it is assumed that the mean polarizability α is independent of the density (expansion) of the material and that the linear expansion coefficient γ does not vary with temperature. Two mechanisms are responsible for the change in refractive index: First the change of polarizability with temperature and second, the decrease of density with thermal expansion. For polymers, the term 3γ is dominantly larger than $(1/\alpha)d\alpha/dT$ and the latter can be neglected [13]. This leads to a negative refractive index change with increasing temperature. For a typical polymer with thermal expansion coefficient $\gamma = 9\cdot10^{-5}$ K^{-1} and refractive index $n= 1.5$, this gives $C_{to} = -1.6 \times 10^{-4}$ K^{-1}. In silica, which has low thermal expansion coefficient, $(1/\alpha)d\alpha/dT$ is much larger than 3γ, resulting in positive C_{to} with relatively small magnitude, $C_{to}= 1 \times 10^{-5} K^{-1}$.

4.1.2 Design of Heat Electrodes

The switching functionality in TO waveguide devices is accomplished by introducing thermal gradients along waveguide(s). This is usually achieved by *micro heaters* positioned above one of the waveguides. An electrical current is supplied to the micro heater to induce thermal energy. Temperature changes the optical path length in the heated waveguide relative to other (cold) waveguide(s). Proper design of heat electrodes is important. The position and dimension of the micro heaters have significant influence on the *efficiency* of the switch and on its *switching time* (τ). The efficiency of the switch is defined as the reciprocal of the mean power dissipation in the bar and cross states. Switching time is the time it takes the switch to change state. Metal micro heaters are deposited by vapor deposition or sputtering techniques and structured by conventional lithography. The minimum heater width is in the order of several μms and the thickness of the metal film is usually below 1 μm.

A heat sink with high thermal conductivity is necessary in the TO switching system. The relative placement of this sink with respect to the electrode affects switching time and temperature distribution in the waveguide. Silicon, which is often used as substrate material for silica and polymer waveguide devices, serves naturally as a heat sink because of its high thermal conductivity. Other advantages of silicon include availability in large quantities, with excellent surface quality and large substrate sizes (up to 12 inches in diameter). Silicon processing technologies, including reactive ion etching, wet etching, thermal oxidation and thin film deposition, are all mature [14].

Better device efficiency can be obtained by maximizing the refractive-index difference induced between the centers of a pair of waveguides ($\Delta n_{1,2}$) due to single micro heater [15]. An approximate model is developed to optimize the position and width (w_{el}) of this micro heater for this purpose. This model is depicted in Figure 4-1. We assume that the lower cladding layer, waveguide core layer, and the upper cladding layer are laid out along the y-direction and are sandwiched between an ideal heat sink (bottom) and a micro heater (top). The core and cladding layers form a homogenous slab with identical thermal and mechanical properties (thermal capacitance, c_{th}, thermal conductivity, σ_{th}, and density ρ). The refractive indices at a reference temperature T_0 and the thermo-optic coefficients of these layers are assumed to be similar too. The refractive index is elevated only in two quadratic regions of the core layer to model the waveguides, as shown in Figure 4-1. These waveguides are positioned at the same height within the slab, and their centers are separated by a distance d along the x-direction. The point in the middle between the two waveguides is referenced as x_0.

To calculate the temperature and refractive index distributions in such a slab the differential equation

$$\nabla(\sigma_{th}\nabla T) = c_{th}\frac{\partial T}{\partial t} \qquad (4.3)$$

must be solved for the static case with appropriate boundary conditions, namely, $dT/dy=0$ at the upper interface without micro heater, and fixed temperatures at the heat sink and the micro heater. For a given supplied thermal power, the largest refractive index difference ($\Delta n_{1,2}$) between the two waveguide centers can be obtained under the following conditions:

1) The micro heater width w_{el} is as small as possible
2) The upper waveguide cladding h_{up} is as thin as possible
3) The lower cladding h_{low} is as thick as possible
4) The mid point between the two waveguides x_0 coincides with the position of maximum change of temperature in the x-direction ($x_0=x|_{max(dT/dx)}$ for $d/2 < x|_{max(dT/dx)}$), or, one waveguide is centered below the micro heater (for $d/2 > x|_{max(dT/dx)}$)

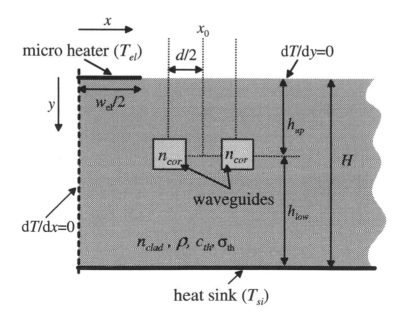

Figure 4-1. Simplified model calculating of the micro-heater induced effective refractive index difference between two waveguides

These conditions are based only on consideration of the thermal behavior of the model. Optical impairments and fabrication techniques and parameters will impact the refractive-index change too. From the optical perspective the model assumes that the waveguides are designed in such a way that ensures single mode operation and ideal guidance of the fundamental mode.

This model can approximate waveguide structures based on buried ridge guides with homogeneous cladding. Other types of waveguides, like polymer rib guides in silica grooves, or polymer guides with silica over-cladding, can still be treated with this model by introducing effective heights and effective micro heater widths. Another important characteristic, apart from the refractive index modulation efficiency, is switching time. The most important parameter to influence the switching time is the overall slab height H. Switching time is proportional to H^2 and can be approximated by:

$$\tau = \frac{c_{th}\rho}{2\sigma_{th}} H^2 \qquad (4.4)$$

assuming that the waveguide is located in the middle of the slab and centered under the micro heater. The thickness of the lower cladding should be made as large as possible for high power-conversion efficiency and as small as possible for high switching speed. Therefore, the design involves a compromise which should be dealt with depending on application.

In the following, we consider two *model examples* of waveguides with identical geometrical dimension and refractive indices, but with different material parameters (for thermal conductivity and thermo-optic coefficient). We use these two examples, throughout the remainder of this chapter, to illustrate the dependence of TO switch performance on the material used to fabricate waveguides (silica and polymers). The overall thickness of the slab in both devices is $H= 30\mu m$, and the waveguides are centered above the thermal sink with $h_{low}= 21$ μm (Figure 4-1). The waveguides are quadratic ($6\mu m \times 6\mu m$) with refractive indices $n_{cor}=1.45$ and $n_{clad}=1.44$. The distance d between them depends on the specific type of switch under investigation and will vary throughout this chapter. However, the absolute position x_0 is chosen to fulfill condition 4) here above. The electrode width is $w_{el}= 15$ μm. Unless otherwise indicated, we assume operation at the wavelength (λ) of 1.55μm. The material properties of the silica switch example are: $C_{to}=0.1$ x $10^{-4}K^{-1}$, $c_{th}=0.8$ $Jg^{-1}K^{-1}$, $\rho=2.2$ gcm^{-3}, and $\sigma_{th}=1.4$ $WK^{-1}m^{-1}$. The corresponding parameters for a typical polymer are $C_{to}= -1\cdot10^{-4}K^{-1}$, $c_{th}=1.5$ $Jg^{-1}K^{-1}$, $\rho=1.2$ gcm^{-3}, and $\sigma_{th}=0.1$ $WK^{-1}m^{-1}$.

Let the micro heater supply 10 W/m of thermal power per unit length (P'_{el}) and let the distance between the waveguides $d= 90$ μm. For these values, and the above material properties, the silica switch example yields a switching time (τ) of approximately 566 μs and the refractive-index difference between waveguide centers becomes 2.86×10^{-5}. The corresponding values for of the polymer switch example are 8.1 ms and 4×10^{-3} respectively. If the waveguide spacing d is reduced to 15 μm, the refractive index difference drops to 1.07 x 10^{-5} for silica and 1.5 x 10^{-3} for polymer. These figures demonstrate the high efficiency in polymer devices compared to silica devices and the high switching speed of silica devices compared to their polymer counterparts. Switching times in the range of a few milliseconds are suitable for many optical-network applications.

4.2   THERMO-OPTIC WAVEGUIDE SWITCHES

Figure 2-4 depicts the most important types of thermo-optic waveguide switches. In the following, we discuss these types in some detail.

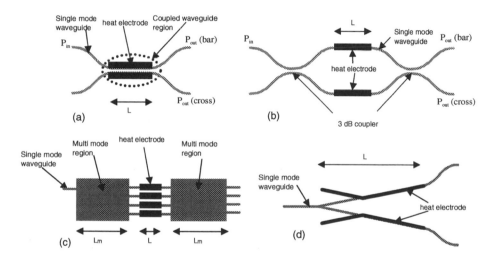

Figure 4-2. Basic configurations of thermo-optic waveguide switches. (a) Directional coupler. (b) Mach-Zehnder interferometer. (c) Generalized Mach-Zehnder interferometer based on multi-mode interference couplers. (d) Y-shaped digital thermo-optic switch.

4.2.1   Directional Coupler Switches

As discussed earlier in chapters 2 and 3, and depicted in Figure 4-2(a), the directional coupler consists of two parallel waveguides brought close to each other over a certain interaction length L. Optical power can be interchanged between the two waveguides. The switch is in an ideal bar state if an optical signal, entering through one waveguide, leaves the switch through the same waveguide. In an ideal cross state, power is completely interchanged between the waveguides.

Two different theories are widely used to describe the behavior of directional couplers: coupled mode theory and normal system mode theory.

We use the coupled mode theory which considers the modes of the two guides separately and assumes a coupling coefficient κ between them. Power transfer among modes is governed by this coefficient [16]. This model is only valid for waveguides that are sufficiently separated by some distance d. Coherent light in the two guides can be represented by two complex amplitudes $a_1(z)$ and $a_2(z)$ propagating in the z-direction according to the coupled wave equations:

$$\frac{da_1}{dz} - j\frac{\Delta\beta}{2}a_1 = -j\kappa a_2 \tag{4.5}$$

$$\frac{da_2}{dz} + j\frac{\Delta\beta}{2}a_2 = -j\kappa a_1$$

Where $\Delta\beta = \beta_1 - \beta_2$ is the propagation constant mismatch between waveguides 1 and 2. In thermo-optic devices, this mismatch is controlled by temperature. The complex output amplitudes can be obtained from the input amplitudes using a transfer matrix:

$$\begin{pmatrix} a_1(z=L) \\ a_2(z=L) \end{pmatrix} = \begin{pmatrix} A_{11} & -jA_{12} \\ -jA_{12}{}^* & A_{11}{}^* \end{pmatrix} \begin{pmatrix} a_1(z=0) \\ a_2(z=0) \end{pmatrix}. \tag{4.6}$$

where matrix coefficients are

$$A_{11} = \cos\left(L\sqrt{\kappa^2 + \left(\frac{\Delta\beta}{2}\right)^2}\right) + j\frac{\Delta\beta}{2}\frac{\sin\left(L\sqrt{\kappa^2 + \left(\frac{\Delta\beta}{2}\right)^2}\right)}{\sqrt{\kappa^2 + \left(\frac{\Delta\beta}{2}\right)^2}} \tag{4.7}$$

$$A_{12} = \kappa\frac{\sin\left(L\sqrt{\kappa^2 + \left(\frac{\Delta\beta}{2}\right)^2}\right)}{\sqrt{\kappa^2 + \left(\frac{\Delta\beta}{2}\right)^2}} \tag{4.8}$$

The ideal cross state, where complete exchange of power takes place between waveguides, requires that $A_{11} = 0$. Thus, the cross state takes place if

$$\frac{\Delta\beta}{2}L = 0 \quad \text{and} \quad \kappa L = (2i - 1)\frac{\pi}{2}$$

(4.9)

where i is an integer. For ideal bar state, where no power exchange takes place, the condition $A_{12} = 0$ is required. Hence, this state occurs for

$$(\kappa L)^2 + \left(\frac{\Delta\beta}{2}L\right)^2 = (i\pi)^2$$

(4.10)

To a first order of approximation, κ is considered as independent of $\Delta\beta$. Therefore, care must be taken to guarantee the second condition for the cross state during device fabrication, Equation (4.9). For any value of κL the switch can be brought into the bar state by adjusting $\Delta\beta$. The minimum $\Delta\beta$ required for switching from cross state to bar state ($\kappa L = \pi/2$ and $i = 1$) is given by:

$$\Delta\beta L = \sqrt{3}\pi$$

(4.11)

The device length required for complete change from constructive to destructive interference is defined as the coupling length L_c. This length is given by:

$$L_c = \frac{\pi}{2\kappa}$$

(4.12)

A device designed with this length using identical waveguides has a default perfect cross state. To illustrate the behavior of a TO directional-coupler switch we use the model examples introduced in section 4.1.2. To simplify calculations, the thermally-induced refractive index change $\Delta n_{1,2}$ is assumed to be homogeneous across the core region of one waveguide. With increasing $|\Delta n_{1,2}|$ the propagation constant difference between the two waveguides ($\Delta\beta$) increases and the switch assumes the bar state. For large d the waveguides are decoupled and κ approaches zero. If d is made smaller, coupling increases and, consequently, L_c decreases. The optimum length L_c of a directional-coupler switch is largely determined by the waveguide separation d. L_c can vary from sub millimeters to 10 mm as d is changed by few microns, as shown in Figure 4-3(a). This imposes very tight design tolerances, as the crosstalk (10 $\log(P_{out,off}/P_{out,on})$) depends critically on the proper choice of $L = L_c$. In Figure 4-3(b), the bar and cross transmittance of the device are plotted against $\Delta n_{1,2}$ for an ideal switch with $d = 11\mu m$ and $L = L_c$. The dotted line corresponds to a deviation of 100 nm in d. All other parameters and dimensions are as per our model example of section 4.1.2. Note that Figure 4-3 holds for both silica and polymers.

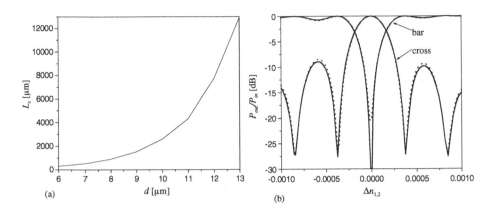

Figure 4-3. (a) The coupling length L_c as a function of the waveguide separation d. (b) Transmittance of an ideal directional coupler ($L = L_c$). The dotted line indicates the results for a deviation of 100 nm from ideal waveguide separation.

So far we discussed ideal directional-coupler devices with two parallel waveguides. Practical devices, however, involve waveguide bends at input/output ports. In these regions, significant amount of pre and post coupling occurs. Approximation methods can be used to take the effect of these bends into consideration. The device can be divided into incremental segments of length Δz and output power can be calculated recursively by multiplying the transfer matrices of all segments as $\Delta z \rightarrow 0$. This does not account for field deformations due to bending and gives reasonable results only for large bend radii.

Although the crosstalk and insertion loss of an ideal TO directional coupler are very low, precise thermal power control is required because of the small thermo-optic switching windows. Design tolerances in this respect are very tight. Since directional-coupler TO switches rely on mode interference, they are wavelength and polarization dependent.

4.2.2 Mach-Zehnder Interferometer Switches

Like directional couplers, the Mach-Zehnder Interferometer (MZI) is an interferometric device [17]. It consists of a 3 dB coupler which splits light into two decoupled waveguides, a phase tuning section of length L in one of the waveguides, and a second 3 dB coupler acting as an output combiner, as shown in Figure 4-2(b). In the thermal tuning section the phase of one of the two split signals is adjusted to give either constructive or destructive interference with the other signal and the combined output signal is routed to the cross or the bar port depending on phase change.

If the 3 dB couplers are realized by directional couplers, the transmission characteristics of the MZI can be calculated by multiplication of their transfer matrices ($L= L_c/2$ for 3-dB splitting) and the transfer matrix of the phase tuning section. The latter is given by:

$$\begin{pmatrix} \exp(-j\Delta\phi) & 0 \\ 0 & 1 \end{pmatrix}$$
(4.13)

where $\Delta\phi = L\Delta\beta$. L is the length of the phase tuning section and $\Delta\beta$ is the change of the propagation constant due to the thermo-optic effect. The transmission characteristics of the MZI switch is given by:

$$\frac{P_{out}}{P_{in}} = \cos^2 \frac{\Delta\beta L}{2} \quad \text{(cross)}$$
(4.14)

$$\frac{P_{out}}{P_{in}} = \sin^2 \frac{\Delta\beta L}{2} \quad \text{(bar)}$$
(4.15)

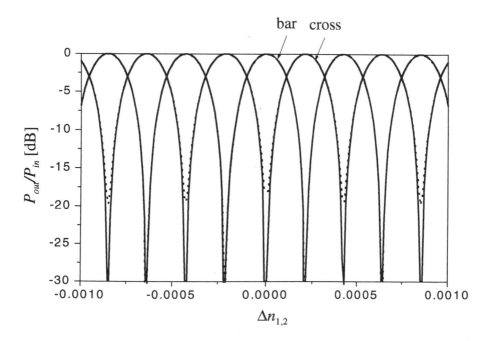

Figure 4-4. Switching characteristic of a MZI

Since the transmission performance is periodic and only dependent on $\Delta\beta$, thermal adjustment of switching states is possible. Furthermore, the required phase change is only $\Delta\beta L=\pi$, which is $\sqrt{3}$ times smaller than that of the directional coupler. However, ideal switching states depend on the critical 3-dB splitting ratio within the couplers and on having equal loss in both waveguide arms. Figure 4-4 shows the calculated transmission characteristics of a MZI based on our model example of section 4.1.2. The phase tuning section in this example has a length of 4583 µm. Crosstalk is critically dependent on the 3-dB splitting ratio of the coupler. With a deviation of, say, 0.4 dB in the 3 dB splitting ratio of the couplers, the crosstalk can not be better than -20 dB (dotted line). Like directional-couplers based TO switches the MZI-based TO switches are wavelength and polarization dependent (due to waveguide birefringence). Since the thermo-optic switching window is narrow, precise temperature control is necessary.

4.2.3 Multi-Mode Interference Switches

The Multi-Mode Interference (MMI) switch, or generalized MZI switch, consists of multi-mode waveguide regions and one or more single mode input/output waveguides, as depicted in the 1 x4 switch of Figure 4-2(c). A multi-mode region can act as an $N \times N$ coupler switch by virtue of a phenomenon called *self imaging* [18]. In this phenomenon, a field pattern produced at the input of a waveguide is reproduced periodically as a single or multiple image(s) in the direction of propagation. If the length of the multi-mode region and the positions of the input and output waveguides are properly chosen, an $N \times N$ coupler is obtained. The length of the MMI section required for an N-fold image, with equal amplitudes and well defined phase differences between images, is given approximately by:

$$L = \frac{3L_\pi}{N} \qquad (4.16)$$

with the beat length of the two lowest order modes $L_\pi=\pi/(\beta_0-\beta_1)$. The transversal spacing between the output waveguides (images) is

$$\Delta x = \frac{w_{eff}}{N}. \qquad (4.17)$$

where w_{eff} is the effective width of the multi-mode region, which is given approximately by:

$$w_{eff} = w_m + \frac{\lambda_0}{\pi}\left(\frac{n_{clad}}{n_{eff,cor}}\right)^{2\sigma}\frac{1}{\sqrt{n^2_{eff,cor} - n^2_{clad}}} \qquad (4.18)$$

and σ is 0 for TE polarization and 1 for TM polarization. The effective refractive index $n_{eff,cor}$ corresponds to the fundamental slab mode of the multi-mode region in the y-direction. The phase relation between images is given by:

$$\varphi_q = (N - q)\frac{q\pi}{N} \qquad (4.19)$$

where q, an integer that is $\leq N$, refers to each of the N images along the x-direction. Due to reciprocity, it is possible to combine the coherent optical field intensities of N waveguides in a single waveguide if the relative input phases are chosen according to Equation (4.19). Hence, generalized MZI switches with N phase tuning elements can be constructed with two of these MMI couplers.

Compact $1xN$ switches can be fabricated as shown in Figure 4-2(c) where $N = 4$. Using the parameters of our model examples, and assuming $w_m = 64$ μm, L need to be 4212 μm to produce a 4-fold image. The waveguides are spaced by $\Delta x = 16.7$ μm and arranged as shown in Figure 4-5(a), where simulations are carried out by normal system mode analysis taking all guided modes into account. The field intensity distribution within the MMI, with equal field amplitudes in all inputs and phases arranged for constructive interference at output 1, is shown. By rearranging the phases of input signals, they can be switched to any output port.

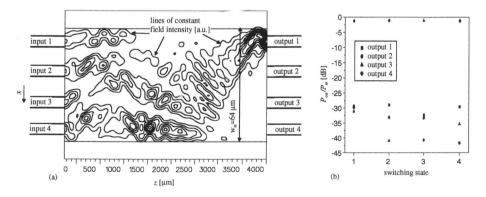

Figure 4-5. (a) Top view of a 4x4 MMI TO switch with lines of constant field intensity. (b) Output power at the four output ports of a 1x4 MZI switch for all four switching states

The transmission performance through the MMI switch is shown in Figure 4-5(b). The minimum accumulated phase change of all phase tuning elements, which is required to tune to all switching states, is only $3\pi/2$. This is half the value required for a 1x4 switch constructed by cascading MZI devices. Therefore, MMI TO switches consumes less power. Switching devices based on MMIs are also generally robust against processing intolerances compared to their MZI and directional coupler counterparts. Nevertheless, they are wavelength dependent and require tight temperature control.

4.2.4 Digital Thermo-Optic Switches

In contrast to interferometric thermo-optic (TO) switches discussed so far, digital TO switches rely on mode sorting and control. The geometry of a single mode waveguide is adiabatically, and gradually, changed to a multi-mode structure. This preserves the power of the fundamental mode where signals are carried [19]. The simplest form of a TO digital optical switch is the 1x2 Y-shaped switch, shown in Figure 4-2(d). In this switch, a single mode waveguide is split with a small angle α (usually $0.1° < \alpha < 0.15°$). Micro heaters are deposited on top of each of the two Y-branches to modulate the refractive indices. The device is examined in more detail in Figure 4-6. Switching is accomplished by heating one of the Y-branches (lowering its refractive index if polymer waveguide). This forces the fundamental system mode (even mode) into the unheated arm, Figure 4-6(a). If the geometric waveguide change along the z direction (from single mode to multi-mode) is gradual, the power of the fundamental mode is conserved and not coupled to higher-order modes. The crosstalk in the heated arm can be very low (off port in Figure 4-6(a)).

Modeling of this switching device is usually done by local system mode analysis, taking mode coupling into account [20]. Modeling is also possible by beam propagation simulations. Figure 4-6(a) shows the normalized modal field distribution (dominant field component) of the even and odd modes at different positions along the device. If the device is perfectly symmetric ($\Delta n_{1,2} = 0$) the fundamental mode is perfectly even and the second mode is perfectly odd with respect to the midpoint between waveguides. Losses due to radiation modes are usually neglected.

When temperature is applied, the fundamental mode is directed to the arm with the higher index and the second mode is directed into the arm with the lower index. This trend is further enforced as the refractive-index difference and waveguide separation increase until the modes are almost totally split among the two arms. Due to the thermally induced asymmetry in the device, input power excites both modes. However, with moderate refractive index difference and small α, the second mode excitation is weak.

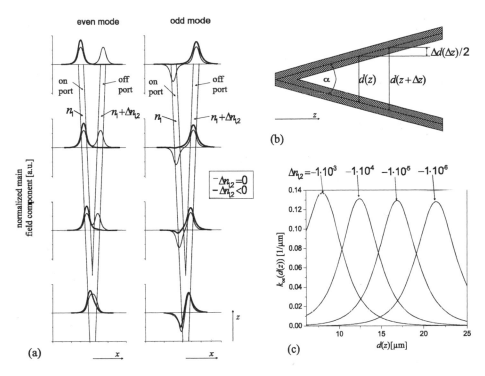

Figure 4-6. (a) Field distribution of the even and odd modes along a y-shaped digital optical switch (DOC). (b) Schematic layout of a y-shaped DOS. (c) Mode conversion coefficient between even and odd mode.

Mode coupling can be described by the conversion coefficients [19] between the even and the odd mode ($k_{o,e}$, $k_{e,o}$) which depend on the distance d between waveguides as follows:

$$k_{o,e} = -k_{o,e} = \cot(\alpha) \lim_{\Delta z \to 0} \left(\frac{\iint_A \psi_e(z)\psi_o(z+\Delta z)^* \, dA}{\Delta z \cdot \sqrt{\iint_A |\psi_e(z)|^2 \, dA \cdot \iint_A |\psi_o(z+\Delta z)|^2 \, dA}} \right) \approx$$

$$\frac{\iint_A \psi_e(z) \frac{\partial \psi_o(z)^*}{\partial d} \, dA}{\sqrt{\iint_A |\psi_e(z)|^2 \, dA \cdot \iint_A |\psi_o(z)|^2 \, dA}} \qquad (4.20)$$

where ψ_e and ψ_o are the transverse field distributions of the even and odd mode, and A is the cross sectional area of the waveguide system. The conversion coefficients $k_{o,e}$ and $k_{e,o}$ are plotted in Figure 4-6(c) as function of the waveguide center separation d using the model example of section 4.1.2. The refractive index difference $\Delta n_{1,2}$ affects Equation 4.20 through ψ_e and ψ_o. The phase relation between the modes plays a very important role. If the modes are in phase, energy is coupled constructively from the even into the odd mode. Coupling is reversed for a 180° phase shift. The relative phase between the modes changes faster with increasing $\Delta n_{1,2}$ ($|\beta_e-\beta_o|$ increases). This leads to alternating power gain and loss of the odd mode in the region of strong coupling. Therefore, the larger $\Delta n_{1,2}$ is, the weaker the second mode excitation becomes.

Figure 4-7(a) shows the crosstalk ($10 \log(P_{out,off}/P_{out,on})$) in a digital optical switch with fixed $\alpha = 0.143°$, different device lengths L, and varying refractive index difference $\Delta n_{1,2}$. Waveguide parameters are again similar to those introduced in our model example. Roughly, three length regimes (indicated by zones 1 to 3 in Figure 4-7) with fundamentally different crosstalk behavior can be identified. In the first zone (short devices), crosstalk is high and is dominated by the proportion of the even mode in the off output port (waveguide with the lower refractive index). The dependence of crosstalk on $\Delta n_{1,2}$ is weak.

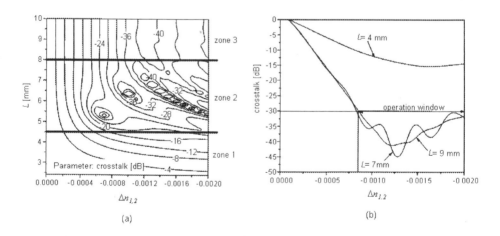

Figure 4-7. Crosstalk in digital optical switches

For relatively longer devices, the second zone in Figure 4-7(a), crosstalk declines, but starts to oscillate with increasing $\Delta n_{1,2}$. This is caused by similar intensity levels of the even and odd modes at the off output port. They add up constructively or destructively depending on the exact value of $\Delta n_{1,2}$ and L. Moreover, the power in the odd mode oscillates due to coupling with the fundamental mode. The crosstalk can approach very low figures, below -44 dB in Figure 4-7(b). However, this interferometric effect is not utilized in digital TO switches, which are meant to be robust against parameter deviations.

The crosstalk behavior of a digital switch is described by the third zone where the device length exceeds a certain limit ($L = 8$ mm in Figure 4-7). The oscillating behavior with varying $\Delta n_{1,2}$ is no longer observed. The power of the even mode in the off output port becomes insignificant for long devices due to mode sorting and crosstalk decreases rapidly with increasing $\Delta n_{1,2}$. The crosstalk saturates for large values of $\Delta n_{1,2}$ and its value is identical to the relative intensity of the odd mode in the off output port. Only marginal change in crosstalk can take place with further increase in the device length since the two waveguide arms become completely isolated at the end of the switch.

The ideal digital switching point in this example (for a device with $L > 8$ mm) would be $\Delta n_{1,2} = 0.0008$ which yields crosstalk below -30 dB even for large parameter deviations. The operation window is usually large and allows digital TO switching performance without tight thermal power control. The influence of the splitting angle α on crosstalk is very important. For smaller angles, the second zone gets larger and embraces longer devices. Meanwhile, the optimum $\Delta n_{1,2}$ in the third zone decrease. This leads to longer devices with better crosstalk performance. However, small opening angles are difficult to fabricate and long switching devices are not favorable. Switching matrices require that unit devices be as short as possible to save substrate space.

The insertion loss in a digital TO switch for $0.1° < \alpha < 0.15°$ can be lower than 0.2 dB, excluding material loss. However, when the refractive-index change is large, insertion loss can increase at high temperatures. Since digital TO switches are not based on mode interference, they are wavelength and polarization independent. They require high thermal power compared to interferometric switches. Therefore, they are usually built on polymers with high thermo-optic coefficients and low thermal conductivity.

4.3 DEVICE FABRICATION AND OTHER SWITCH CONFIGURATIONS

4.3.1 Silica Based Thermo-Optic Switches

Silica based integrated optical circuits are generally fabricated on silicon or quartz substrates. Silica layers are deposited on these substrates by either chemical vapor deposition (CVD) or flame hydrolysis deposition (FHD).

The latter can produce low-loss waveguides (as low as 0.02 dB/cm). The core-cladding refractive index step is about 0.4 %. Higher index contrasts for high density integrated circuits are also possible (1.5 %) but this increases loss to about 0.07 dB/cm [21]. The birefringence in these devices is very low. In the FHD process, fine glass particles are chemically generated in an oxy-hydrogen flame and deposited on the substrate. Then, the substrate is heated at about 1000 °C to form a transparent glass layer. The core layer is usually created by Germanium doping and waveguides are defined by conventional lithography and reactive ion etching.

Among silica-based interferometric devices, the MZIs and MMIs devices exhibit the highest TO device efficiencies and are almost exclusively used. The performance of these devices can be further improved by etching trenches between the two branches of the interferometer in the phase tuning section. This enhances thermal isolation between the waveguides [22]. Device efficiency can also be improved by broadening the under cladding (SiO_2 grooves in the substrate) [23]. These techniques are shown in Figure 4-8(c).

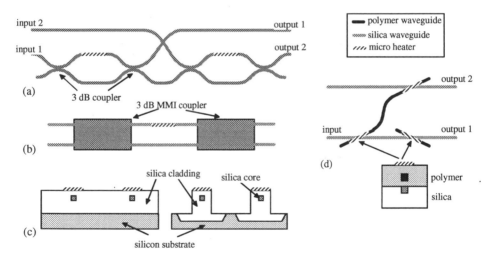

Figure 4-8. Modified interferometric waveguide switches. (a) Double gate MZI. (b) MZI with MMI couplers instead of directional couplers. (c) Cross sectional area of the phase shift section in a MZI for a conventional and thermally optimized design in silica on silicon. (d) Vertical directional coupler cascade in a hybrid polymer/silica device.

An interesting version of the MZI is the so called double gate MZI, which is shown in Figure 4-8(a) [24]. It is basically a 1x2 switch where input 1 can be switched between output 1 and output 2. A signal entering through input port 2 can either be routed to output port 2 or sent to the dead arm (optical sink) of the second MZI. The switch is designed to minimize the crosstalk at output port 2. This is used in certain matrix switch architectures. In the bar state, input 1 is routed to output 1, input 2 is routed to output 2, and crosstalk suppression is doubled. In the cross state, when input 1 is routed to output 2, input 2 carries no signal. Only crosstalk from previous switch stages may exist at this input and is then coupled to the dead arm. In this latter state, both MZIs are in the cross states with large crosstalk suppression. The switch require no thermal energy in the bar state where both MZIs are in the bar state too. The length of the two arms in the MZI must be adjusted for precise phase shift of π. This can be done by trimming the refractive index profile by UV exposure [25] or heat treatment [24].

The MMI 2x2 coupler switch, as depicted in Figure 4-8(b), has been realized too. It has higher fabrication tolerances for the 3 dB splitting ratio. The switch in [22], for example, has a crosstalk of -21 dB and an insertion loss of 1 dB, power consumption of 110 mW and a switching time of 180 μs.

Some efforts have been made to build silica digital thermo-optic switches [26] for protection switching application. This is the x-switch design with asymmetric waveguide branches shown in Figure 4-9(a). It is passive (require no thermal energy) in one state, but the thermal power required for the other state is as large as 1.3 W. The crosstalk and insertion loss are about -15 dB and 2.5 dB, respectively.

4.3.2 Polymer Based Thermo-Optic Switches

Polymer materials for integrated optics describe a whole family of organic materials ranging from acrylates, polyimides to polycarbonates and olefins. Only very few of them are commercially available and meet the desired criteria for integrated optics such as low optical loss at the telecommunications transmission windows (around 1310nm and 1550nm) and thermal and environmental stability. Therefore, production of high performance devices usually requires close cooperation between chemists and component designers. Strongly cross linked fluorinated polymer complexes are usually used for TO devices and switches. In these materials, losses of 0.1 dB/cm can be obtained with index contrasts between core and cladding ranging from 0 % to 35 %. It is possible to precisely trim the refractive index profile by blending and copolymerizing selected monomers [27].

Thin films of the pre polymer material are spin coated on silicon, glass or polymer substrates. The Film thickness varies from the sub micrometer range to several tenths of micrometers. Cladding and core layers are produced in this manner. If silicon is used as a substrate, silica is also a good alternative for the lower cladding layer especially in connection with rib guides and high index contrast waveguides. Waveguides in the core layer are often formed by direct photo patterning, which keeps side wall roughness to a minimum and allows for rapid and inexpensive production. Alternatively reactive ion etching, in combination with thermal curing and conventional photo lithography, is used. The side wall roughness is higher in this case. Therefore, the etch mask and fabrication parameters are critical and must be carefully adjusted.

As a base material for thermo-optic waveguide switches, polymers exhibit excellent performance due to their high thermo-optic coefficient and low thermal conductivity. It is especially useful to use silicon as a substrate because it serves as a good heat sink. Due to the different thermal characteristics of polymers and silicon, stress is induced in the waveguide films. This can lead to significant birefringence, which is not harmful in digital optical switches since they are based on mode sorting. In interferometric switches, based on mode interference, this birefringence would lead to polarization dependent switching characteristic. By using highly cross linked polymers operated above their glass transition temperature, birefringence can be reduced to $\Delta n_{TE-TM} = 10^{-6}$ and polarization dependence can be avoided [27].

The main focus of research in recent years has been on digital thermo-optic switches. They have better crosstalk performance than interferometric devices. Digital TO devices are robust against birefringence, wavelength independent, and do not need complex control circuitry. A challenging task in the fabrication of Y-shaped digital optical switches is producing the sharp vertex at the Y-splitting point, as the splitting angle should be in the order of only 0.1° to secure low insertion loss and low crosstalk. The quality of the vertex strongly depends on the fabrication process. If the waveguides are formed by etching grooves in the lower cladding layer, which are then filled with the core material, any kind of under etching sharpens the vertex [28]. On the contrary, the vertex becomes blunt if the core material is directly etched to form the waveguides.

Some theoretical studies modifies the Y-shape of the digital optical switch by introducing a local opening angle, which varies along the switch leading to curved branches [29]. Theoretically, this approach can almost double crosstalk suppression for a given wavelength. However, the improvement is due to interferometric effects and the resulting switching characteristic is strongly wavelength dependent. Therefore, Y-shaped switches with curved branches are inadequate for digital optical switching.

Practical modified versions of digital thermo-optic switches are shown in Figure 4-9. Best results in crosstalk suppression are achieved by placing an adjustable thermo-optic attenuator behind each arm of the Y-switch as indicated in Figure 4-9(c). These devices further reduce the output power in the off port. The attenuators are sometimes integrated into the Y-branches to shorten the device length. Different attenuators have been proposed in this regard [30]-[32]. Most of them are based on disturbance of lightwave guidance by heating. Crosstalk figures of TO digital switches with attenuators can be in the order of -45 dB and thermal power dissipation is in the range between 200 mW and 400 mW.

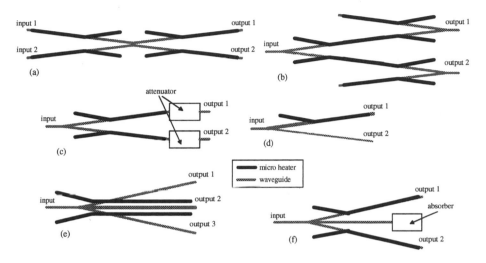

Figure 4-9. Examples for modified digital optical switches. (a) X-shaped 2x2 switch. (b) Cascade with partially merged switching stages. (c) Switch with attenuators. (d) Asymmetric switch. (e) Three branch 1x3 switch. (f) W-shaped switch.

Crosstalk can be reduced as well by using W-shaped three branched digital switches as shown in Figure 4-9(f) [33]. If the relative index distribution between the branches is properly adjusted, the middle arm reduces the crosstalk by attracting optical power from the off port and leading it to an absorber. Other methods to suppress crosstalk include cascading two stages of digital optical switches, but this can be space consuming and doubles the required thermal power. A more elegant solution is obtained by turning the second stage upside down and shifting it into the first stage as shown in Figure 4-9(b) [34]. This can be done while keeping device length and power consumption under control. Using this method, -40 dB crosstalk has been reported. Other modified digital TO switches include the 2 × 2 switch shown in Figure 4-9(a) [35]. This utilizes the even and the odd modes to carry the signal. It is basically an X-crossing waveguide structure with electrodes on each branch. The crossing region must by large enough to support two modes. A crosstalk of -25 dB was reported [35].

Asymmetric passive switches (the term passive means that they have a defined switching state at zero thermal power, whereas the other state is activated by heat, as indicated earlier) were also reported [26], [36]. The effective index of one waveguide branch is lowered by reducing its cross sectional area. In the passive state the light is guided to the larger arm. The power required to tune the device to the other state is approximately double the corresponding power in conventional symmetric switch. This switch can be advantageous in applications like protection switching with preferred paths.

1×3 or 1×4 switches are either realized by three fold and four fold waveguide splitting in a single stage [37]-[38] as shown in Figure 4-9(e), or by cascading several switching stages. For the single-stage multifold splitter the device length of a 1×N switch increases by approximately a factor of $\log_2(N)$ compared to a 1×2 switch (to maintain reasonable crosstalk values). Devices based on multifold splitting regions have crosstalk in the range from -15 dB to -20 dB. No advantage over multi-stage switch solutions is identified.

Interferometric devices, like MZIs, have been also fabricated with polymer [39]-[40]. Very low power consumption (4.8 mW) and low crosstalk (-40 dB) were achieved. Polymer devices cover a wide range of optical applications apart from switches. These include lasers, amplifiers, couplers, modulators, filters, attenuators, polarization controllers and dispersion compensators. The potential of integrating these functionalities and even building three-dimensional multi-layer polymer structures has been demonstrated [41]-[42].

4.3.3 Polymer/Silica Hybrid Devices

Recent research has attempted to combine the advantages of silica and polymer materials in a single thermo-optic switching device. The low loss of silica waveguides allows for large integrated circuits with large number of ports and other optical components, such as arrayed waveguide gratings (AWG). Polymer waveguides, on the other hand, offer efficient digital switching devices. The two materials may complement each other. Therefore, integration of their features into hybrid TO switches has been reported [43].

Figure 4-8(d) illustrates an approach to this integration using vertical directional couplers [43]. The vertical coupler is similar to the conventional directional coupler, except that waveguides are not positioned side by side. Instead, they are placed on top of each other with a thin isolation layer in between. Silica waveguides may be produced in the conventional manner, with or without the upper cladding. Then polymer films are processed on top. The thickness of the isolation layer should be adjusted carefully. The switch is passive in the bar state. By applying thermal actuation, the refractive index of the polymer waveguide is decreased until it matches that of the silica guide. The light tunnels through the polymer guides into the neighboring silica guide. Crosstalk can be reduced by adding a dummy directional coupler in the path to output port 1 (Figure 4-8(d)).

To realize this device, the refractive index of the polymer waveguides must be tailored to be slightly higher than that of silica. Since the thermal conductivity of silica is higher and its thermo-optic coefficient is lower, the device is relatively efficient. Crosstalk is below -35 dB, insertion loss is 1.5 dB, and power consumption is 80 mW (to actuate all couplers) [43]. 1×8 switches were also demonstrated by interconnecting 1x2 units.

4.4   THERMO-OPTIC MATRIX SWITCHES

Thermo-optic switching devices can be interconnected to form switching matrices with several inputs/outputs. Many matrix architectures can be built using 1×2 or 2×2 TO switches as building blocks. Matrix switch architectures are discussed in chapter 9. In the following, we only highlight certain details pertinent to TO waveguide matrix switches. It is desirable to realize large switching matrix on a single chip to minimize insertion loss and cost. We discuss single-substrate TO binary-tree structures and modified crossbar structures. Both structures are strictly/wide-sense non-blocking.

For relatively large port count TO matrix switches (up to 16 x 16) the crosstalk in single substrate binary trees is dominated by waveguide crossings, not by switches. Figures 4-10(a) and 4-10(b) give schematics of single substrate binary tree matrix switches. Waveguides are laid out in such a way to minimize crossings. In contrast to binary trees, the crossbar architecture minimizes power consumption since only one switching device is actuated per lightpath, assuming passive devices. The classical crossbar architecture may be modified as shown in Figure 4-10(c) to obtain equal path lengths (homogenous loss) for all lightpaths. This switch is not dilated, but dilation can be obtained by using double gate MZIs switches as depicted in Figure 4-10(d). However, this increases the depth[1] of the matrix and adds extra waveguide crossings.

The dominant architecture for polymer matrix switches is the binary tree. This is due to two main reasons. First, losses in polymer waveguides are higher than their silica counterparts. Therefore, it is favorable to minimize matrix depth in order to minimize its insertion loss. Binary trees are characterized by short depths. Second, polymer switches exhibit high efficiency, but their crosstalk is relatively large. Hence, it is more critical to optimize the switch matrix design for minimal crosstalk rather than for

---

[1] The number of switching devices encountered by the lightpath through the matrix

minimal power consumption. Redundant switches can also be introduced in binary trees to bypass crosstalk and reduce its effect [44].

The largest thermo-optic matrix switch demonstrated so far with polymer devices is a 16×16 switch on a chip of 4×10.4 cm^2 [45]. The waveguides have a refractive index contrast of 1.5 % and are fabricated by photo patterning. The switch is recursively build from 4×4 and 8×8 binary tree switches based on Y-shaped digital thermo-optic switches as shown in Figures 4-10(a) and 4-10(b). Insertion loss is about 6 dB, crosstalk is below -30 dB and the total power consumption is 6.4 W. The main source of crosstalk is waveguide intersections. All heat electrodes can be driven by a single power source and no complex bias point control is necessary.

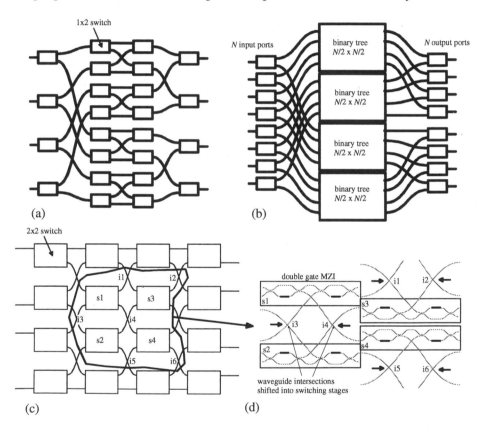

Figure 4-10. Optimal architectures for TO waveguide matrix switches. (a) 4x4 binary tree. (b) Large binary tree built by recursion from smaller binary trees. (c) Modified cross-bar architecture. (d) Modified cross bar architecture with double gate MZIs.

Smaller polymer switching matrices based on the binary tree architecture have also been demonstrated, including 8×8 [46] and 4×4 [47]-[48]. In [48], the worst-case insertion loss and crosstalk are 4 dB and -50 dB respectively, taking into consideration operation at 1310 nm and 1550 nm wavelengths. The power consumption is 2.1W and the switching time is below 3 ms.

The largest TO switch matrix developed with silica waveguides is also 16×16. This is realized with the modified cross bar architecture, using double gate MZIs, as shown in Figures 4-10(c) and 4-10(d) [24]. This architecture has moderate power consumption despite the inefficiency of silica TO switches. The matrix was fabricated on 6 inch substrates with Ge doped waveguides (flame hydrolysis). The overall waveguide length for single path approaches 66 cm, with 0.03-dB/cm loss. The worst case insertion loss and single-assignment[2] crosstalk is 8 dB and -40 dB respectively, in a wavelength range of 1530 nm to 1560 nm (for TE and TM polarization). The maximal assignment crosstalk is estimated to be -33 dB. The required switching power is 17 W. Excellent crosstalk performance is achieved in this matrix by alternating the orientation of the double gate MZIs from row to row and from column to column as shown in Figure 4-10(d) (compare the switches labeled s1-s4 to the corresponding switches in Figure 4-10(c)). In the cross state, crosstalk is led to the optical sink of the double-gate MZIs. Only the crosstalk in the bar state is routed to a signal output port. However, the crosstalk in the bar state is very low, as described earlier in section 4.3.

In this matrix design, substrate space can be saved by shifting the waveguide intersections (i1-i4 in Figure 4-10) into the switching stages, as indicated in Figure 4-10(d). This also enlarges intersection angles and reduces their contribution to crosstalk. 8×8 matrix switches based on this design have been demonstrated with a worst-case insertion loss and crosstalk of 6 dB and -50 dB, respectively. Power consumption was 16 W [49], which is high compared to the 16×16 matrix switch described above. This is due to using non-passive MZIs.

Binary tree architectures have been also realized in silica technology. An 8×8 switch is built with 1×2 and 1×4 MMIs/MZIs [50]. The insertion loss and crosstalk is about 6 dB and -35 dB, respectively. Substrate area is as small as 65×13 mm^2. 4×4 switches are fabricated using the same design, with similar crosstalk and lower insertion loss (2.8 dB) [51].

Thermo-optic matrix switches are suitable for low to medium switch sizes and applications, including small optical cross-connects, optical add/drop multiplexers, and protection switching applications. By comparing the characteristics of polymer and silica 16×16 matrices, it becomes apparent that power consumption is significantly lower in the case of polymer matrices. Although waveguide loss in polymer matrices is higher, their insertion loss is lower. This is due to shorter path lengths in polymer binary

---

[2] A single assignment crosstalk originates from a single active input port, whereas a maximal assignment crosstalk originates from all input ports.

trees compared to the modified crossbar silica architecture. Crosstalk in matrix switches based on the two material systems is comparable (in the order of -30dB). For higher port counts single-chip matrix switches, crosstalk is limited by inevitable waveguide crossings. For the time being, crosstalk in thermo-optic switching fabrics that are larger than 16×16 is too high to make them competitive with other optical switching technologies.

REFERENCES

[1] M. Haruna and J. Koyama, "Thermooptic deflection and switching in glass," Appl. Optics Lett., Vol. 21, No. 19, pp. 3461–5, October 1982.

[2] M. Haruna and J. Koyama, "Thermo-optic effect in LiNbO3 for light deflection and switching," Electron. Lett., Vol. 17, No. 22, pp. 842–4, October 1981.

[3] J. M. Cariou, J. Dugas, L. Martin, and P. Michel, "Refractive-index variations with temperature of PMMA and polycarbonate," Appl. Optics, Vol. 25, No. 3, pp. 334-6, February 1986.

[4] M. B. J. Diemeer, J. J. Brons, and E. S. Trommel, "Polymeric optical waveguide switch using the thermooptic effect," J. Lightwave Technol., Vol. 7, No. 3, pp. 449–53, March 1989.

[5] Y. Silberberg, P. Perlmutter, and J. E. Baran, "Digital optical switch," Appl. Physics Lett., Vol. 51, No. 16, pp. 1230–2, October 1987.

[6] J. F. Nye, "Physical Properties of Crystals," Oxford University Press, Oxford, 1979.

[7] R. C. Weast, "Handbook of Chemistry and Physics," CRC Press, Cleveland, Ohio, 1976.

[8] E. D. Palik, "Handbook of Optical Constants of Solids," Academic Press, New York, 1985.

[9] H. Borchers, H. Hausen, K.-H. Hellwege, K. Schafer, and E. Schmidt, "Eigenschaften der Materie in ihren Agregatzuständen," Springer-Verlag, New York, 1971.

[10] A. Frank and K. Biederbick, "Kunststoff-Kompendium," Vogel Verlag und Druck KG, Würzburg, 1990.

[11] C. Z. Tan, "Review and analysis of refractive index temperature dependence in amorphous SiO2," Journal of Non-Crystalline Solids, Vol. 238, pp. 30-6, 1998.

[12] R. S. Moshrefzadeh, M. D. Radcliffe, T. C. Lee, and S. K. Mohapatra, "Temperature dependence of index of refraction of polymeric waveguides," IEEE J. Lightwave Technol., Vol. 10, No. 4, pp. 420–5, April 1992.

[13] R. M. Waxler, D. Horowitz, and A. Feldman, "Optical and physical parameters of plexiglas 55 and lexan," Appl. Optics, Vol. 18, No. 1, pp. 101–4, January 1979.

[14] J. D. Plummer, M. D. Deal, and P. B. Griffin, "Silicon VLSI Technology," Prentice Hall, Inc., New York, 2000.

[15] O. G. Ramer, "Integrated optic electro optic modulator electrode analysis," IEEE J. Quantum Electron., Vol. 18, No. 3, pp. 386–92, March 1982.

[16] A. Yariv, "Coupled mode theory for guided wave optics," IEEE J. Quantum Electron., Vol. 9, No. 9, pp. 919–33, September 1973.

[17] R. C. Alferness, "Guided-wave devices for optical communication," IEEE J. Quantum Electron., Vol. 17, No. 6, pp. 946–59, June 1981.

[18] L. B. Soldano and E. C. M. Pennings, "Optical multi-mode interference devices based on self-imaging: Principles and applications," IEEE J. Lightwave Technol., Vol. 13, No. 4, pp. 615–27, April 1995.

[19] W. K. Burns and A. F. Milton, "Mode conversion in planar-dielectric separating waveguides," IEEE J. Quantum Electron., Vol. 11, No. 1, pp. 32–9, January 1975.

[20] D. Marcuse, "Theory of Dielectric Optical Waveguides," Academic Press, Inc., New York, 1991.

[21] T. Miya, "Silica-based planar lightwave circuits: Passive and thermally active devices," IEEE J. Select. Topics in Quantum Electron., Vol. 6, No. 1, pp. 38–45, January 2000.

[22] Q. Lai, W. Hunziker, and H. Melchior, "Low-power compact 2x2 thermooptic silica-on-silicon waveguide switch with fast response," IEEE Photon. Technol. Lett., Vol. 10, No. 5, pp. 681–682, May 1998.

[23] R. Kasahara, M. Yanagisawa, A. Sugita, T. Goh, M. Yasu, A. Himeno, and S. Matsui, "Low-power consumption silica-based 2x2 thermooptic switch using trenched silicon substrate," IEEE Photon. Technol. Lett., Vol. 11, No. 9, pp. 1132–4, September 1999.

[24] T. Goh, M. Yasu, K. Hattori, A. Himeno, M. Okuno, and Y. Ohmori, "Low loss and high extinction ratio strictly non blocking 16x16 thermo-optic matrix switch on 6-in wafer using silica-based planar lightwave circuit technology," IEEE J. Lightwave Technol., Vol. 19, No. 3, pp. 371–9, March 2001.

[25] L. Leick, M. Svalgaard, and J. H. Povlsen, "Impact of weak index perturbations on 3 dB multi mode interference couplers," In Proc. Integrated Photonics Research, pp. IMC3–1, Monterey, 2001.

[26] M. Hoffmann, P. Kopka, and E. Voges, "Thermooptical digital switch arrays in silica-on-silicon with defined zero-voltage state," IEEE J. Lightwave Technol., Vol. 16, No. 3, pp. 395–400, March 1998.

[27] L. Eldada and L. W. Shacklette, "Advances in polymer integrated optics," IEEE J. Select. Topics in Quantum Electron., Vol. 6, No. 1, pp. 54–68, January 2000.

[28] R. Moosburger, G. Fischbeck, C. Kostrzewa, and K. Petermann, "Digital optical switch based on oversized polymer rib waveguides," Electron. Lett., Vol. 32, No. 6, pp. 544–5, March 1996.

[29] R. Moosburger, C. Kostrzewa, G. Fischbeck, and K. Petermann, "Shaping the digital optical switch using evolution strategies and BPM," IEEE Photon. Technol. Lett., Vol. 9, No. 11, pp. 1484–6, November 1997.

[30] K. Sakuma, D. Fujita, S. Ishikawa, T. Sekiguchi, and H. Hosoya, "Low insertion-loss and high isolation polymeric y-branching thermooptic switch with partitioned heater," In Proc. Integrated Photonics Research, pp. WR3–1, Monterey, 2001.

[31] U. Siebel, R. Hauffe, J. Bruns, and K. Petermann, "Polymer digital optical switch with an integrated attenuator," IEEE Photon. Technol. Lett., Vol. 13, No. 9, pp. 957-9, September 2001.

[32] M.-S. Yang, Y. O. Noh, Y. H. Won, and W.-Y. Hwang, "Very low crosstalk 1x2 digital optical switch integrated with variable optical attenuators," Electron. Lett., Vol. 37, No. 9, pp. 587–8, April 2001.

[33] U. Siebel, R. Hauffe, and K. Petermann, "Crosstalk-enhanced polymer digital optical switch based on a w-shape," IEEE Photon. Technol. Lett., Vol. 12, No. 1, pp. 40–1, January 2000.

[34] R. Hauffe, F. Kerbstadt, U. Siebel, J. Bruns, and K. Petermann, "Reliable low cross talk digital optical switch based on a cascade like structure with small dimensions," In Proc. Integrated Photonics Research, page ITuH2, Monterey, 2001.

[35] N. Keil, H. Yao, and C. Zawadzki, "A novel type of 2x2 digital optical switch realized by polymer waveguide technology," In the Proc. European Conference on Optical Communication (ECOC'96), page TuC.1.6, Oslo, 1996.

[36] S. Lee, J. Bu, S. Lee, K. Song, C. Park, and T. Kim, "Low-power consumption polymeric attenuator using a micro machined membrane type waveguide," IEEE Photon. Technol. Lett., Vol. 12, No. 4, pp. 407–9, April 2000.

[37] T. W. Oh and S.-Y. Shin, "Polymeric 1x3 thermo-optic switch," In Proc. Integrated Photonics Research, pp. RTuD2–1, Santa Barbara, 1999.

[38] H.-C. Song, S.-Y. Shin, W.-H. Jang, and T. H. Rhee, "1×4 thermo-optic switch based on four branch waveguide," Electron. Lett., Vol. 35, No 18, pp.1546–8, September 1999.

[39] Min-Choel Oh, Hyung-Jong Lee, Myung-Hyun Lee, Joo-Heon Ahn, and Seon Gyu Han, "Asymmetric x-junction thermooptic switches based on fluorinated polymer waveguides," IEEE Photon. Technol. Lett., Vol. 10, No. 6, pp. 813–5, June 1998.

[40] Y. Hida, H. Onose, and S. Imamura, "Polymer waveguide thermooptic switch with low electric power consumption at 1.3µm," IEEE Photon. Technol. Lett., Vol. 5, No. 7, pp. 782–4, July 1993.

[41] R. T. Chen, "Polymer-based photonic integrated circuits," Optics and Laser Technology, Vol. 25, No. 6, pp. 347–65, June 1993.

[42] S. M. Garner, S. Lee, V. Chuyanov, A Chen, A. Yacoubian, W. H. Steier, and L. R. Dalton, "Three-dimensional integrated optics using polymers," IEEE J. Quantum Electron., Vol. 35, No. 8, pp. 1146–55, August 1999.

[43] N. Keil, H. H. Yao, C. Zawadzki, K. Losch, K. Satzke, W. Wischmann, J.V. Wirth, J. Schneider, J. Bauer, and M. Bauer, "Hybrid polymer/silica vertical coupler switch with <-32 db polarization independent crosstalk," Electron. Lett., Vol. 37, No. 2, pp. 89–90, January 2001.

[44] R. Hauffe, U. Siebel, and K. Petermann, "Crosstalk-optimized integrated optical switching matrices in polymers by use of redundant switch elements," IEEE Photon. Technol. Lett., Vol. 13, No 3, pp. 200–2, March 2001.

[45] F. L. W. Rabbering, J. F. P. van Nunen, and L. Eldada, "Polymeric 16×16 digital optical switch matrix," In Proc. European Conference on Optical Communication, Amsterdam, 2001.

[46] A. Borreman, T. Hoekstra, M. Diemeer, H. Hoekstra, and P. Lambeck, "Polymeric 8×8 digital optical switch matrix," In Proc. European Conference on Optical Communication (ECOC'96), ThD.3.2, Oslo, 1996.

[47] R. Moosburger and K. Petermann, "4x4 digital optical matrix switch using polymeric oversized rib waveguides," IEEE Photon. Technol. Lett., Vol. 10, No. 5, pp. 684–6, May 1998.

[48] L. Guiziou, P. Ferm, J.-M. Jouanno, and L. Shacklette, "Low-loss and high extinction ratio 4x4 polymer thermo-optical switch," In Proc. European Conference on Optical Communication (ECOC'96), Tu.L.1.4, Amsterdam, 2001.

[49] T. Goh, A. Himeno, M. Okuno H. Takahashi, and K. Hattori, "High-extinction ratio and low-loss silica-based 8x8 strictly nonblocking thermooptic matrix switch," IEEE J. Lightwave Technol., Vol. 17, No. 7, pp. 1192–8, July 1999.

[50] M. P. Earnshaw and J. B. D. Soole, "8x8 optical switch matrix in silica-on-silicon," In Proc. Integrated Photonics Research, IMC2, Monterey, 2001.

[51] M. P. Earnshaw, J. B. D. Soole, M. Cappuzzo, L. Gomez, E. Laskowski, and A. Paunescu, "Compact, low-loss 4x4 optical switch matrix using multimode interferometers," Electron. Lett., Vol. 37, No. 2, pp. 115–6, January 2001.

Chapter 5

LIQUID-CRYSTAL BASED OPTICAL SWITCHING

J. Michael Harris
Robert Lindquist
JuneKoo Rhee
James A. Webb

5.1 LIQUID CRYSTALS

Liquid crystals are a class of materials that are both wonderful and mysterious and have been the focus of studies for long time [1]-[2]. The name is an oxymoron in that no material can simultaneously be in both a crystalline solid state and a liquid state. Rather, liquid crystals exhibit intermediate phases often referred to as mesomorphic phases in which the state of matter exhibits some mechanical and symmetry properties of both crystals and liquids. For example, many liquid crystals flow like a liquid while at the same time showing birefringence and other anisotropic[1] characteristics of a crystal. Liquid crystals can be classified into one of the three types, thermotropic[2], polymeric and lyotropic[3] depending on the physical parameters controlling the liquid crystalline phases. The most widely studied and used type of liquid crystals in electronic applications is thermotropic in which the mesomorphic phases appear in distinct temperature ranges defined by phase transition temperatures. In this chapter, only thermotropic liquid crystals will be considered and discussed. For information on lytropic and polymeric the reader is referred to [3]-[4].

---

[1] Material characteristics vary depending on the direction along which they are measured
[2] A type where phase is dependent on the temperature
[3] A type where phase is dependent on material concentration in a solvent

The product applications of liquid crystals are vast and include thermometers, novelty items such as mood rings, many types of displays (laptop, desktop, projection, instrument, gas pump, etc.), telecommunication components and more. In the design and engineering of these products an understanding of the basic properties and chemical structure of the liquid crystal (LC) material is critical. Thus, we start by attempting to provide a brief review of the basic chemical structure, the liquid crystal phases, basic electrical and optical properties, and basic cell construction. Then, we discuss the use of LCs in optical switching, investigate several issues of reliability in LC-based switches and discuss system and network considerations.

5.1.1 Chemical Structure

A liquid crystalline phase is typically observed in organic substances having highly anisotropic molecules. These molecules are often flat (disk-like) or more commonly elongated (rod-like). Our main focus is on small elongated organic molecules that form the thermotropic type nematogen or smectogens. Nematogens and smectogens are liquid crystals material that exhibit nematic and smectic phases respectively.

The molecular structure of a typical small rod-like liquid crystal molecule is shown in Figure 5-1. The building blocks for the rigid rod include two or more aromatic rings bonded by linkage group X and terminated at each end by terminal groups R and R`. One of the terminal groups, R, is also called the side chain. The physical and optical properties of the liquid crystal are governed by these basic constituent groups and the manner in which they are synthesized.

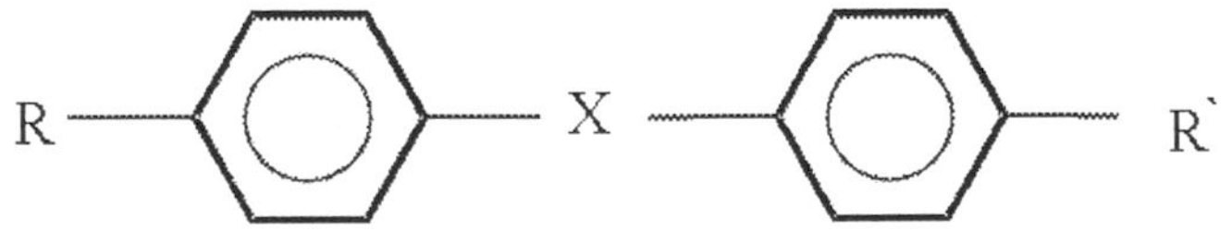

Figure 5-1. Structure of common thermotropic liquid crystal.

Although Figure 5-1 depicts benzene rings, the rings can be either benzene, cyclohexanes, or a combination of both. These rings provide a rigid core and the intermolecular forces that are needed to form the nematic phase. They affect absorption particularly in the ultraviolet region and have significant influence on dielectric anisotropy, birefringence, elastic constants, and viscosity.

The linkage group X can be as simple as a single bond between the two rings like the biphenyl or linking groups such as diphenylethane, stilbene, tolane, an ester, or a Schiff's base. These linkage groups have a strong influence on the chemical stability of the liquid crystal as well as its susceptibility to moisture, temperature change, and ultraviolet radiation.

The side chain R is typically an alkyl chain C_nH_{2n+1}, an alkoxy chain $C_nH_{2n+1}O$, or an alkenyl chain. The length of the chain will influence phase transition temperatures and the elastic constants of the nematic phase. Short chains that have less than two carbon atoms are often too rigid to exhibit a nematic phase. Medium chains of 3-8 carbon atoms provide for appropriate nematic phase. In general, the longer chain molecules exhibit a lower melting point and a potentially higher viscosity.

The terminal group R` can be split into two categories; polar and nonpolar. They play a role in the dielectric constants and the dielectric anisotropy. The nonpolar terminal group which is similar to the side chains does not significantly influence the dielectric anisotropy. The polar group, however, can. These groups commonly include cyano (CN), flourine (F), and chlorine (Cl). Those with CN offer the highest polarity and thus yield high dielectric anisotropy and high birefringence. CN terminal groups also have high viscosity and poor ultra-violet (UV) stability. The compounds with F and Cl in the terminal groups offer lower polarity and thus lower birefringence and dielectric, but higher stability and lower viscosity.

Clearly, the process to synthesize the ideal liquid crystal molecule is complex and highly dependent on application. In general, a single liquid crystal compound cannot meet all the characteristics required for an application. As a result, most commercially available liquid crystals are a mixture of multiple liquid crystal compounds. Significantly larger operating temperature ranges are possible using mixtures and many optimum physical parameters can be engineered through appropriate mixtures.

5.1.2 Liquid Crystalline Phases

The mesomorphoric phases of liquid crystals are best understood by looking at the ordering of the molecules. For simplicity and instructive purposes, the liquid crystal molecules are assumed to be rigid rods. As the temperature rises, the degree of ordering of these rods decreases. An illustrative example of a phase sequence for a common thermotropic liquid crystal is shown in Figure 5-2.

Starting in the liquid state, often referred to as isotropic, the rods are completely disordered. The random molecular orientations and positions result in macroscopic properties that are isotropic. Liquid crystals in the isotopic state are typically clear. As the liquid crystal is cooled from a liquid state to the nematic liquid crystal state, the molecules begin to show rotational ordering. Although the molecules exhibit no positional ordering, the rods are generally pointed in the same direction. As the LC is cooled further, a change from a nematic state to a smectic state takes place. The smectic state exhibits both rotational and positional ordering as shown by the layered structure. The difference between smectic A and smectic C phase is the tilt rotational ordering within the layers. Finally, completely ordered state of a crystalline solid is reached upon further cooling of the smectic C state. Each phase has unique mechanical, optical, electrical, and chemical properties that can and are exploited for numerous applications.

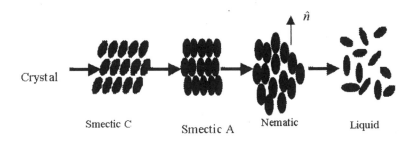

Figure 5-2. Common phase transition sequence.

5.1.2.1 Nematics

Nematic is the most widely used liquid crystal phase for electro-optic applications. Generally speaking, the use of the term liquid crystal in the literature and in products most often implies nematic liquid crystal. Due to the elongated nature of the molecules, the nematic state exhibits rotational ordering as illustrated in Figure 5-2. The molecules align approximately parallel to one another and are directionally correlated. Yet, positional order is not apparent and molecules can flow along one another. The average direction of the aligned molecule is defined by a unit vector $\hat{n}$ along the director axis. The degree of rotational ordering is represented by the order parameter, S, which is defined as

$$S = \frac{1}{2}\left\langle 3\cos^2\theta - 1\right\rangle \tag{5.1}$$

where θ is the angle between the long axis of an individual molecule and the director axis, and "< >" implies an ensemble average. S approaches 1 for perfectly aligned molecules and approaches 0 for disordered states like isotropic. Typical values for the order parameter (S) in the middle temperature range of a nematic phase are between 0.4 and 0.6.

5.1.2.2   Electrical Properties

With rotational ordering, the nematics can be treated at least locally as uniaxial with the axis of symmetry being the director axis $\hat{n}$. Thus, the dielectric tensor can be written as

$$\varepsilon = \begin{pmatrix} \varepsilon_\perp & 0 & 0 \\ 0 & \varepsilon_\perp & 0 \\ 0 & 0 & \varepsilon_\parallel \end{pmatrix} \tag{5.2}$$

in which $\varepsilon_\perp$ and $\varepsilon_\parallel$ are the perpendicular and parallel dielectric constants, respectively. The dielectric anisotropy which is defined as

$$\Delta\varepsilon = \varepsilon_\parallel - \varepsilon_\perp \tag{5.3}$$

plays a critical role in the electro-optic properties of the liquid crystal. As a result of anisotropy the induced dipole moment of the molecules will not be aligned to an electric field unless the director axis is perpendicular or parallel to the electric field. As a result, a torque force is exerted on the molecules. The sign and magnitude of the dielectric anisotropy dictates in part the direction and strength of these electric field induced torque forces. A positive dielectric anisotropy will tend to align the molecules parallel to the electric field while a negative dielectric anisotropy will tend to align the molecules perpendicular to the electric field. Overall, the torque forces are sufficient to rotate the liquid crystal molecule with a voltage of less than 5 volts.

The electric field induced torque is not the only force on the molecules in nematics. Nematics transmit a restoring torque resulting from perturbations of an equilibrium condition such as an electric field. The deformations fall into 3 basic types:

Splay: $F_1 = \dfrac{1}{2} K_1 (\nabla \cdot \hat{n})^2$, conformation with div $\hat{n} \neq 0$ (5.4)

Twist: $F_2 = \dfrac{1}{2} K_2 (\hat{n} \cdot \nabla \times \hat{n})^2$, conformation with $\hat{n} \cdot$ curl $\hat{n} \neq 0$ (5.5)

Bend: $F_3 = \dfrac{1}{2} K_3 (\hat{n} \times (\nabla \times \hat{n}))^2$, conformation with $\hat{n} \times$ curl $\hat{n} \neq 0$ (5.6)

where F_1, F_2, F_3 and K_1, K_2, K_3 represent the free energy and elastic constants for splay, twist and bend respectively. By minimizing the free energy, the director axis orientation in the equilibrium state can be calculated. More details about this are provided in [1]-[2].

In addition to the field-induced and elastic forces, boundary condition and surface forces must be taken into account to calculate the director axis orientation. Liquid crystal components typically have a thin layer of liquid crystalline material sandwiched between two substrates. The boundary conditions are dictated by an alignment layer. The two alignment types are homeotropic and homogenous as shown in Figure 5-3. Homogeneous alignment, in which the liquid crystal molecules align parallel to the substrate, is typically created by physically rubbing a baked polyimide. Homeotropic alignment, in which the liquid crystal molecules align perpendicular to the substrate, is typically created using chemical surface treatment.

(a)   (b)

Figure 5-3. Molecule alignment: (a) Homogeneous alignment, (b) Homeotropic alignment.

The rubbing process determines the director axis orientation at the surface as well as provides a pretilt. Pretilt is a small tilt angle between the substrate surface and the liquid crystal director and is used to provide a preferential direction for the rotation of the LC molecules [5]. Most alignment layers are assumed to be hard anchoring. The surface forces are so strong that the LC molecules at the surface do not respond to applied field perturbations. Thus, the alignment layers determine boundary conditions, but do not enter into the dynamic equation describing the field-induced effects in nematic liquid crystals.

5.1.2.3 Optical Properties

Similar to the dielectric tensor, the refractive index tensor of a nematic can also be described locally as a uniaxial crystal with the axis of symmetry being the director axis, $\hat{n}$. The index tensor can be written as

$$n = \begin{pmatrix} n_o & 0 & 0 \\ 0 & n_o & 0 \\ 0 & 0 & n_e \end{pmatrix} \tag{5.7}$$

where n_o and n_e are the ordinary and extraordinary indices of refraction, respectively. The birefringence is then defined as

$$\Delta n = n_e - n_o \tag{5.8}$$

and can range from 0.01 to ~ 0.3. This large birefringence coupled with the low voltage rotation of molecules provide the opportunities to engineer some very unique solutions to optical systems requiring polarization management.

Since nematics are locally uniaxial, the polarization state of light propagation can be tracked through a liquid crystal cell by dividing the cell into a series of thin layers. The director axis is fixed within a layer. For each layer, the polarization is mapped into the eigenpolarization states which are linear for uniaxial mediums and are referred to as the ordinary and extraordinary rays. The ordinary ray propagates with an index of refraction of n_o and the extraordinary ray propagates with an index of refraction of n_{eff} given by

$$n_{eff} = \sqrt{\frac{n_e^2 n_o^2}{n_e^2 \cos^2 \theta + n_o^2 \sin^2 \theta}} \tag{5.9}$$

where θ is the angle between the director axis and the wave vector of the extraordinary wave. By tracking the propagation of the eigenpolarization state through each layer the polarization effect can be modeled. Techniques such as Jones matrices or extended Jones matrices are commonly used to describe the polarization staes of light propagation in a liquid crystal device. A more thorough description of these methods is given in [6].

5.1.3   Twisted Nematic Cells

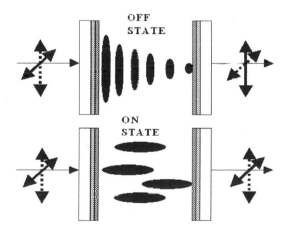

Figure 5-4. The powered and un-powered states of a TN cell.

A common LC cell used for display and telecommunication switches is a twisted nematic (TN). The basic cell structure is shown in Figure 5-4 where molecular alignment is homogeneous at each substrate surface. However, the direction of the director alignment is rotated typically 90° between the substrate. Thus, the molecules align such that the director axis rotate or twist as one propagates through the cell. A right-hand or left-hand twist is determined partially by the pre-tilt angle and any chiral additive that is added to the LC mixture. By applying a voltage across the cell the molecules will align with the field thereby eliminating any twist. Polarization management can be achieved by switching between the twist and untwisted state of the LC.

In the presence of a high voltage ($\sim> 6V_{rms}$), the polarization state of an optical beam propagating through the cell will remain unaffected. Thus linearly polarized light will exit the cell unmodified and will pass through a second parallel polarizer. The degree to which the LC cell affects the polarization state can be measured by placing the cell between crossed polarizers. The extinction ratio (ER) of the signal, which is given by Equation (5.10), can be greater than 100,000:1 or -50 dB:

$$ER = 10\log\frac{I(output)}{I(input)} \qquad (5.10)$$

In this equation, the intensity I is measured in any consistent unit of measure and the loss through the setup is considered small in relationship to the ER.

In the twist state linearly polarized light incident on the pixel is rotated 90° by means of "adiabatic" following in the limit of a slow twist, which is also known as the Mauguin condition. This condition exists when

$$u = 2\frac{d}{\lambda}\Delta n \gg 1 \qquad (5.11)$$

where u is the Mauguin parameter, d is the cell thickness, λ is the wavelength, and Δn is the LC birefringence. However, most twisted nematic cells do not satisfy the Mauguin condition, because the cells would have to be too thick which degrades response time. As a result, linearly polarized incident light does not rotate by exactly 90° and this leads to performance degradation.

In telecommunication switching applications where both short response time (<10ms) and high extinction ratios (<-40dB)[4] are required, a TN cell should be designed to achieve high extinction in the absence of the Mauguin

---

[4] Nomenclature of extinction ratio (ER) often imply the sign. As the ratio of the output to the input becomes smaller, the ER becomes a more negative number, but is generally described as a higher ER

condition. The degree to which the rotation is a perfect 90° can be obtained by measuring the contrast ratio η[5]. For a 90° TN LC cell sandwiched between parallel polarizer, η is given by:

$$\eta = \frac{I_{out}}{I_{in}} = \frac{\sin^2\left(\frac{\pi}{2}\sqrt{1+u^2}\right)}{1+u^2} \qquad (5.12)$$

Hence, the contrast ratio becomes smaller as u increases. However, it only becomes zero under discrete conditions that satisfy:

$$\left(\frac{\pi}{2}\sqrt{1+u^2}\right) = m\pi \quad \text{where } m = 1,2,3... \qquad (5.13)$$

To achieve the extinction required in a telecommunication switch with a single TN cell, the pixel must satisfy Equation (5.13). Since it is fairly difficult to obtain precise cell thickness with current manufacturing capability, the TN cell must be tuned into a minimum operating temperature, voltage, and/or a thickness wedge. Without such compensation techniques, the contrast will be limited to > -30dB except for a few discrete wavelengths. With compensation methods extinction ratios of less than −40 dB can be achieved.

The twisted nematic cell is just one of many configurations that enable polarization management. Additional examples are electrically controlled birefringence (ECB) cells, super twist nematics (STN) cells, and many more. Fundamentally, the fabrication of these cells is very similar with varying alignment layers and LC additives. Again, an excellent review of some common cells is given in [6].

5.2   LIQUID-CRYSTAL OPTICAL SWITCHES

The optical properties of LC materials, where the refractive index and birefringence is determined by molecular alignment, can be exploited to build optical switches. Molecular alignment can easily be controlled by applying relatively modest electric fields across the LC cell. There are switch designs that are as simple as 2x2 space switches, more complex MxN switches, and wavelength-selective switches that are used to control spectral

---

[5] We use the terms contrast and contrast ratio in defining the performance of the LC cell, and use the terms extinction and extinction ratio in describing the LC switch. Both ratios can be expressed in terms of a linear ratio or in decibels. In practice, these terms are often used interchangeably which can contribute to some confusion.

slices or channels. The latter are suitable for wavelength division multiplexing (WDM) systems. Examples of each are given below.

5.2.1 Switches Based on Polarization Management

The first complete description of a polarization management switch that was polarization independent was reported in 1980 [7]. In this switch, light of unknown polarization enters the switch from a multimode fiber to a polarization beam splitting cube. The light is divided into two beamlets which have orthogonal linear polarization states. If these beamlets are then passed through a polarization rotator and recombined at the output of the rotator via a second polarization beam splitting cube, the direction of the light through the second cube can be controlled via the amount of rotation applied by the rotator. A schematic of this type of switch is shown in Figure 5-5. An experimental loss of only 1.6 dB, an extinction ratio of ~20 dB, a switching speed of 140 msec, and operating voltages of 0.8-2.5 volts (rms) have been reported.

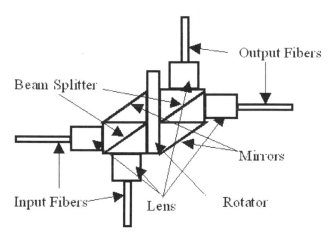

Figure 5-5. Schematic of a 2x2 switch.

Since the principle of operation of this switch provides the fundamentals of most switches in use today, it is reasonable to take time to discuss it in some detail. Figure 5-6 shows the switch with lightpaths in both the bar (through) state and the cross state. In Figure 5-6(a), the light passes from Fiber 1 to Fiber 3 (bar state) when the LC or rotator is unpowered. This is due to the changing of the polarization state of the light as indicated by the arrows and circles in the drawing, where the arrows indicate the vertical polarization component and the circles represent the horizontal component. By tracing the light paths and polarization states, it is possible to see that input light coupled to fiber 2 with the rotator in this state will pass through to fiber 4. The alternate state of the switch is shown in Figure 5-6(b) where the polarization state is unmodified by the rotator since the LC cell is now powered and the light passes from fiber 1 to fiber 4. Input to fiber 2 would be forwarded to fiber 3. Polarization states and optics are discussed in more detail in literature (see [8] for example).

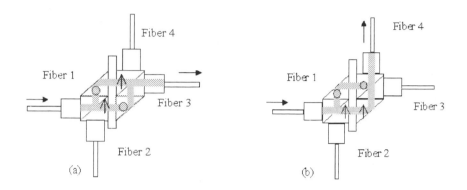

Figure 5-6. Lightpath through an LC switch: (a) bar state with the LC cell un-powered (b) cross state with the cell powered.

This fundamental design forms the basis of many implementations, but it is not without concern. While the overall design of the switch is polarization independent, this is accomplished by separating the beam into two beamlets which trace different paths through the switch. Any difference in optical loss for the two beamlets will result in polarization dependent loss (PDL). Any misalignment in the optical paths through the switch will result in non-symmetric coupling at the output fiber, which leads to additional PDL. This creates significant concern about manufacturing and optical-alignment tolerences. Also the device must exhibit mechanical and thermal stability over extended periods of time. While the crosstalk was measured for a prototype device at ~20 dB [7], which is not an acceptable level, this was indeed due to the coatings on the beam splitters and not due to performance of the LC cell. Thus, careful design and deposition of coatings is important. Testing of the device is essential to reveal performance changes during thermal excursions of both coatings and certainly the LC cell.

Many of these issues and concerns have been overcome. Operation from $-5\ ^0C$ to $70\ ^0C$ with less than 0.2 dB change in loss, over -45 dB of extinction ratio, insertion losses of ~1 dB and switching times of less than 1 msec have all been reported [9]. This high level of performance led to development of numerous practical products. With such characteristics there has also been significant work to expand the overall functionality of this structure into MxN switches.

A free-space MxN switch based on using LC arrays and routing layers has been demonstrated. The switch design uses diverse routing of the polarization similar to the 2x2 example, but relies on sandwiches of a birefringent material such as calcite or precise polarization beam splitters and LC cells to route lightpaths [10]-[11]. In this configuration a 2^n x 2^n switch requires $2n$-2 routing sandwiches to create a Benes network. To date, this and similar structures have not proven to be manufacturable.

5.2.2 Switches Based on Refractive-Index Change

For larger systems, another type of LC switching devices has been proposed for commercial use. This relies on using the birefringence of LC crystals not to manipulate the polarization state, but to control the reflection and refraction of the light off a surface. As discussed earlier, the refractive index of an LC material is anisotropic and can be controlled by controlling the alignment of the molecules relative to the direction of the light propagating through the material. This can be used to control the amount of light that is either refracted at a surface or reflected from it. Since the reflectivity is determined by the angle of incidence upon the surface, the relative indexes of the transmission medium and the reflecting surface, the polarization of the light and of course its wavelength [7], this approach offers many opportunities for design optimization. In a carefully controlled design, the change in index is enough to cause refraction of the light beam through the surface or total internal reflection (TIR) of this beam away from the surface. Figure 5-7 depicts this scheme.

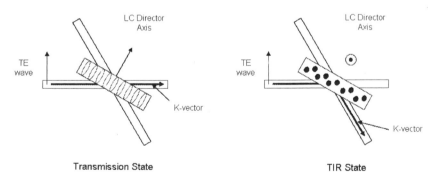

Figure 5-7. Detail showing LC state for transmission and TIR

While this type of system can be used in free space designs, as discussed in chapter 8, it is most useful when coupled with planar waveguide structures to build switching matrices. An example is shown in Figure 5-8. The promise of such designs is to be able to make MxN switches that can be fabricated using silica, silicon, or polymer planar devices. The transmission waveguides are formed using one of the many techniques available, depending on the material used for waveguide. Many of these fabrication methods have been discussed in chapters 2, 3 and 4. The LC switch points (crosspoints) are formed by deep etching trenches within the planar substrates at the junctions of the waveguides. The electrodes necessary to control the orientation of the molecules are deposited across the trenches. Small fill channels can be etched into the planar device to allow for filling the trenches with liquid crystal material.

These designs have the potential to realize medium size switching matrices using photolithic techniques and wafer processing. This enables the production of low cost devices. However, there are several requirements to be taken into consideration. The materials have to be carefully chosen to ensure that the proper index mismatch is achieved for TIR conditions. The walls of the trenches must have high surface quality and the trenches must be very narrow to reduce coupling losses. High loss materials or high loss trenches will prevent fabrication of high performance matrices. Of course, to maintain high isolation, the dimensions of the trenches must be carefully controlled. Techniques must be employed to ensure that the LC molecules achieve proper alignment in the trenches. Finally, planar devices have general issues concerning PDL. Diverse polarization routing must be employed to control the isolation, which adds to the burden of keeping PDL low. Since LC devices are generally small in size, environmental concerns can be reduced by controlling temperatures and enclosing the entire device in hermetic packaging [12].

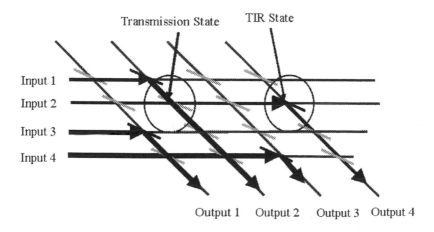

Figure 5-8. Schematic of a 2-d LC cross-connect.

Another way to implement a switch using LCs is to create a switch point by bringing two optical waveguides into proximity with each other, but not so close as to cause optical coupling between the guides. By placing an LC material between the guides and carefully choosing the indexes of the materials and the spacing of guides, light can coupled between the guides in one state and blocked in the other state [12]-[15]. While these have been demonstrated in several materials, they have not moved forward into commercial applications as yet.

5.2.3 Wavelength-Selective Switches

LCs can be utilized to build wavelength selective switching systems for WDM applications. There are three general classes of optical designs for this purpose. The first is based around adding diffractive optics to the optical path to separate the channels spatially and then switch each channel individually. The second uses gratings within guiding structures to create reflections of specific spectral slices, and the last is based on etalons.

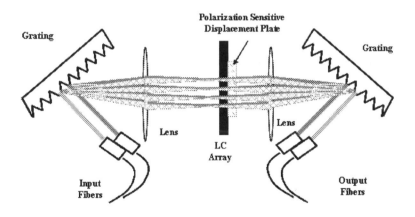

Figure 5-9. Optical schematic for free space wavelength dependent switch.

Figure 5-9 shows a schematic of a switch that uses diffraction gratings to separate the optical channels in a WDM system and uses an array of pixels in a liquid crystal cell to switch each channel independently. This switch is based upon the same polarization management techniques used in the basic 2x2 switch of Figure 5-5, except that the spatial separation of the light from the gratings allows channel beamlets to be focused on an array of pixels in the LC cell. To achieve high extinction, the optical paths for the two polarization beamlets for each channel should pass through one pixel on the LC so that both polarization beamlets experience the same rotation. This also ensures that both beamlets experience the same loss and reduces PDL. Also, an alignment offset must be established between input and output fibers so that when the beamlets are switched to the desired output depending on the polarization state of the LC pixels [16].

Early demonstrations of this design yielded approximately 30 dB of contrast with about 4 nm channels' spacing. Recently, this type of switches has been demonstrated with more than 80 wavelength channels, 0.4 nm channel spacing, >40 dB of isolation and approximately 5 dB of total loss [17]. Again, alignment and quality of optics are important issues taking into consideration the dimensions of the LC pixels and the requirement of centering the beams on the pixels. The accuracy of the alignment of these beams against pixel arrays and the stability of the setup over ranges of temperature and time determines not only loss and PDL of the device, but also the spectral stability and efficiency of the systems. There have been proposals to relax alignment constraints by using larger numbers of pixels and dynamically controlling them. In this concept, channel allocation docs not need to be pre-specified [18].

Another wavelength-selective switching concept uses a resonant cavity as a filter. These switches depend on tuning the resonant cavity (etalon) to the wavelength of interest reflecting it back to the input of the switch where it is collected and redirected by either a circulator or an optical coupler. Such a scheme has been demonstrated with 0.24 nm filter bandwidth, ~12 nm tuning range and an insertion loss of approximately 4.1 dB [19]-[20]. Switching (tuning) times of ~30 microseconds were demonstrated too. One of the disadvantages of this approach is that the filter needs to 'park' in an unused slot of the spectrum adjacent to the channel of interest or be tuned through active channels to reach the channel of interest. These methods either waste useful spectrum or disrupt adjacent channels. Additionally, monitoring the central wavelength of the filter is important to ensure that there is no wavelength drifts over ranges of temperature and time. This increases the cost of this kind of devices.

LC wavelength selective switches can also be realized by using Bragg diffraction gratings instead of Fabry-Perot filters. It has been shown that these gratings can be created with periodic polymer-dispersed liquid crystal (PDLC) planes. By applying voltage across the planes, the switch can be toggled on or off. This work is in its early stages, but shows promises for not only switches, but other optical devices as well [21]-[22].

5.3   RELIABILITY CONSIDERATIONS FOR LIQUID-CRYSTAL COMPONENTS

Deployment of LC based devices in communication networks depends not only on the ability of these devices to meet certain performance specifications, but also on their fulfillment of the reliability requirements of communication networks. Most telecom LC devices are currently designed for installation in a central office (CO) environment. The operating conditions recommended by Telecordia are 5 to 40 ^{0}C and 5% to 85% relative humidity (RH) continuous, with short term excursions of –5 to 50 ^{0}C and 5% to 95% RH. Of course, depending on deployment conditions, additional requirements may be added. For example, there are requirements concerning flammability, vibration and shock resistance that are recommended. It is also possible to require that devices function in accordance to specification within systems that are installed at diverse altitudes, ranging from 60 meters below sea level to 1800 meters above sea level [23].

The study of reliability of LC components for optical networks starts with a look at reliability of traditional liquid crystal devices for use in information display applications. The widespread use of LCs for information display over the past few decades has resulted in a considerable understanding of their reliability issues [24]-[28]. Failure characteristics of these devices are determined by criteria such as cosmetic appearance, void formation, electrical current consumption, contrast ratio and response time [24]. Environments known to accelerate failure in LC devices include exposure to UV radiations, temperature, humidity, and electrical fields (AC voltage and DC offset) [24], [26]. With careful design; proper choice of materials and processing technologies; and reasonably controlled exposure environments, lifetimes of 20 years can be expected from LC devices [25].

In the following we survey some potential concerns specific to telecommunications systems and discuss some examples of particular tests and performance measures.

5.3.1 Material and Fabrication Techniques

Selection and processing of the various elements and materials that make up LC components naturally have a large impact on reliability. As depicted in Figure 5-10, substrates from the basis of a series of films to be deposited, including anti-reflective or reflective coatings, various conductive electrode materials, and rubbing or alignment layers. Additionally, barrier layers can be added at various locations in the basic design to provide mechanical or chemical protection to certain films. Material choices made during the design phase will have major impact not only to the proper operation of the device, but its cost and the long term reliability.

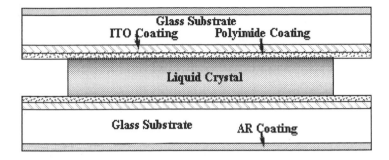

Figure 5-10. Schematic of a reference cell design.

5.3.1.1   Substrate

An LC device is generally manufactured by building the various coatings up on the substrates and then joining the two substrates to form the cavity into which the LC material is filled. A variety of substrate materials can be used for LC devices; however, for high reliability devices such as would be applicable to telecommunications, glass and silicon are the materials of choice. Silicon is often selected for reflective devices while glass can be used for either configuration. There are mechanical characteristics such as surface smoothness, flatness and thickness that impact performance, cost and yields. Chemical composition of the substrate and cleanliness of the substrates can significantly impact the long term performance and reliability of the device. Additionally, the substrates should be matched to the films in terms of the coefficient of thermal expansion (CTE) and to each other as they will see significant changes in temperature during the manufacturing process and will also be exposed to milder temperature exposure in operation.

5.3.1.2   Coatings

There are two types of thin film coatings found in most LC cells. They are transparent conductors, usually indium tin oxide (ITO), and antireflective (AR) coatings. Sometimes coatings which reflect or absorb light are also found in LC cells to improve the handling of stray light or to be able to use the cell in reflective mode. These coatings may fail through mechanical or electro-chemical means and therefore must be carefully chosen. As mentioned above, a catastrophic failure can be induced in the manufacturing of the cell if a film is stressed to rupture during its application or in subsequent manufacturing steps. For this reason, any design requires a systematic study of the specific manufacturing processes which will be used and the material selections.

For AR and reflective coatings there are two main failure modes. Many of the coatings nominally used in these applications are very soft. A common AR coating will contain magnesium fluoride which is very easily scratched. These films almost always require a protective coating such as silicon dioxide to be applied over them to protect them from abrasion. If gold is used as a reflective coating, it will require similar protection. Since most of these films are deposited on oxide glasses, adhesion of metals can also be a significant problem. Often a few angstroms of chrome can be used as an adhesion layer to form a strong chemical bond to the substrate and act as an anchor for the metal layers. Of course, each barrier layer or adhesion layer applied to the substrate will add manufacturing costs to the production of the cell and should be used judiciously.

The most commonly used transparent conductive film is ITO. While ITO is generally used in display applications, it is an extremely difficult film to use in practice. This is due to the fact that its conductivity is a function of not only the thickness of the film, but also the oxygen content and oxidation state of the film. This is confounded by the fact that the transparency is also a function of the oxidation state of the film. By increasing the oxidation state, the conductivity is reduce while the transparency is increased. Additionally the location of the ITO just beneath the alignment layer puts it at high risk of being abraded during any rubbing process which occurs later in the manufacturing process. Therefore, to increase the durability of the film and to increase its inertness this layer is often protected by a barrier layer such as silicon dioxide. Of course to be able to protect ITO from a change in oxidation state due to electro-chemical effects, the barrier film must be nearly perfect with no pin holes or voids. If a reflective metal is used in the cell, such as gold, the problem becomes a little less complex since the conductivity and reflectivity can both be enhanced by using thicker films. However, adhesion and protection of these types of materials must still be dealt with as discussed above.

5.3.1.3   Alignment Layer

The alignment layers in an LC cell help define the initial alignment of the crystals and anchoring forces (Sec. 5.1.2.1). A number of techniques can be used to create the alignment layer. These include high angle deposition of inorganics such as SiO_2, rubbed coatings such as polyimides, and even rubbing of bare substrates. The range of alignment technologies available also result in a range of reliability issues to understand. Here we assume that rubbing of bare substrates is not a serious option for high reliability liquid crystal devices. Thus, all other technologies involve the use of an alignment layer coating. Reliability concerns any changes in anchoring forces that will be exhibited over time. Changes or degradation of the alignment quality either along the director or the pretilt is highly dependent upon the material chosen and its processing. The most commonly used alignment layers in the industry are polyimides and these materials generally have high T_g's[6] compared to most polymeric materials and are quite stable. However, the performance can be significantly impacted by the quality of the processing of this coating.

5.3.1.4   Liquid Crystal

Two specific failure modes that the LC itself is at risk for are Ultra-Violate (U-V) degradation and voiding. UV degradation is generally not a risk for the telecommunications operational environment, but could be for the processing of the cells as many adhesives used in LC assembly are UV curable. UV adhesives are often used due to their advantage in manufacturing. The adhesive can be applied to parts, but does not cure until initiated by high fluence UV radiation exposure. Curing can be accomplished without exposing the adhesive to high thermal excersions and in a relatively short time period. Voiding of the LC is caused by liberation of gases dissolved into the LC. Voiding of this type can be eliminated or minimized by degassing of the LC prior to use. UV radiation exposure and mechanical impact have also been shown to be contributors to voiding [28]. Voiding can also be the result of outgassing from contamination, therefore, if voiding is found in cells, the source of the defect is important to realize corrective action and adequate reliability.

5.3.1.5   Adhesives

Adhesives used in cell fabrication essentially provide both mechanical alignment of the substrates and form part of the containment for the LC material. Generally, two adhesives are used in a cell. One of them makes up the majority of the perimeter length and is cured prior to filling the cell with LC. The second is a plug adhesive which completes the sealing of the vessel after it has been filled with LC material.

There are several reliability concerns regarding adhesives. These include contamination of the LC by the adhesives themselves, change in distance between the two substrates (creation of gaps), permeability of the adhesive which allows other contaminants such as water vapor to mix with the LC, and catastrophic failures such as delamination. Choosing a perimeter adhesive has fewer constraints than the plug since it is applied and cured prior to contact with the LC material. The choice of perimeter adhesive and its cure schedule must be optimized with respect to dimensional stability,

---

[6] T_g is the glass transition temperature of a polymer, below which it is brittle and above which it is elastic.

permeability, adhesion and to minimize any contamination of the LC material. The choice of a plug adhesive is much more limited in that it is placed in direct contact with the LC prior to curing. This liquid to liquid contact increases the risk of LC contamination and poor adhesion. After the LC is present, thermal processing is limited which is why UV curable adhesives are preferable. Due to the small perimeter length of this adhesive, dimensional stability is not as critical as for perimeter adhesive. However, protecting the LC from external environmental effects are critical requirements.

Another adhesion system that is mentioned in the literature uses a glass frit perimeter with an indium plug. This clearly presents a major change in processing. Such inorganic materials will have the benefits of being more dimensionally stable and more effective as environmental barriers. There may also be additional reliability issues that do not apply to organic adhesives such as development of fractures due to thermal expansion and mismatches between the frit and substrate.

5.3.2 High Optical Power

Optical power is a particular concern because of the much higher power densities typical of lasers and optical amplifiers used in communications networks compared to those typical of display applications. Another key difference is the wavelength. Optical network devices will almost exclusively be exposed to wavelengths in the Infra-Red (IR) spectrum with the exception of a few pump wavelengths. The major concern of working with IR wavelengths for these devices is absorption and subsequent localized heating which can accelerate failure modes.

Optical power testing has been conducted on a candidate LC by sandwiching it between two single mode optical fibers inside a capillary tube. The gap between the fibers, where the LC was placed, was estimated to be approximately 2 μm. Optical power was supplied by a 4 watt tunable laser at a wavelength of 1550nm. Both the input and output powers of the setup were measured. Input power was changed at a rate of approximately 2 watts/minute with a maximum of 3.5 watts.

Figure 5-11 plots the output optical power through the LC-filled capillary joint as a function of the input power. Experimental measurements are given by the solid line whereas the dashed line describes the ideal case with no loss (output power equals input power). The observed loss is probably due to three sources; misalignment between the two fibers, light scattering by the liquid crystal itself and an index mismatch between the fibers and the LC. The most important observation of this experiment is that the plot is linear. This observation indicates that no damage was done to the LC material up to the maximum optical-power exposure. If damage were to occur to the liquid crystal, a nonlinear curve would be expected.

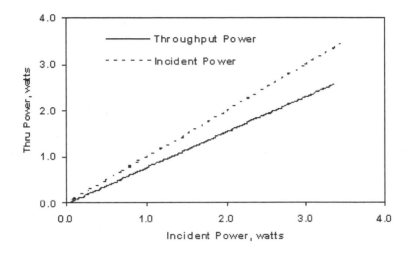

Figure 5-11. Observed losses due to optical power in LC material.

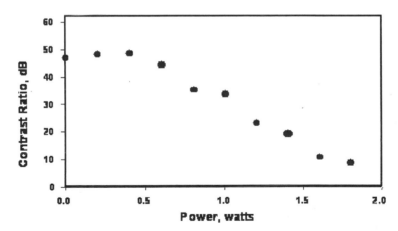

Figure 5-12. Contrast ratio degradation as a function of optical power though a completed LC cell.

Optical power experiments have also been conducted on completed LC cells of the typical transmissive twisted nematic type. The cells were constructed as shown in Figure 5-10. Optical power was delivered through a beam collimator and lens producing a spot size of approximately 85 μm. This size was estimated by microscopy of regions exposed to the high power. The LC cell was exposed to various optical-power levels. Exposure was for a period of 5 minutes in each case. Immediately after each exposure, the contrast ratio of the LC cell was measured.

Figure 5-12 plots the contrast ratio as a function of optical power for each of the 5 minute exposures in this experiment. The contrast ratio degrades progressively as optical power increases. The damage threshold is about 500 mW (27 dBm) and the contrast ratio drops below 40 dB at exposures greater than 800 mW (29 dBm). Optical microscopy of these regions showed that damage to the polyimide coating and the ITO layer were first observed at powers of 800 mW and 2.0 W respectively. The damage to the polyimide layer is consistent with thermal damage observed in other exposure tests. The alignment characteristic of the polyimide coating has been shown to degrade above 200°C.

5.3.3   Electrochemical Stability

In an LC cell, the LC material performs the function of an electrolyte while the conductive coatings play the roles of the electrodes. Selection of the materials and voltages that minimize complementary oxidation and reduction reactions is desirable in LC devices. This is because these reactions can have many unfavorable effects including changes in insertion loss and changes in conductivity of the electrodes. A simple method of assessing the electrochemical stability of LC cells is to apply a DC voltage across the cell and measure the leakage current through it. The magnitude of this current is an indication of the quality of the LC cell.

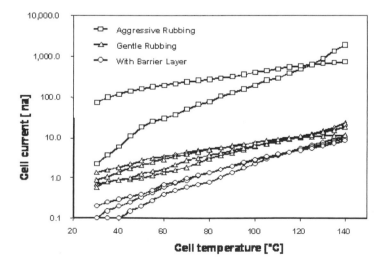

Figure 5-13. Evaluation of various manufacturing methods used to protect LC cells from electrochemical instabilities.

This method was used with a 10 V DC voltage across transmissive LC cells with polyimide alignment layers designed for optical network applications. Figure 5-13 plots the leakage current as a function of temperature. These LC cells were manufactured with aggressive and gentle rubbing techniques, as well as with a protective barrier coating on the active substrate. Significant decreases in electrochemical activity are seen by improving the rubbing process. Additional improvements are obtained by further adding the barrier layer. Note that controlling the rubbing process alone can make a marked improvement in reducing the electrochemical effects.

5.3.4 Environmental Testing

As with any commercialized telecommunications product, LC based optical devices are expected to successfully survive a variety of standardized environmental exposures [23]. Carrier-grade components are normally required to survive exposure to 85% relative humidity and 85°C temperature. Figure 5-14 shows the average contrast ratio of seven transmissive TN LC cells using organic adhesives. Measurements are carried out under the two polarization conditions, as shown in figure. After 1200 hours of exposure there is no clear trend, suggesting that these LC cells are performing quite well.

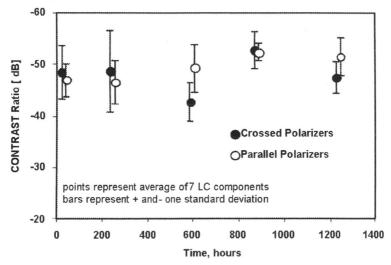

Figure 5-14. Variation of extinction ratio over time with exposure to 85% relative humidity and 85°C temperature.

On the other hand, Figure 5.15 shows the results of an experiment where a TN LC cell was placed in a freezer at -180 degrees centigrade for one hour. After this time, the cell was returned to room temperature and the extinction ratio was measured using both parallel and crossed polarizers. The temperature was increased by 10 degrees and the test repeated. This continued until the temperature reached 260 degrees. The experiment was duplicated on a second cell. Degradation at high temperatures was linked to deterioration of the polyimide coating's ability to maintain alignment of the LC at this level of heat.

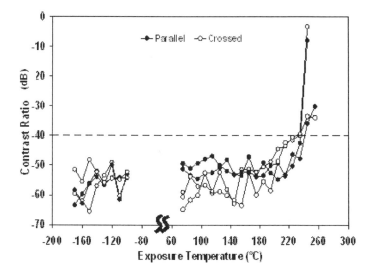

Figure 5-15. Impact on contrast ratio of exposure to extreme temperatures.

5.4   SYSTEM AND NETWORK PERFORMANCE CONSIDERATIONS

Liquid crystal based optical switches have several qualities and features including, high extinction ratio, low power consumption, and high reliability. Wavelength selective LC switches can also be realized. These features make LC switches desirable for numerous optical network applications.

Protection switching is one of the main areas of application where LC switches can be utilized in optical communication networks. Protection switching requires high reliability, uninterrupted service with low FIT[7], and

---

[7] FIT is a unit for measuring failure rates. The acronym comes from the phrase Failure unIT. Although it is an accurate description, it is not from the phrase "Failure In Time". The FIT

moderate switching times (in the range of milli-seconds). With no moving parts, LC switches have all these feature requirements and their switching time can be engineered to be shorter than 1 msec.

LC-based optical switches can also be deployed in reconfigurable optical add/drop multiplexers (OADM). Applications of LC switches in network environments also demands low insertion loss and high extinction ratio. Low insertion loss is important since considerable loss is introduced by mux/demux pairs and other optical components. With currently available mux/demux technologies, based on either thin-film dielectric filters or arrayed waveguide gratings, the total insertion loss of the pass-through traffic can be greater than 6 dB for four channels, or 12 dB for forty channels [29]. LC based switches with losses of < 1dB and isolation of better than 50 dB are commercially available. LC-based wavelength selective switches offer lower losses. Commercially available 80-channel systems have been demonstrated with total pass-through losses of < 6 dB per channel.

In a network environment crosstalk is affected by the extinction ratio of the switches. In order to guarantee 99.999% availability, it is required that cumulated crosstalk does not exceed - 40 dB [12]. This level of performance has not only been reported in several research papers, but is also achieved in commercial products.

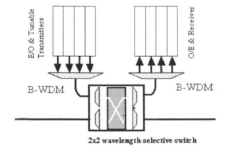 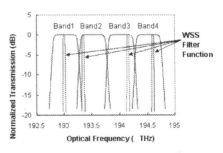

Figure 5-16. Schematic diagrams of a partially tunable OADM design with band WDMs that uses a 2x2 wavelength-selective LC switch (a), and its pictorial explanation of the tuning function in the drop path for 8-skip-2 banded channel plan. B-WDM stands for broad band mux/demux.

The LC-based OADM design shown in Figure 5-16 is characterized by low loss and inherently low out-of-band crosstalk. The wavelength-selective switches deployed in this system need to match the transmission plan of the network. Switches of this type have been reported to have flat top channels of 25 GHz, spaced at 50 GHz spectral intervals [17]. For more details, and discussion of key requirements and experimental results of this application, the reader is referred to references [29]-[33].

REFERENCES

[1] P. G. deGennes, "The Physic of Liquid Crystals," Clarendon Press, Oxford, 1974.

[2] S. Chandrasekhar, "Liquid Crystal," 2nd ed., Cambridge University Press, New York, 1992.

[3] P. Mariani, F. Rustichelli, and G. Torquati, "Physics of Liquid Crystalline Material," (I. C. Khoo and F. Simoni, Eds.), Gordon &Breach, Philadelphia, 1991.

[4] A. Ciferri, W. Krigbaum, and R. B. Meyer (Eds.), "Polymer Liquid Crystals," Academic Press, New York, 1982.

[5] R. P. Raynes, "Optically Active Additives in Twisted Nematic Devices," Revue de Physique Appliquee, Vol. 10, No. 3, pp. 117-20, 1975.

[6] P. Yeh and C. Gu, "Optics of Liquid Crystal Displays," John Wiley and Sons, Inc., New York, 1999.

[7] R. E. Wagner and J. Cheng, "Electrically Controlled Optical switch for Multimode Fiber Applications," Applied Optics, Vol. 19, No. 17, pp. 2921-5, September 1, 1980.

[8] E. Hecht, "Optics," Second Edition, Reading, MA, Addison-Wesley Publishing Company, pp. 270-326, 1987.

[9] J-C. Chiao, K-Y. Wu, and J-Y. Liu, "Liquid-Crystal WDM Optical Signal Processors," BroadBand Communications for The Internet Era Symposium Digest, 2001 IEEE Emerging Technologies Symposium, pp. 53-7, 2001.

[10] K. Noguchi, T. Sakano, and T. Matsumoto, "A Rearrangeable Multichannel Free-Space Optical Switch Based on Multistage Network Configuration," IEEE J. Lightwave Technol., Vol. 9, No. 12, pp. 1726-32, December 1991.

[11] K. Noguchi, "Optical Free-Space Multichannel Switches Composed of Liquid-Crystal Light-Modulator Arrays and Birefringent Crystals," IEEE J. Lightwave Technol., Vol 16, No. 8, pp. 1473-81, August, 1998.

[12] Corning Incorporated, unpublished internal research (2001).

[13] R. A. Betts, F. Lui, and E. Gauja, "Tunable Couplers Fabricated in K+/Na+ Ion Exchanged Glass," Electron. Lett., Vol. 26, No. 12, pp. 786-8, June 7, 1990.

[14] J. Zubia, U. Irusta, A. Aguirre, and J. Arrue, "Design and Measurment of POF Active Couplers," Proc. IEEE 10th Annual LEOS Meeting (LEOS '97), Vol. 2, pp. 48-9, 10-13 Nov., 1996.

[15] R. Asquini and A. d'Alessandro, "A Bistable Optical Waveguided Switch Using a Ferroelectric Liquid Crystal Layer," Proc. IEEE 13th Annual LEOS Meeting, LEOS'2000, Vol. 1, pp.119-20, 2000.

[16] J. S. Patel and Y. Silberberg, "Liquid Crystal and Grating-Based Multiple-Wavelength Cross-Connect Switch," IEEE Photon. Technol. Lett., Vol. 7, No. 5, pp. 514-6, May 1995.

[17] J. Kondis, B. A. Scott, A. Ranalli, and R. Lindquist, "Liquid Crystals in Bulk Optics-Based DWDM Optical Switches and Spectral Equalizers," Proc. of the IEEE 14th Annual Meeting of the IEEE Lasers and Electro-Optics Society (LEOS'2001), Vol. 1, pp. 292-3, 2001.

[18] J.-K. Rhee, F. Garcia, A. Ellis, A. Hallock, T. Kennedy, T. Lackey, R. G. Lindquist, J. P. Kondis, B. A. Scott, J. M. Harris, D. Wolf, M. Dugan, "Variable Passband Optical Add-Drop Multiplexer Using Wavelength Selective Switch," the 2001 European Conference on Optical Communications (ECOC'2001), Vol. 4, pp. 550-1, 2001.

[19] Y. Bao, A. Sneh, K. Hsu, K. M. Johnson, J-Y Liu, C. M. Miller, Y. Monta, and M. B. McClain, "High-Speed Liquid Crystal Fiber Fabry-Perot Tunable Filter," IEEE Photon. Technol. Lett., Vol. 8, No. 9, pp. 1190-2, September 1996.

[20] A. Sneh and D. Johnson, "High-Speed Continuously tunable Liquid Crystal Filter for WDM Networks," IEEE J. Lightwave Technol., Vol. 14, No. 6, pp. 1067-80, June 1996.

[21] L. V. Natarajan, R. L. Sutherland, V. P. Tondiglia, and T. Bunning, "Electro-optical Switching of Volume Holographv Gratings Recorded in Polymer Dispersed Liquid Crystals," Proceedings of the IEEE Aerospace and Electronics Conference, Vol. 2, pp. 744-9, 1996.

[22] L. H. Domash, Y-M Chen, P.Haugsjaa, and M. Oren, "Electronically Switchable Waveguide Bragg Gratings for WDM Routing," Digest of the IEEE/LEOS Summer Topical Meeting, Vertical Cavity Lasers, Technologies for a Global Information Infrastructure and Applications, Gallium Nitride Materials, Processing and Devices, pp. 34-5, 1997.

[23] Telcordia, Network Equipment-Building Systems (NEBS) Requirements: Physical Protection, GR-63-CORE, Issue 2, April 2002.

[24] T. S. Chang, "Evolution of Liquid Crystal Display Reliability," Electrochemical Society Fall Meeting, Vol. 76-2, pp. 507-9, 1976.

[25] J. A. Castellano, "Reliability and Standards in the U.S.A.," Molecular Crystal Liquid Crystal, Vol. 4, No. 63, pp. 265-80, 1981.

[26] K. Kitagawa, K. Toriyama, and Y. Kanuma, "Reliability of Liquid Crystal Displays," IEEE Transactions on Reliability, R-33, No. 3, pp. 213-8, 1984.

[27] J. Roman, "A New Approach to LCD Reliability," Electronic Packaging and Production, Vol. 9, pp. 108-11, 1986.

[28] A. R. Revels, "Void Formations in Liquid Crystal Displays," AIAA/IEEE Dig. Avionics System, pp. 243-8, 1994.

[29] A. Boskovic, M. Sharma, N. Antoniades, and M. Lee, "Broadcast and Select OADM Nodes Applications and Performance Trade-off," OFC'2002, Tech. Digest, p.158, Anaheim, CA, 2002.

[30] J.-K. Rhee, M-J Li, P. I ydroose, M. Zhao, B. Hallock, I. Tomkos, and M. Ajgaonkar, "A Novel 240-Gbps Channel-by-channel Dedicated Optical Protection Ring Network Using Wavelength Selective Switches," OFC'2001 Postdeadline Papers, Tech. Digest, Vol. 4 pp. PD38-1-PD38-3, Anaheim, CA, 2001.

[31] I. Tomkos, M. Vasilyev, J-K Rhee, M. Mehendale, B. Hallock, B. Szalabofka, M. Williams, S. Tsuda, and M. Sharma, "80/spl times 10.7 Gb/s Ultra-long-haul (+4200 km) DWDM Network with Reconfigurable Broadcast & Select OADMs," OFC'2002 Postdeadline Papers Tech. Digest, pp. FC1-1-FC-3, Anaheim, CA, 2002.

[32] H. Kim and S. Chandrasekhar, "Dependence of In-band Crosstalk Penalty on the Signal Quality in Optical Network Systems," IEEE Photonics Tech. Lett., Vol. 12, No. 9, pp. 1273 –4, Sept. 2000 (and references therein).

[33] J.-K. Rhee, F. Garcia, B. Hallock, T. Kenneday, T. Lackey, R. G. Lindquist, J. P. Kondis, B. A. Scott, J. M. Harris, D. Wolf, M. Dugan, "Variable Passband Optical Add-Drop Multiplexer Using Wavelength Selective Switch," Proc. 27th European Conference on Optical Communications (ECOC'01), Amsterdam, Vol. 4, pp. 550-1, 2001.

Chapter 6

MEMS BASED OPTICAL SWITCHING

Roland Ryf
David T. Neilson
Vladimir A. Aksyuk

Micro electro mechanical systems (MEMS) technology [1] plays an important role in the optical switching space. It brings many of the advantages of macroscopic opto-mechanical switching and adds many desirable features [2]. Conventional mechanical optical switches, which are discussed in Chapter 8, such as movable mirrors and shutters, provide a versatile way of switching light with very good optical performance characteristics such as low loss and high contrast. However, the size, cost, speed, and reliability associated with mechanical parts limit their practical use.

Using MEMS technology, optical elements and functions have been demonstrated, including mirrors, shutters, filters, lenses, and polarizers [3]-[12]. They maintain the excellent optical performance characteristics of their macroscopic counterparts, but are much smaller, faster and can be fabricated using very large scale integration (VLSI) fabrication processes, similar to those of the electronics industry. Additionally, such optical MEMS can easily be batch-fabricated in arrays of hundreds or thousands of elements, which can be put in a single compact package. This makes them suitable for the demanding high-complexity optical networking applications, which need high levels of functionality per module. Integration is a key enabler of high-complexity subsystems, allowing a new application to be addressed by providing novel functions, impractical to achieve otherwise. An example of this is enabling large strictly non-blocking NxN transparent optical cross-connects (OXC). This system connects any of the N optical input ports to any of the N optical output ports. It has N^2 connections and the number of different states the system can be in is $N!$.

This Chapter will start with a general overview of optical MEMS and the basic principles of free-space optical switches. Then, we discuss numerous types of optical switches based on optical MEMS. We cover different switch designs, sizes (8x8 to 1296x1296 fabrics), and various WDM switching systems.

6.1   THEORY AND PRINCIPLES OF BASIC MEMS SWITCH ELEMENTS

All micro mechanical optical switching devices may be roughly classified into two distinct classes. Devices in the first class achieve the switching function by manipulating the intensity or the direction of propagation of a light beam via reflective or refractive structures. In the second class the steering or switching function is achieved by diffractive or interference effects, where the mechanical motion adjusts the phase of light. Examples of the first class are: a mirror of a fixed orientation inserted into the path of a light beam to redirect all or part of the beam into a different output; a mirror permanently located in the path of the beam and can change orientation to steer the beam to the desired output. Some examples from the second class are: a tunable cavity formed by a membrane such as the mechanical anti-reflection switch (MARS) [6]; switches based on tunable gratings [13]-[15], which are revealed by moving a grating structure out of plane.

Two devices operating by motion of a mirror are shown schematically in Figure 6-1(a) and (b). One uses linear translation of the mirror into the path (b). The other employs rotational motion of a mirror in the path (a) to switch light between output paths. Two devices utilizing interference processes are also illustrated in Figure 6-1(c) and (d). One uses a grating structure and is a wavefront splitting interferometer (c). The other uses a thin film structure and is an amplitude splitting interferometer (d).

The systems in the second class rely on the wave properties of the electromagnetic field and are fundamentally wavelength dependent. Achieving high contrast switching relies on the precision of the phase control to cancel out the unwanted signal(s) via destructive interference. Consequently achieving high contrast over a wide wavelengths range can be difficult. However, one benefit of this kind of devices is that the small range of motion required, of the order of half a wavelength, which can allow for low actuation forces, or high modulation [6] and switching rates. The first class has the advantage of wavelength independent operation over a broad spectrum. This can be achieved by appropriate mirror coatings. This approach is therefore much more suitable for building large scale, low loss, wavelength independent optical switching fabrics.

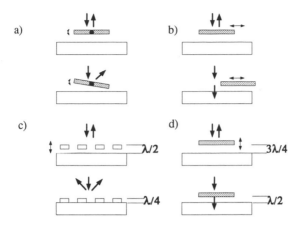

Figure 6-1. Basic switching elements obtained by micromachining (a) Tilt motion of fixed location mirror, (b) Insertion of fixed angle mirror [16]-[17], (c) Tunable diffraction grating [13], and (d) Thin film interference device

The requirements in terms of size and motion of the switching elements will depend upon the type of optical system in which they are to be used. For a large, low-loss free-space switch fabric the beam is typically several hundreds of microns in diameter, dictating large mirror size and mirror stroke. Alternatively, for a waveguide based system, switching may be accomplished by inserting a mirror in a narrow gap between fixed guides, where the beam size is comparable to the mode size. In this case, strokes and mirror sizes of a couple of tens of microns may be sufficient.

An additional consideration for designing switching elements is whether the motion is to be digital or analog in function. We define digital switching elements as capable of selecting from fixed position states, and analog switch elements as capable of selecting from a continuous range of positions. Digital devices may provide simpler actuation and control, since the states are known. However, digital is typically limited in the number of states that can be addressed, with many of the devices switching between only two states. Analog motion allows more states to be addressed and so is more appropriate for switches with high numbers of ports. It is also useful where intermediate states are to be used to provide variable attenuation of signals. This functionality is often desirable to equalize power among lightpaths. Analog devices must be designed to be stable with either open loop or closed loop controllers.

A final consideration is the number and type of degrees of freedom the device will have. Single axis motion of a rotational type is usually the simplest to implement since the motion in this case is around a fixed point and generally the support structure and the actuators can remain fixed. Pure translation is typically more challenging since the actuation and support may be required to move with the element. Motion about two axes adds significant complexity. Two-axis rotational elements based on gimbal structures have been demonstrated. Implementation of robust two-axis translation still poses significant challenges.

6.1.1 Mechanics of Micromachines

The purpose of the mechanics of any switching device is to allow the optical element to be adjusted accurately with respect to the optical beam in response to a control signal. Part of the device structure comprises an actuator that uses some physical principle to convert the applied control signal into an output mechanical force and displacement – the output work. In addition to the actuator and the optical element, the switching device includes the support structure, which is composed of multiple mechanical structures. These perform many functions including: connecting or linking the optical element to the actuator; correctly pre-positioning self-assembled parts of the structure; enabling the desired motion trajectory while restricting undesired displacements; and providing well-defined hard stops and restoring forces.

Regardless of the actuation principle, micromechanical design plays a critical role in achieving the desired function. Most effective mechanical design approaches on the micro-scale are often different from our macroscopic-based intuition. This is caused by the way various physical properties, such as mass and inertia, friction forces, stresses and strains scale with decreasing size. For example, mass and inertia decrease much faster than surface friction forces. If conventional mechanical links were used between moving micro-parts, friction in the joints would dominate. This would introduce hysteresis, decrease accuracy and possibly lead to wear. Therefore, micro flexure hinges and compliant mechanisms are preferred to conventional kinematic joints. In MEMS it is not typically the moving parts that are a reliability risk, but friction contacts between them. Fortunately, the scaling also allows the design and fabrication of small compliant flexure hinges without exceeding elastic limits of the material, which enable large angular rotation. Micro-parts operate in a regime much like a quartz resonator in a watch. Devices can operate in the pure flexure regime by suspending them completely on the appropriate torsional or linear springs. Devices with elastic links have been driven through billions of full amplitude cycles and prolonged set and hold periods, without any noticeable change in performance characteristics.

Finally, since the mass of the moving parts is so low, the mechanical resonance frequencies of these devices are typically high. This decouples them from external mechanical vibrations and noise sources. Unless sophisticated electronic feedback control circuits are used, the device response time is defined by its mechanical resonance frequencies, which depend on spring stiffness and the device mass. The mass is dictated by the device size, often dominated by the optical element (e.g. mirror reflector) and often not easily changed. Micro-fabrication techniques are available to make springs with a very wide range of stiffness. The stiffness is typically chosen to accommodate the actuator output force and the required amplitude of motion. The response times (switching time) for most optical MEMS based switching elements that are powered by electrostatic actuators lies between 10 μs and 10 ms. To achieve the shortest response time, not only must the device resonance frequency be maximized, but also the damping condition should be close to optimal. This requires that the mirror moves into position quickly and settles without excess oscillation. Since springs are typically designed to work in the elastic regime, there is negligible mechanical energy loss in them. Most often energy dissipation, resulting in the appropriate damping coefficient, is achieved by engineering the interaction of the mirror with the air around it. As the mirror moves, it has to push the air out of its way, and the amount of energy required to do that can be controlled to some extent by introducing appropriate features, such as gaps or walls, to control the flow.

6.1.2   Fabrication of Micromachines

A wide variety of micro-fabrication techniques and materials are available for making these devices. For example, silicon surface micro-machining techniques rely on processes developed for manufacturing of electronic components. This includes: thin film deposition (silicon, silicon dioxide, silicon nitride, metals); chemical mechanical polishing (CMP); micro-lithographical patterning and etching.

Starting with a silicon wafer, several layers of different materials are sequentially deposited, photolithographically patterned and etched to define parts of the micro-system, such as reflectors, springs, electrodes, electrical wiring, mechanical attachment points and support structures. Then, layers made of so-called "sacrificial material" (typically Si dioxide) are removed by selective etching to free or "release" the mechanically movable parts which are made of other materials, e.g. silicon, silicon nitride and metals [18]. These remaining parts constitute the final MEMS device. For more complex devices, some parts may be "spring-loaded" during fabrication, such that after the release they push themselves and other parts into the desired position automatically to self-assemble the devices. An example of such a self assembly surface micromachined structure is shown in Figure 6-2(a).

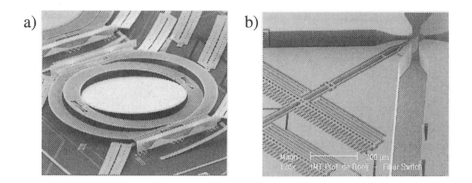

Figure 6-2. Micro fabricated structures: (a) Two axis tilt mirror obtained by surface micromachining [11]. (b) Shutter switch fabricated with DRIE technique [10].

A variety of other processes and techniques have been adapted or specifically developed for fabrication of MEMS devices [19]. These include deep reactive ion etching (DRIE) of Si and various types of wafer or flip-chip bonding [20]. DRIE allows the creation of structures out of a silicon wafer by etching lithographically defined deep cuts with vertical walls into it (see Figure 6-2(b) for an example of a structure obtained by DRIE). Wafer bonding allows joining together silicon wafers with various micro-fabricated structures to form a complete set of devices. For example, a wafer containing multiple micro-fabricated mirrors suspended from it on springs may be joined with a wafer containing multiple wires and electrodes used to apply electrostatic force and move the mirrors.

In the majority of cases the fabrication processes involved are batch-fabrication processes where thousands or millions of individual devices (e.g. mirrors or switches) are fabricated simultaneously on a single silicon wafer. While similar to the way electronics are fabricated, MEMS fabrication is significantly different in multiple respects. It often involves unconventional materials, such as gold for optical elements and nickel for magnetic actuation. Unconventional processes, such deposition of thick layers of polysilicon and special annealing and doping cycles for residual mechanical stress control are also employed. Control of the mechanical properties of materials is much more important in MEMS than in electronics. In particular, residual stress control is critical in MEMS fabrication. Residual stress engineering can also be used to achieve a desired function, such as self-assembly of MEMS devices [11] or to ensure flatness of metalized silicon mirrors [21].

Since many of the optical MEMS devices work in reflection it is necessary to provide highly reflective surfaces. Thin metal coating (such as aluminium or gold) is most often used for making reflector surfaces. The underlying layer supporting the coating is typically single crystal or polycrystalline silicon. Management of the difference in the material properties between the metal and silicon is critical to achieving useful devices. Typically, $\lambda/20$ flatness over the operating temperature range is required. By using appropriate fabrication processes, choosing the correct layer thicknesses and careful design, both surface roughness and deviation from flatness can be reduced to below the desired value. Alternatively, athermal, dielectric stack mirrors with tailored curvature (including flat) have been experimentally demonstrated [22]. Such mirrors can often be made thinner and lighter, while still maintaining the desired flatness. They have the disadvantage of requiring a more complex multilayer fabrication sequence.

6.1.3 Actuation Methods for Micromachines

In order to provide switching functionality it is necessary to be able to move the micro-optical elements. The microstructures are moved to the desired position by applying external forces. The process is often referred to as actuation. Many actuation principles based on electrostatic, magnetic, piezoelectric and thermal expansion forces have been demonstrated. However, electrostatics emerged as a preferred technology for optical switching applications particularly for large port-count switch fabrics. There are several reasons for this choice. Electrostatic micro actuators can be built with materials, processes and techniques available from microelectronics manufacturing. They do not require exotic materials and do not generally suffer from hysteresis associated with materials properties, as common in magnetic and piezoelectric materials. More importantly, electrostatic actuators are uniquely suitable for dense integration for two reasons – they dissipate very little power and electrostatic fields can be shielded very effectively with conventional materials such as doped polysilicon or metal. Heat dissipation is a challenge for integrating thermal actuators, while the need for effective shielding to prevent magnetic crosstalk limits the integration density for actuators using magnetic forces.

Electrostatic actuators generally use forces generated between two or more conductors at dissimilar electrical potentials. Such forces are defined by the conductor geometry and the resulting position-dependent capacitance matrix between these conductors, as well as the applied voltages. The applied forces can be changed at electrical-charging and discharging time scales, which are determined by the actuator capacitances C and resistance

R, or its electrical RC time constant. This time constant is almost always much shorter than the mechanical response time of the actuator. In the majority of the optical switching applications, electrostatic actuators are used to move an element into a desired position and maintain that position for a prolonged period. For such steady state condition, the actuator presents a voltage-dependent capacitive load and dissipates virtually zero power. The dissipation is also near zero when the actuator is switched in a quasi-stationary manner, i.e. at frequencies much lower than the mechanical resonance frequencies of the actuator structure. The change in voltage and capacitance is associated mainly with a change in the stored energy.

Although the actuator dissipates very little power (even at high voltage), the power dissipated and the complexity of the control electronics increases with increasing voltage. It is also advantageous to decrease the actuation voltage as much as possible to avoid electric breakdowns. However, for a given geometry, the electrostatic force is proportional to the square of the applied voltage. If the actuator output force is reduced, the spring stiffness has to be reduced to maintain the device range of motion. If the spring stiffness is reduced too much, the device will be too slow and possibly too susceptible to gravity, low-frequency inertial forces and mechanical vibrations.

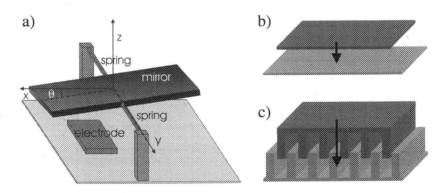

Figure 6-3. (a) Schematic of electrostatic actuated tilt mirror. (b) Parallel plate actuation. (c) Comb drive actuator for increased electrostatic force.

The geometry of the actuator conductors can be changed to obtain higher force. Let us consider for example a simple single axis tilt mirror as represented in Figure 6-3(a). It incorporates one of the most common actuators using an attractive force between two nearly parallel conductive plates (Figure 6-3(b)) – beam steering micromirrors are often actuated by applying voltage to plane electrodes located directly beneath the mirror. In this geometry the mechanical torque τ produced on the mirror will be related to the electric capacity C of the mirror by the relation

$$\tau(V,\theta) = \frac{1}{2}V^2\frac{\partial C}{\partial \theta} \tag{6.1}$$

where V is the voltage applied between the mirror and the electrode and θ is the mirror tilt angle. This torque can be increased by more than an order of magnitude by changing the actuator geometry as shown on Figure 6-3(c). This is because the capacitance C between the two conductors of the actuator is much larger and has a much stronger variation as a function of the actuator angle θ. Although this more complex geometry works quite well for a single tilt, a similar geometry suitable for a two axis tilt mirror (tip-tilt mirror) would contain curved walls that are beyond current microfabrication capabilities.

Another obvious way of increasing the actuator force at a given applied voltage is to increase the capacitance, and its rate of change with mechanical motion, by bringing the two conductors, e.g. the two plates, closer together. However, this results in a decrease of the total linear or angular range of motion the actuator can achieve. In Figure 6-3(a), the angle at which the mirror touches the substrate decreases with reducing the initial mirror-electrode gap. Moreover, when the position of the mirror is maintained by keeping it at a constant potential difference with respect to the electrode, only a fraction of the total mechanical angular range can be stably achieved. As the voltage is increased, the electrostatic force grows more rapidly than the restoring mechanical force and the mirror "snaps down"– rotates uncontrollably until it mechanically touches down. The snap-down, or the loss of stability of voltage-controlled electrostatic actuators is generally due to the nonlinear nature of the electrostatic force. A force of attraction between two such conductors at any given voltage difference is given by the derivative of the capacitance between the conductors with respect to the mechanical displacement, and generally depends on their relative positions. In equilibrium the electrostatic force is balanced by the elastic mechanical restoring force. To maintain equilibrium and stability, as the positions of electrodes change, the restoring force should increase faster than the electrostatic force. While the rate of change of the restoring force is constant, depending on spring stiffness, the rate of change of the electrostatic force for many electrode geometries, especially for nearly-parallel plates, is inversely proportional to the distance between electrodes, in most cases. This results in loss of stability at short distances, thus limiting the actuator range.

The device shown in Figure 6-4 overcomes this limitation by using some mechanical transmission mechanism. A parallel plate actuator with a small gap between its plates produces high output work while tilting by a small angle. This work, in turn, can be converted via angle-amplification mechanism to achieve high angle of rotation for the beam steering mirror. Conceptually similar approaches were used in several other devices, including linear [23] and vertical (e.g. Figure 6-1(a)) translation actuators.

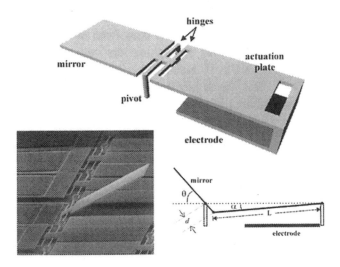

Figure 6-4. Transmission mechanism to increase mirror deflection angle using electrostatic actuation.

Finally, using electronic closed-loop control circuit it is possible to further improve the actuator performance by stabilizing the unstable part of the actuator range [24]-[26]. For example, for a parallel plate actuator the stroke can be increased, theoretically by up to a factor of three, without increasing the maximum applied voltage. This improvement is achieved by locally sensing and controlling the actuator capacitance via a feed-back loop. Since the capacitance is a monotonic function of the actuator degree of freedom for most actuator geometries, the whole range of possible motion is reachable. By changing actuator design parameters this gain can be translated into a decrease in voltage at a fixed range, or an increase in the output force.

For this scheme to work, the sensing part of the electronic circuit has to be located as close as possible to the MEMS device and both MEMS and electronics have to be properly designed to minimize stray capacitance and improve sensitivity. The sensor has to be able to resolve the actuator capacitance with the required accuracy and have a bandwidth, which is large compared to the mechanical response frequency. For arrays of optical MEMS devices this can be practically achieved only when the electronics is either monolithically integrated with the device chip or connected with extremely small stray capacitance, e.g. via flip-chip attachment.

6.1.4 Packaging of Micromachines

Although micromechanical structures obtained by silicon micromachining are very robust against mechanical shocks, their performance can degrade rapidly if they are not protected from environmental contamination, like dust, moisture, chemical elements, or in some cases even simple airflows. Protection is achieved by a package, which is essential for the reliable operation of the device. In optical MEMS, the requirements on the package are very challenging. The package has to keep the microstructure at a precise and defined location. It must provide sealed protection against environmental effects. In order to interact with light, a transparent window with low reflection is required. Finally, the package has to provide the electrical interconnection to activate the microstructures. The last requirement is particularly demanding for complex devices, like a 1000 port cross-connect, where packages with more than 5000 interconnects are used. The preferred materials for optical MEMS packages are ceramics. They can provide hermetic sealing of the MEMS structure in conjunction with seam sealed windows. They support high density and large scale wiring through multilayer structures.

6.2 OPTICS OF MEMS SWITCHES

MEMS switches control light by moving parts, which interact with a light beam in free space. These parts can be reflectors, refractive devices, absorbers, or partially transparent structures. Light is typically guided to and from the switch by mean of optical fibers. A significant issue with interfacing optical switching systems to the fibers is to project and adapt the mode profile at fiber outputs onto moving MEMS parts. Generally, the natural diffraction of a light beam propagating in free space dictates the optimal size of the moving part and the overall size of the optical system. Since the intensity profile of the guided mode in the fiber is very similar to a fundamental Gaussian function and since the Gaussian intensity profile exhibits simple diffraction in free space, we start this section by reviewing the fundamentals of Gaussian beams. Then, we discuss how beam parameters can be changed with lenses and study the impact of optical design on switching performance.

6.2.1 Gaussian Beams

In the following, we discuss the theory of the evolution of Gaussian beams during propagation. This powerful theory consists of a simple set of algebraic expressions, which allow us to identify optimal beam diameters in free-space optical switch designs. The Gaussian beam is a solution of the wave equation in paraxial approximation that exhibits an intensity $I(x,y,z)$ and complex amplitude $A(x,y,z)$. The amplitude is a zero order Hermite-Gaussian function in the x and y direction at any position along the propagation direction z [27]:

$$A_{Gauss} = A_0(z)\exp\left(-\frac{x^2+y^2}{w^2(z)} + i\frac{x^2+y^2}{R^2} + i\phi(z)\right) \tag{6.2}$$

where $i=\sqrt{-1}$, A_0 is the on-axis amplitude, $\phi(z)$ is the on-axis phase shift, and R is the radius of curvature of the phase front, which is given by

$$R(z) = z\left(1+(z_0/z)^2\right) \tag{6.3}$$

$2w$ is the 1/e amplitude width of the beam which is evolving as

$$w(z) = w_0\sqrt{1+(z/z_0)^2} \tag{6.4}$$

$e= \exp(0)$ is the base of the exponential function, and w_0 is the minimal beam half diameter or beam "waist" at $z = 0$, and is related to the Rayleigh length z_0 by

$$z_0 = \pi\frac{w_0^2}{\lambda} \tag{6.5}$$

where λ is the wavelength. The Rayleigh length is a characteristic length, which defines a Gaussian beam and describes its diffraction properties. If the propagation distance from the waist location z is small compared to the Rayleigh length, the beam will have an almost flat phase front and behave like a plain wave. If the propagation distance is much larger than the Rayleigh length, the phase front will be curved with a radius proportional to the propagation distance z, resembling a spherical wave. In both extremes the propagation can be modeled using rays that are perpendicular to the wavefronts.

The Gaussian beam which minimizes the optical spot size in between two apertures separated by distance L is of particular interest for free-space optical switching systems. The optimal beam can be derived from Equations (6.4) and (6.5), where the waist w_0 is substituted by the Rayleigh length z_0:

$$w(L, z_0) = \sqrt{\frac{\lambda}{\pi} z_0 + \frac{\lambda}{\pi} \frac{L^2}{4} \frac{1}{z_0}} \tag{6.6}$$

where we assumed that the beam waist is located in the middle of the distance L and thus the aperture are located at a distance $z = \pm L/2$. This is justified because the width of the Gaussian beam along the propagation direction is symmetrical with respect to the waist position, and grows monotonically around this position. The beam with minimal spot size can then be identified by evaluating the derivative of the term under the square root in Equation (6.6) with respect to z_0. We obtain:

$$L = 2z_0 \tag{6.7}$$

and the beam half diameter at the apertures is

$$w_{min, spot} = \sqrt{\frac{\lambda}{\pi} L} = \sqrt{2} w_0 \tag{6.8}$$

The minimal spot size $w_{min,spot}$ is proportional to the square root of the distance and this property is essential for enabling the scaling to large port counts in an optical switch. For example, consider an existing switch and scale each of its geometrical dimensions by a factor of four. The distance between the switching planes will be increased by a factor of four, whereas the beam size will grow only by a factor of two, thus allowing more beams (lightpaths) to be accommodated by the switch and thus enables larger port counts. Equations (6.7) and (6.8) can be considered as the design rules for selection of the optimal Gaussian beam for a given optical arrangement.

6.2.2 Collimation and Imaging Optics

In free-space optical switches, light is often brought into the switch by optical fibers, where it is strongly confined to fiber medium. The light coupled off a standard single mode fiber, can be described in good approximation by a Gaussian beam which at 1.5-µm wavelength has a waist of $w_0 = 5$ µm that is located at the edge of the fiber, and a corresponding Rayleigh length of 50.7 µm. The beam is strongly diverging with an angle of ±6.5°, and its optimal working distance would be just 0.1 mm. Some applications exist, like simple 2x2 switches or optical variable attenuators, where output light from a fiber can be directly manipulated by the MEMS device. However, in most cases additional lenses are required either to collimate the light beam or to relay it from the fiber to a desired location. Collimation is the process to form the beam, make it less divergent, and expand its reach with a desired width. The term collimator is used to refer to

one or more lenses which performs this task for light emerging out of an
optical fiber [28].

6.2.2.1 Collimators

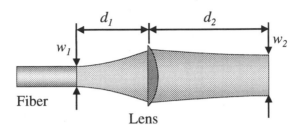

Figure 6-5. Collimation of light from a fiber by mean of a lens

Consider first a configuration where a single lens is used to collimate the
light coupled out from a fiber as depicted in Figure 6-5. After propagating
the distance, d_1 the phase front of the diverging beam is flattened by a lens
with focal length $f \approx d_1$. The new beam will form a waist with half
diameter w_2 at the distance d_2 from the lens. The new beam parameters are
described by

$$w_2 = w_1 M \tag{6.9}$$

and

$$d_2 = M^2(d_1 - f) + f \tag{6.10}$$

where the magnification factor M is given by

$$M = \frac{|f|}{\sqrt{(d_1 - f)^2 + \pi^2 w_1^4 / \lambda^2}}. \tag{6.11}$$

Two special cases of Equations (6.9) and (6.10) are of particular interest.
The first is the Fourier transform configuration defined by the condition
$d_1 = d_2 = f$ [27]. In this case, we get the following relation for the waist
radii:

$$w_1 w_2 = \frac{f \lambda}{\pi}, \quad \text{or} \quad z_1 z_2 = f^2 \tag{6.12}$$

where z_1 and z_2 are the Rayleigh length of the corresponding beams. Thus for
a given focal length the waist radii and Rayleigh lengths are inversely
proportional. The Fourier transform configuration is cornerstone for the
imaging system configuration, which is discussed soon in the following

section. The second configuration of interest exhibits a maximum distance d_2 for a given focal length f where, using Equations (6.10) and (6.11) and we have:

$$d_1 = f + \pi w_1^2 / \lambda \qquad (6.13)$$

$$d_2 = \frac{\lambda f^2}{2\pi w_1^2} + f \qquad (6.14)$$

and

$$w_1 w_2 = \frac{f\lambda}{\pi\sqrt{2}}. \qquad (6.15)$$

Equation (6.14) gives that maximal working distance for a pair of collimators for a given focal length f. Planoconvex lenses (i.e lenses with a first a flat surface followed by a second curved convex surface) as represented in Figure 6-5 generally gives a good optical performance for micro-optics [29]. In the situation where $w_1 << w_2$ the optical wavefront is almost identical to a spherical wave, and thus the lens shape can be optimised by traditional ray trace techniques.

Pointing accuracy is important in geometries with large working distance and particularly for large-scale NxN switches. In the collimation geometry, an axial alignment error Δx between the fiber and the lens produces a pointing angle error $\Delta\theta$ of the collimated beam, given by

$$\Delta\theta = \Delta x / f . \qquad (6.16)$$

Collimator assemblies, where the fiber and the lens combined are available as individual component, are commercially available. Different manufacturing techniques and lens technologies such as gradient index lens or laser machined aspheric lens are used. Recently, compact collimators where fiber is fused directly to the lens have been introduced.

6.2.2.2 Imaging Optics

Imaging systems are an alternative way to expand and project the light coupled out from the fiber onto a desired location. In this section, we limit the discussion to a double telecentric imaging system, which in addition to reproducing the intensity pattern of the object plane onto the image plane, also maintains the incidence angle of the beam between the planes. Such a system is obtained by cascading two lenses in Fourier transform configuration as depicted in Figure 6.6. In this system, the waist size w_2 depends on the focal lengths f_1 and f_2 according to the relation

$$w_2 / w_1 = f_2 / f_1 \qquad (6.17)$$

and thus multiple sets of lenses can produce the same spot size w_2.

The overall distance between the object and image plane will be given by

$$d = 2(f_1 + f_2)$$ (6.18)

which determines the absolute focal lengths of the required lenses.

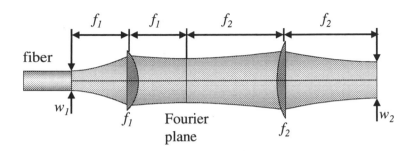

Figure 6-6. Two lens imaging system for beam expansion

The functionality of the single lens collimation and the imaging system are the same when applied to Gaussian input beam. Note however, that the imaging system will also work for input beams that are non-Gaussian, including output beams from multimode fibers.

6.2.3   Design Considerations for Optical MEMS Switches

The task of designing Optical-MEMS based switching system is sophisticated. This is due to the non-static nature of the lightpaths. Free-space beam steering switches are based on deflection of lightpaths (beams) from an input port to the desired output ports. This involves angular and displacement changes in positions of optical elements, such as lenses, mirrors and apertures, each of which has different optical properties. Depending on the location of the input and output ports and the angle of incidence of the beam on a deflector, the total traverse path length of the beam may also change. This affects insertion loss, which must be as low as possible.

Insertion loss within the optical switch can be attributed to several effects, the most important of which are clipping at apertures (for example, due to a limited mirror size) and coupling with fiber. Coupling loss can be caused by any distortion of the beam shape or its phase front. We now analyze the main contributions to insertion loss using a general device-independent approach based on Gaussian beams and beam-propagation numerical analysis.

6.2.3.1 Gaussian Beams and Coupling Loss

Whenever an optical beam is launched into a fiber, any difference between the beam and the mode field will result in power loss. Differences in beam width, position, angle of incidence, curvature or aberration of the incoming phase, all contribute to this loss. The amount of loss introduced by each of these factors can be calculated by a method known as Gaussian beam decomposition [30]. Here we present the results of applying this method.

Consider a Gaussian beam, which is centered on the fiber and propagates in parallel to the fiber. The power coupling efficiency η_0 is given by

$$\eta_0 = \frac{4}{\left(\dfrac{w}{w_0} + \dfrac{w_0}{w}\right)^2 + \left(\dfrac{\pi w_0 w}{\lambda R}\right)^2} \tag{6.19}$$

where w_0 is the mode field half diameter of the fiber, w is the half diameter of the incoming Gaussian beam, and R is the radius of curvature of its phase front. The coupling efficiency exhibits a maximum of 1 for $w = w_0$ and $R \to \infty$ as expected, and is only weakly dependent on the ratio between the mode field half diameter and the beam half diameter. A variation of ±20% in this ratio, for example, will produce a coupling of >97%, or a loss of less than 0.15 dB. The phase front curvature effect is also small as long as

$$R << \frac{\pi w_0^2}{\lambda} = z_0 \tag{6.20}$$

i.e. R larger than the Rayleigh range of the corresponding mode profile.

Now, let us consider the impact of introducing a lateral position error Δx in the position of the incidence light beam. According to [30], we obtain

$$\eta_{\Delta x} = \eta_0 \exp\left[\frac{-2\Delta x^2}{w_0^2} \cdot \frac{\dfrac{1}{w^2}\left(\dfrac{1}{w^2} + \dfrac{1}{w_0^2}\right) + \left(\dfrac{k}{2R}\right)^2}{\left(\dfrac{1}{w^2} + \dfrac{1}{w_0^2}\right)^2 + \left(\dfrac{k}{2R}\right)^2}\right] \tag{6.21}$$

where $k = 2\pi/\lambda$. Equation (6.21) simplifies to

$$\eta_{\Delta x} = \exp\left[\frac{-\Delta x^2}{w_0^2}\right] \tag{6.22}$$

in the case where $w = w_0$ and $R \to \infty$, thus the Gaussian beam has the exact amplitude to match the mode but is shifted laterally. Equation (6.22) is a

Gaussian function with a $1/e^2$ width of $2w_0$, giving the required position requirement of ± 2 μm for a single mode fiber for less than 1dB loss. This degree of precision requires active alignment and athermal mount designs. Therefore, collimators or collimator arrays are usually packaged in high-precision subassemblies, which are then used as part of optical setups with much more relaxed positioning requirements.

The impact of an error $\Delta\theta$ in the angle of incidence can be calculated similarly [30] and given by

$$\eta_{\Delta\theta} = \eta_0 \exp\left[\frac{-k^2}{2}\Delta\theta^2 \cdot \frac{\frac{1}{w^2} + \frac{1}{w_0^2}}{\left(\frac{1}{w^2} + \frac{1}{w_0^2}\right)^2 + \left(\frac{k}{2R}\right)^2}\right] \tag{6.23}$$

which can be simplified to

$$\eta_{\Delta\theta} = \exp\left[\frac{-k^2 w_0^2}{4}\Delta\theta^2\right] \tag{6.24}$$

in the case where $w = w_0$ and $R \rightarrow \infty$. Also Equation (6.24) is a Gaussian function with $1/e^2$ width of $\lambda/(\pi w_0) = 2.8°$ for a standard single mode fiber at a wavelength of 1.5 μm. The angular requirements for the fiber alignment are therefore relatively easy to achieve with standard mechanical mounts.

In the above discussion, we have considered the positional errors Δx and $\Delta\theta$ separately. In a practice, however, this is not the case. For example the effect of a tilt error $\Delta\theta$ leaning towards the direction of a displacement error Δx will have a different effect than a tilt error that is orthogonal to Δx [30].

6.2.3.2 Clipping of Gaussian Beams

Clipping of a Gaussian beam at any aperture inside an optical switch will cause loss. Generally, the goal will be to make the loss as low as possible. In most optical systems, there will be some apertures, which contribute most to the beam clipping loss, and we refer to them as limiting apertures. In MEMS switches the limiting apertures are often the micromirrors. This is because typically the smaller the MEMS mirror the better it functions and it allows the mirrors to be tightly packed to achieve a small system size and a large number of ports.

The amount of optical power lost by traversing an aperture is not equal to the power of the light blocked by the aperture, but twice as much as explain in the following. Also, the amount of loss will depend on the location and the number of apertures and on beam waist. In the following, we consider the situation where light from a fiber is coupled into another fiber by means of a collimator pair, pointing at each other. To get a first understanding of the issue we consider the power in a Gaussian beam of diameter d_0 that passes through an aperture of diameter D_A located at the beam waist ($d_0 = 2w_0$). We define the transmitted power fraction, η_P, as the proportion of input power, P_0, that is transmitted through the aperture:

$$\eta_P = \frac{P}{P_0} = (1 - \exp(-2D_A^2 / d_0^2)) = (1 - \exp(-2\kappa^2)) \tag{6.25}$$

where $\kappa = D_A/d_0$ is the ratio of aperture size to beam size. For $\kappa = 1$ the power fraction is $\eta = 0.63$dB and the loss does not seem excessively high. However, this is simply the power that propagates through an aperture and we are interested in how much of that power actually couples into the output fiber. The clipping of the beam not only removes energy from it but also distorts the mode shape and so results in a reduced fiber coupling efficiency. For a clipped Gaussian beam the coupling efficiency is given by the expression

$$\eta_C = (1 - \exp(-2\kappa^2)) \tag{6.26}$$

which is an identical additional loss to the transmitted power. The resulting coupling efficiency to the fiber is then given by

$$\eta = \eta_P \eta_C = (1 - \exp(-2\kappa^2))^2 . \tag{6.27}$$

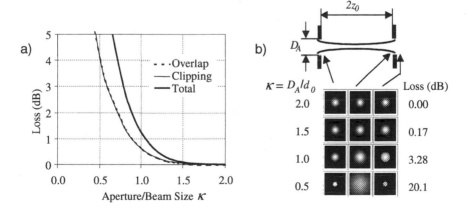

Figure 6-7. Fiber coupling loss as a function of relative aperture size κ for (a) light clipped at a single aperture (b) clipped between two apertures (see text for description)

These results are illustrated numerically in Figure 6-7(a) from which we can conclude that there is little advantage to going to κ values greater than 2 unless large misalignment tolerances are required. However, the condition of $\kappa > 1$ is necessary to obtain low insertion loss. While the above equation and Figure 6-7(a) give a good indication of some design constraints with respect to insertion loss, they do not fully reflect the system aspects of clipping. In general, there will be at least two limiting apertures in the switching system: usually the MEMS mirrors and the coupling micro-lenses. These will be separated in order to allow beam steering and to place optical components in between them, if needed. Hence, the effect of beam diffraction between these apertures must be taken into consideration too.

The clipping of the beam at the first aperture will cause diffraction and results in greater clipping and loss at the second aperture than predicted by Equation (6.27). Numerical analysis of this is shown in Figure 6-7(b) for two apertures spaced by two Rayleigh ranges, which is the condition for minimum loss of unclipped beams as derived above. The losses and beam profiles indicate that a clipping ratio of 1 results in significant loss (> 3dB) and distortion of the propagating beam. It is also clear that for a clipping ratio of 1.5 the beam is not obviously distorted and the loss contribution is sufficiently small to contemplate low-loss switching designs. While κ ratios of 1.3 to 1.7 are reasonable, we should note that this is a relatively small range of the order of ±12% and so there is little design freedom in this if a low-loss system is desirable.

6.2.3.3 Impact of Path Length Variation

In optical switches the beam path for a particular connection depends in general on the input and output port locations and is controlled and aligned by the deflective elements. The total optical path length may be different for each connection and may be expressed as

$$L_{InPort,OutPort} = \int_{InPort}^{OutPort} \frac{dL}{n(L)} \tag{6.28}$$

where n(L) is the refractive index at position L. The variation of a path length is given by

$$\Delta L_{InPort,OutPort} = L_{InPort,OutPort} - \frac{1}{n} \sum_{All\,connections} L_{Nin,Nout} \tag{6.29}$$

where n is the total number of connections.

The impact of the optical path length difference on insertion loss, for an optimized system, depends only on the Rayleigh length z_0 of the beam in the switching section, and is given by

$$\eta_{\Delta L} = \frac{1}{1 + \left(\dfrac{\Delta L}{2z_0}\right)^2}. \tag{6.30}$$

Note that Equation (6.30) is symmetric with respect to $\pm \Delta L$ and does not depend on the type of optics used to produce the Gaussian beam. The only assumptions made are that the beam is not clipped at any aperture and that the aberrations of the lens system are negligible. The loss introduced by the path length variation is smaller than 1 dB provided that $\Delta L < z_0$. In general, path length variations are undesired because they introduce connection dependent loss, which may require an additional equalization step when the switch is operated in an optical network.

6.2.3.4 Origin of Polarization Dependent Loss

Polarization Dependent Loss (PDL) is also an important contributor to loss variability in switching elements. PDL arises from any optical component part which exhibits polarization dependent transmission or reflection. The main contributors in this respect are the beam deflecting components, and we focus here on micromirrors. To achieve high reflectivity on the mirrors either multilayer coatings or metal coating are used. Both typically exhibit polarization dependent reflectivity when the angle of incidence on the mirror is non-zero. In Figure 6-8 the reflection loss of a mirror with a gold reflector is plotted as a function of the angle of incidence on the mirror for the two linear polarization components. PDL is measured as the difference in path loss between these two components. For an incidence angle $\alpha < 50°$ PDL is <0.1 dB, and grows rapidly for larger angles. In the case of beam steering cross-connects, usually more than one mirror is involved and PDL takes effect multiple times.

We have thus far reviewed some of the most important design issues and guidelines for free-space optical MEMS switches. Note that in general, the implications of these issues and design rules are not independent, i.e. by improving one characteristic, another may get worse and a merit weighting of the desired overall performance is required. This process needs to encompass the optical design, the design of the microstructure, packaging, physical design of device housing, and design of driver electronics and electrical interconnects. The task of development of these switches is therefore a truly interdisciplinary activity.

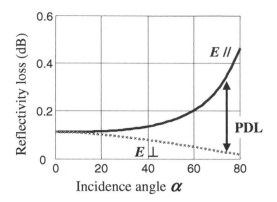

Figure 6-8. Loss after reflection from a gold mirror for two polarization states, where *E//* designate the polarization with electrical field parallel to the plain of incidence, and *E⊥* the electric field perpendicular to the plain of incidence. The loss difference between the two components gives the Polarization Dependent Loss (PDL).

6.3 N × N OPTICAL MEMS SWITCHES

N×*N* switches are expected to be capable of interconnecting *N* input fibers with *N* output fibers in any desired combination. Two general MEMS approaches for building *N*x*N* switches have been used. The first is to use individual switching elements (mainly mirrors) that are binary, i.e., capable of being in one of only two discrete states [12], [31]-[33]. In this case, the complexity of the system (the total number of switching elements) is proportional to N^2. This is commonly referred to as a digital architecture since the mirrors have binary states. Digital architectures are more attractive for small switches ($N \leq 32$).

In the second approach, which is known as beam steering or analog architecture, each element is capable of being in at least *N* distinct states, and then the total number of switching elements required is 2*N* [4], [5], [34]-[40]. In this approach, each mirror acts as a 1x*N* switch, where two sets of such 1x*N* switches are interconnected in free space. Reducing the number of switching elements enables the system to scale to much larger port counts than possible with digital mirrors[1].

There is often some confusion about how the switching system scales, depending on whether it is based on digital/analog mirrors and planar/volume configurations. This is because planar systems are usually

---

[1] In effect, *O(N)* of the system complexity is undertaken by the individual MEMS elements which simplifies the rest of the system complexity to O(N). The split of complexity in the beam steering regime boosts system scalability.

made of digital mirrors and three-dimensional (volume) systems are usually made of analog mirrors. Indeed, both digital and analog architectures can be implemented in a planar configuration (2D-MEMS). Taking into consideration the effects of beam diffraction, the physical dimensions of a planar switch scales with N^2 whether it is made of digital or analog mirrors. For volume configuration (3D-MEMS), only analog beam steering has been demonstrated. However, the physical size of a three-dimensional switch scales proportional to N whether it is based on digital or analog mirrors.

6.3.1 Planar Configurations

In planar configurations (2D MEMS) the light beams are steered within a single plane. This can be advantageous because the system can be manufactured over a single substrate. Unlike several other optical switching technologies which are based on planar waveguide geometries, planar optical MEMS switches are based on free-space beam deflection.

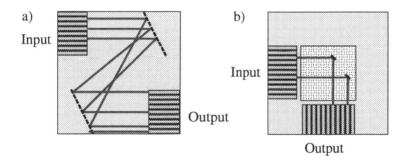

Figure 6-9. Planar free-space switch configurations: (a) Analog beam steering switch (b) digitally activated popup mirrors switch matrix

Two types of free-space planar configurations are possible using MEMS. The first type, depicted in Figure 6-9(a), uses micromirrors which can tilt around an axes perpendicular to the switching plane with analog actuation [33]. The key advantage in this configuration is that only $2N$ switching elements are required. The elements however must be able to precisely rotate to N separate angles. The details of this configuration are described in section 6.3.2.

The second configuration, Figure 6-9(b), uses digitally actuated mirrors. In this case, the switching element places its mirror either in the beam path or away from it. These elements are used as basic 2x2 optical switches. The number of switching elements in this matrix switch is proportional to the square of the number of ports. The number of elements crossed by the lightpath will depend on the connection and will vary from a minimum of one 2x2 element to a maximum of *2N* of these elements. This imposes stringent requirements on insertion loss of the elementary 2x2 switch [41]. Meanwhile, the overall size of the switch scales unfavorably fast because the diffraction of the optical beams requires the use of larger beam diameter and therefore larger micromirrors. We have seen that the minimal width of a Gaussian beam over a propagation length L is given by $z_0 = L/2$. Introducing the ratio $\kappa = D_A/d_0$ between the mirror diameter D_A and the beam diameter d_0 and the mirror fill ratio $\mu = D_A/p$, where p is the spacing between the mirrors, we obtain a relation between the mirror diameter D_A and the number of ports N as follows:

$$N = \frac{\pi}{8\lambda_0} \frac{1}{\kappa^2} \mu D_A \tag{6.31}$$

where λ_0 is the central wavelength of the light, and we made use of the longest propagation distance $2pN$ in the planar switch. If we consider for example a 16-port switch, with $\mu = 0.5$, $\kappa = 1.6$, we obtain a mirror size $D_A \approx$ 320 μm. We can also express the lateral length L_a of the device as function of the number of port by:

$$L_a = \frac{8\lambda_0 \kappa^2}{\pi\mu^2} N^2 \tag{6.32}$$

and for the same 16-port switch this would lead to lateral length of 8 mm. Equation (6.32) implies that the length of the device, which scales linearly with N in the case of planar waveguide devices, will scale as N^2 in free-space devices. This means that if our exemplary 16x16 device would be expanded to a 32x32 switch, its lateral size would grow to 32 mm. Therefore planar MEMS switches will not scale much larger than 32x32.

Although a switching fabric with larger port count can be built by cascading multiple units in a Clos arrangement [42]-[44], the achievable switch size is severely limited by the number of required interconnects between these units. The usefulness of the Clos arrangement is also limited by the significant total insertion loss of the resulting switch, including fiber connectors and interconnects.

An example of a 16x16 switch based on a digital MEMS matrix is shown in Figure 6-10. These switches have been commercialized by several vendors with up to 16x16 ports. They exhibit small footprints and typical mean insertion loss of 2-3 dB. Note that because of the digital activation of micromirrors, the loss variability strongly depends on the pointing accuracy of the coupling collimators and on the end angle of the activated mirror. A loss variability of ±2 dB is typical for 16x16 systems, with a worst case of 4-5 dB. PDL is typically <0.4 dB and mostly resulting from the 45° incidence angle on the mirrors. The switching time is typically 10 ms.

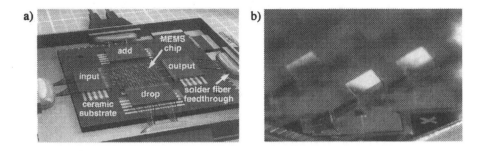

Figure 6-10. Planar 16x16 optical MEMS switch based on popup mirrors: (a) Package
switching system (b) Popup mirrors [32].

These planar MEMS switches have been proposed to realize
reconfigurable Optical Add/Drop Multiplexers (OADM). One useful feature
they have in this respect is the possibility to provide multiple input and
output collimator arrays around the switch matrix. An example for such a
configuration is shown in Figure 6-10. When no mirror is activated, the
beams propagate straight through the switching matrix from the input
collimator array to output collimator. By activating a particular mirror, an
optical signal can be redirected to the drop fibers and/or added via the add
port. The integration of the add and drop functions in a single module results
in a compact systems but may compromise robustness.

6.3.2 Free-Space Beam Steering Configurations

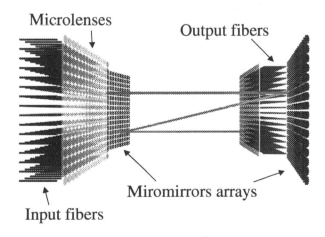

Figure 6-11. Schematic view of a beam steering cross-connect (see text for description)

In free-space beam steering MEMS switches, light beams are steered by micromirrors, which in general tilt about one or two axes. We describe the two-axis case since the one-axis case is a simpler special case. In order to connect a lightpath through the switch, a micromirror pair with each mirror pointing at the other will be required. The interconnection principle is shown in Figure 6-11. The light from a two-dimensional fiber bundle is collimated by a microlens array, and each collimated beam is pointed at one particular micromirror of a two dimensional mirror array. These micromirrors, referred to hereafter as input mirrors, redirect the beams onto a second two dimensional array of micromirrors, referred to as output mirrors. The output mirrors will each have a corresponding collimator microlens and output fiber. The output mirror will deflect the incidence beam in the direction of the output collimator where it is collimated by microlenses and coupled into to the output fiber. In a 3-D MEMS architecture, $2N$ micromirrors are used to accomplish the $N \times N$ cross-connection functionality.

The two main limiting apertures in a beam-steering optical switch are the micromirrors and the collimating microlenses. Since the microlenses lie further from the beam waist, they might be expected to be the limiting aperture of the system [45]. However, the fill factor of the micromirror, $\mu = D_A/p$, is much smaller than the corresponding fill factor of the microlens. This is because the need for a tilting element with support structures, electronics and wiring makes achieving high fill factors for the MEMS mirror more problematic. Hence, the micromirror is generally the limiting aperture, as long the distance separating it from the microlens is approximately one Rayleigh length of the beam (z_0) or less.

The optical system then consists of two arrays of MEMS mirrors spaced by a distance of two Rayleigh ranges $2z_0$ since this is the minimum clipping condition derived earlier. The beam diameter at each aperture is given by

$$d_0 = 2\sqrt{2}w_0. \tag{6.33}$$

The ratio of aperture diameter to beam diameter or beam factor is $\kappa = D_A/d_0$, with κ typically around 1.5 (Section 6.2.3.2). This gives the optimal Rayleigh length as

$$z_0 = \frac{\pi D_A^2}{8\lambda_0 \kappa^2} = \frac{\mu^2 \pi p^2}{8\lambda_0 \kappa^2}. \tag{6.34}$$

For a two-dimensional array of N mirrors the linear number of mirrors is $n = N^{1/2}$. If we consider a planar beam steering configuration as shown in Figure 6-9(a) then $n = N$. The maximum optical beam deflection angle range α, is thus given by

$$\tan(\alpha) = \frac{(n-1)p}{z_0} = (n-1)\frac{4\lambda_0 \kappa^2}{\mu^2 \pi p}. \tag{6.35}$$

The maximal mechanical tilt range of the mirror will be half this value, i.e. $\alpha/2$. All the mirrors must sweep this same range in order to address all connections, but for the centre mirrors it is split equally to both sides and at the edge it is single sided. If the mirror array is composed of uniform elements, then they must all be capable of sweeping this angle to both sides. For small angles, where we can take $\alpha = \tan\alpha$, there are some observations. The angular range scales inversely with the mirror pitch, so making the beam system larger reduces the required angular range. The angular range also scales with the square of the ratio of the aperture size to beam size, which implies that the beam should be as large as practically possible relative to the mirror, though this is limited by loss due to clipping. The angular range also scales inversely with the square of the linear fill factor, which is just the relative area fraction of the chip filled with the mirror.

We take the example of a 256 port cross-connect switch ($n = 16$) with a mirror pitch of $p = 1$ mm, a mirror fill factor of $\mu = 0.7$, and a beam factor $\kappa = 1.5$. Assuming the wavelength to be 1.55 μm, this gives $\alpha = 9.9$ degrees and the separation of the chips is 86 mm or a $z_0 = 43$ mm. This configuration is not optimal with respect to the use of the micromirror tilt range. The switch can be improved by introducing a lens in Fourier transform configuration between the micromirror arrays, as shown in Figure 6-12.

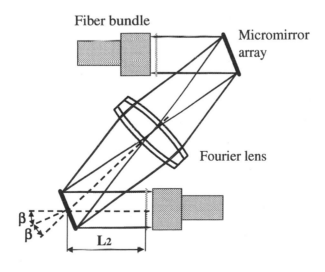

Figure 6-12. Beam steering cross-connect with Fourier lens

The function of the lens is twofold: first, it translates the beam deflection angle after the first micromirror array into a position on the second micromirror array. Thus, all mirrors will have to be steered at the same angle independent of their location on the input array in order to aim at a particular output mirror. This simplifies the steering and testing of the mirrors. The second function of the Fourier lens is to refocus the light beams. In fact, in this geometry the beam waist will be located on the micromirror. The requirement of having the same spot size on input and output mirror will lead to a Rayleigh length of the beam that is equal to the focal length of the Fourier lens. This can be verified by Equation (6.12). For a given beam size on the mirror, the resulting optimal Rayleigh range of the system in the Fourier lens geometry will be twice that of the lensless system. The tilt angle requirement will be half that of the lensless system. In other words, for a given micromirror design the introduction of the field lens can allow a four times larger number of ports. Having longer Rayleigh range brings some additional advantages for the system including more tolerance to micromirror curvature errors and less polarization dependent loss (PDL). The only drawback of the field lens system is that the distance between the micromirror array will be twice as large.

An example of a switch with 256x256 ports (nominal) is shown in Figure 6-13. 256 micromirrors are arranged in a 16x16 square matrix with 1 mm spacing between mirrors [46]. A picture of the mechanical subassembly of the switch module and the packaged MEMS device is shown in Figure 6-13. A detailed picture of a single micromirror is actually shown in Figure 6-2(a). In this switch a Fourier lens (f = 70mm) was used. A low skew angle (β = 15°, Figure 6-11) Z configuration for the optical switch ensures that PDL are <0.1 dB.

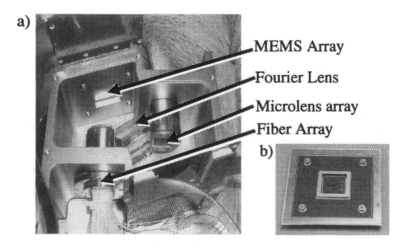

Figure 6-13. 256x256 switch module: (a) Mechanical assembly, (b) Packaged MEMS device

Figure 6-13(b) shows the MEMS mirror array mounted in a hermetic package that includes an antireflection coated sapphire window and a backside electrical pin-grid array. The mirror-to-mirror spacing of the MEMS mirror array is 1 mm with wiring arteries every fourth row or column. The individual mirrors are 600-µm diameter, and the $1/e^2$ beam size on the MEMS mirror is 372 µm. These parameters ensure that >99% of the light will hit the mirror even with a +/-30-µm error in spot position. The light enters the system from a two dimensional fiber array with a 1-mm pitch. An array of the refractive silicon micro lenses is used to relay the beams and produce the correct size of waist at the MEMS mirrors. In order to obtain a low loss system it is necessary to have microlenses with very low aberrations ($\sim\lambda/10$), low wave front errors across the entire array and high focal-length uniformity ($\sim$1%). The silicon microlens array is made by a reflow process and has excess losses ranging from 0.1 dB to 0.3 dB, which is comparable to the best individual collimators available today. The microlenses are aligned and attached to the fiber arrays, which are shown as part of the assembled switch fabric in Figure 6-13(a). The performance of this switch was verified by measuring the insertion loss on 56,644 connections that are supported by the 238x238 implemented ports. The histogram (and cumulative distribution) of the insertion loss is shown in Figure 6-14.

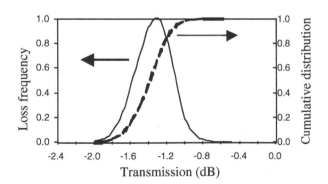

Figure 6-14. Insertion loss distribution for 56,644 connections (solid line), and cumulative distribution (dashed line) in the 238x238 switch. Mean fiber-to-fiber insertion loss is 1.3 dB

Note that the real loss distribution due to microlenses, optical coatings, mirror curvature, and beam clipping is narrower and estimated to be 1.3 ± 0.4 dB. The measured distribution is in fact broadened by the loss variability among different connector pairs. This was confirmed by an accurate calibration of the measurement system, where the loss distribution of reference fibers was compared to that of the switch, and also be repeating the measurements with different test sets. These results show that 3D-MEMS switches represent a viable switching technology and can achieve very good optical performance. The technology is scalable. 1296x1296 port switches have been reported [39] and multi thousand port switches appear to be feasible.

6.4 WAVELENGTH SELECTIVE SWITCHES

Wavelength selective switches are desirable for applications in wavelength division multiplexing (WDM) systems and networks. A wavelength selective switch can be realized by a hybrid system comprising some means of wavelength demultiplexing/separation, a MEMS switching fabric and a means of wavelength multiplexing/combining. We focus our attention in this section on wavelength selective switches that have multiplexers and demultiplexers integrated in the free-space switching system. Multiplexing/demultiplexing is usually carried out by diffraction gratings. We start by presenting a typical configuration of a free-space grating demultiplexer utilizing reflective MEMS. We will describe the dependence of amplitude transmission functions on wavelength. This is often referred to as the pass-band and is of fundamental importance for designing optical transmission systems and networks. We then explore a number of exemplary wavelength-selective MEMS switches.

6.4.1 Diffraction Grating Based Multiplexers

An example of a diffraction-grating based wavelength selective device is shown in Figure 6-15 [47]-[48]. The light from a fiber is collimated by a lens with focal length f and demultiplexed by diffraction off the grating.

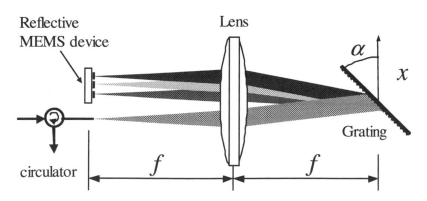

Figure 6-15. Principle of diffraction based wavelength selective device.

The direction of the beam after the grating will depend on the wavelength λ_0 of the beam. The diffracted beams then pass through the lens for a second time, and the spectrally resolved light is focused on the reflective linear MEMS device, which is also referred as 1D MEMS device. The MEMS device then either changes the amplitude or the direction of the beam. The reflected light passes through the lens, is wavelength-multiplexed by diffraction off the grating, and finally the lens couples the light back into the fiber. The output light is separated from the input light by a circulator. This configuration has the significant property that if a flat mirror is introduced in the device plane, the amplitude and phase of the reflected light will be completely independent of the wavelength, and will allow a distortion free demultiplexing and multiplexing of the signal. This property arises from the use of the same lens on multiple passes with the light diffracted twice by the same grating. This ensures that the focal length on each pass is matched and the demultiplexer and multiplexer are perfectly matched.

The reflective grating in Figure 6-15 can be a blazed or sinusoidal relief Echelette gratings or higher order Echelle grating [49]. The Echelette gratings which operate in the first or low (<5) diffraction orders typically have strong polarization properties requiring the use of polarization diversity or tolerance for increased insertion loss by working at a condition of equal polarization loss [48], [50]. The Echelle gratings offer the advantage of a diffraction efficiency η_g, defined as the power of the diffracted beam divided by the incident power, which is independent of the polarization of the incident beam. They have limited free-spectral range (~30 nm), due to their high diffraction (>20) order. For further information on gratings see [49]. The relation between the incidence angle and the diffracted angle is described by

$$\sin(\theta) = \frac{\lambda_0}{\Lambda_g} N_g - \sin(\theta') \qquad (6.36)$$

where λ_0 is the wavelength of the light, Λ_g is the grating period, N_g is the diffraction order of the grating, and θ and θ' are the angle of incidence and the angle of diffraction of the beam respectively.

The grating in Figure 6-15 operates near the Littrow configuration. In this configuration the diffracted light emerges from the grating in the direction opposite to that of the incident light beam ($\theta = \theta'$), and in this situation the incident angle θ on the grating will be equal to the angle of the grating α, which is given by

$$\sin \alpha = \frac{1}{2} \frac{\lambda_0}{\Lambda_g} N_g \qquad (6.37)$$

The dispersion D of the grating, defined as angular change $d\theta$ for a given wavelength change $d\lambda$, will then be given by

$$d\theta = \frac{2}{\lambda_0} \tan(\alpha) d\lambda = D d\lambda. \tag{6.38}$$

A high dispersion grating which has more than 1000 lines/mm, will therefore exhibit a dispersion of more than 0.1°/nm. In the geometry of Figure 6-15 this is translated to a spot separation b given by

$$b = \frac{2}{\lambda_0} \tan(\alpha) f d\lambda = D f \, d\lambda \tag{6.39}$$

which leads to of 87 μm/nm if a lens with focal length of 50 mm is used. This results in a channel spacing of 70 μm for a 100 GHz (0.8 nm) standard Dense WDM (DWDM) system. Micromachining is ideally suited to manufacturing structures of this size and for WDM applications.

6.4.2 Optical Pass-Bands of MEMS Based Switches

The optical pass-band of a device can be defined as the range of the wavelength spectrum that the device can pass at low loss. Usually a specific loss ratio is defined as acceptably low. The pass-band defines the bandwidth available for data transmission and so is critical to achieving the desired performance in optical transmission systems. If certain device has to be traversed multiple times by different wavelengths, then maximizing the flatness of its pass-band is important. For the system shown in Figure 6-15, which uses a MEMS mirror as aperture at the image plane, the pass-band is defined by the coupling efficiency η of the light passing this aperture. For a Gaussian beam of width $2w_0$ passing through an aperture $2a$ wide with a displacement b from the centre of the aperture, as shown in Figure 6-16(a), the coupling efficiency into a fiber is given by

$$\eta = \frac{1}{4} \left\{ \mathrm{erf}\left(\sqrt{2}\left[\frac{a}{w_0} - \frac{b}{w_0} \right] \right) + \mathrm{erf}\left(\sqrt{2}\left[\frac{a}{w_0} + \frac{b}{w_0} \right] \right) \right\}^2 \tag{6.40}$$

where erf(x) is the error function (integral Gaussian function). The wavelength dependence takes effect through the displacement b of the beam, which varies with wavelength. The shape of the coupling efficiency η function with respect to the normalized offset (b/a) is illustrated for various values of a/w_0 in Figure 6-16(b).

As can be seen the pass-bands become flatter as the mirror width to beam size ratio a/w_0 gets larger. Note that the pass-band will be typically flatter than with a multiplexer arrangement based on a single pass on a diffraction grating with the same spectral resolution.

From Equation (6.39) we can rewrite the coupling efficiency as

$$\eta(\delta\lambda) = \frac{1}{4}\left\{\text{erf}\left(\sqrt{2}\left[\frac{a}{w_0} - \frac{fD\delta\lambda}{w_0}\right]\right) + \text{erf}\left(\sqrt{2}\left[\frac{a}{w_0} + \frac{fD\delta\lambda}{w_0}\right]\right)\right\}^2 \quad (6.41)$$

which for $a > 3w_0$ gives the following pass-band $\delta\lambda_{6.02dB}$ (for a level of loss equals to -6.02 dB):

$$\delta\lambda_{6.02dB} \approx \frac{a}{fD} \quad (6.42)$$

This pass-band is independent of the beam size. If we are to consider the limiting case of apertures with zero-gap between them, then the 6.02 dB bandwidth is equal to the channel spacing. The 6.02 dB pass-band is independent of spectral resolution. This is not the case for widths measured at other loss points. For example the 3.01-dB bandwidth is given by

$$\delta\lambda_{3.01dB} \approx \frac{a}{fD} - \frac{0.262w}{fD} \quad \text{or} \quad \frac{\delta\lambda_{3.01dB}}{\delta\lambda_{6.02dB}} \approx 1 - \frac{0.262w}{a} \quad (6.43)$$

Thus the 3.01 dB bandwidth is a function of size of the aperture and the spectral resolution of the system. It can be interpreted that the pass-band is equal to the channel spacing minus the spectral resolution of the system. Therefore, the sharper the spectral resolution of the system, the broader its pass-band. This is in contrast to single pass systems where typically the opposite is true. This property enables the realization of extremely sharp pass-band filters.

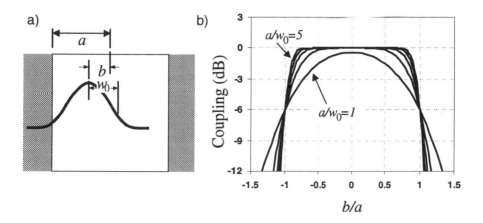

Figure 6-16. (a) Pass-band dependence on spot size and mirror size. (b) The pass-band shapes for a/w_0=1,2,3,4,5, with 5 being the flattest.

In practice, determining the pass-band is generally more complex than this since the adjacent aperture is not totally absent but is simply tilted. This introduces some complications [51]. However, this analysis provides useful insight about the principles governing pass-bands and their determination.

6.4.3 Exemplary MEMS Wavelength Selective Switches

6.4.3.1 The Wavelength Blocking Filter

The wavelength blocking filter [47] is the simplest wavelength selective switch since it is a 1x1 switch. It can be seen as a derivative of a channelized gain-equalizing filter [48], [50], [52] Figure 6-17 shows how this filter can be used to realize a reconfigurable OADM. In this scheme the incoming light is separated by a splitter: one arm, carrying the through traffic, will be filtered by the blocking filter, and the second arm will be used to drop the desired channel, either using a demultiplexer or multiple tuneable filters to access each wavelength. Similarly, the add channels are introduced on the output fiber using a combiner. In this architecture a blocking filter is only required to filter out the dropped channels from the through path to allow new traffic to be added.

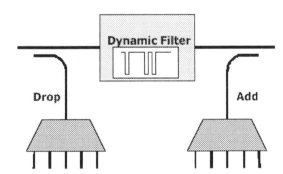

Figure 6-17. Dynamic wavelength blocking filter used in reconfigurable OADM.

Several technologies have been proposed for these high-resolution filters including free-space demultiplexing combined with MEMS tilt mirrors [47], [52] or with diffractive structures [13], [53]. We briefly describe the system in [47]. which is shown in Figure 6-18. In this system the fiber output is imaged and dispersed onto the MEMS tilt mirror array using a 50 mm focal-length lens and a 600 lines/mm grating at the Littrow configuration ($\theta = 27.7°$). The light is then re-imaged and multiplexed by the same optical system onto the same fiber. The 600 lines/mm grating has low PDL, which is further reduced by placing a $\lambda/4$ wave-plate that rotates the polarization of the light by 90° between reflections off the grating [48]. Input and output light are separated by means of a circulator. The channel spacing is 100 GHz which requires the mirrors to be spaced at 27.1 μm in the 1.55 μm wavelength transmission window. Given a spot diameter of 10.6 μm (equal to that of the fiber), this leads to a 3.01 dB bandwidth of 89.7 GHz. In practice a 90 GHz bandwidth was observed at this level of loss. The insertion loss of this device is 5 dB. This is broken down to 1.7 dB due to two diffractions off the grating, I dB arising from two passes through the circulator, 1 dB due to coatings and absorption in the lens, 1 dB because of aberrations, and, finally, 0.3-dB MEMS' reflection loss.

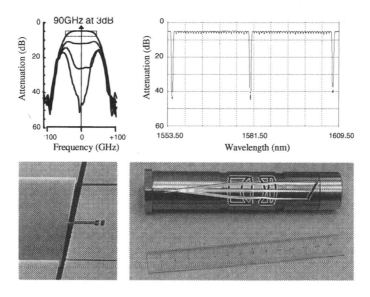

Figure 6-18. Wavelength 1×1 switch or blocker, showing performance and design

6.4.3.2 Optical Add/Drop Multiplexer

In the wavelength blocker the micromirrors are tilted to introduce a coupling loss for the reflected light. Other schemes can also be used to utilize MEMS devices in reconfigurable OADM. In the example shown in Figure 6-19, the tilt is used to separate the reflected beam after diffraction from the grating. This extends the functionality of the MEMS device to that of an elementary 2x2 wavelength-selective switch. As illustrated in Figure 6-19, a fold mirror separates the pass beams from the drop beams, which hit different parts of the grating. In order to use the device as an OADM, both fiber connecting to the device are provided with circulators.

The MEMS mirrors used in this device are 30x50 μm in size and were digitally driven with three different states: undeflected, +3° and -3°. The device described in [54] uses a 600 lines/mm grating with 16 channels at 200 GHz (57 μm) spacing and a −3 dB pass-band of 88 GHz. Due to the small mirror size and electrode design, the switching time was as short as 20 μs and the total fiber-to-fiber insertion loss was less than 4.6 dB.

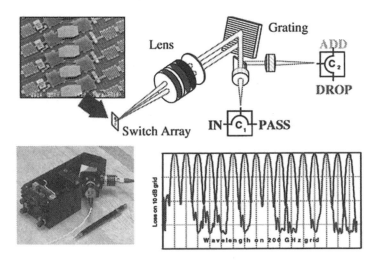

Figure 6-19. Schematic of an OADM based on two-position tilt mirrors

6.4.3.3 The 1 × N Wavelength Selective Switch

The 1xN switch [55]-[57] can be considered as a generalization of the 1x2 switch embedded in the OADM discussed above. Because every wavelength in the 1xN switch can be switched to any one of N output ports, this switch can be used in a fully flexible OADM with multiple add/drop fiber ports, each of which carries single or multiple wavelengths. The 1xN switch can also be used in other applications, such as programmable wavelength channel interleavers and programmable band splitters. 1xN switches can be cascaded to form larger architectures. Finally a NxN wavelength-selective matrix switch can be built by interconnecting back-to-back 1xN switches. Switching architectures are discussed in chapter 9.

The optical design of the 1xN switch is shown in Figure 6-20 [55]. Compared to the design of the OADM discussed in section 6.4.3.2, 1xN switch uses an additional lens in Fourier transform configuration to perform a space to angle conversion in the first stage of the switch. Also the 1xN switch will require tilt mirrors with N different tilt angles. These are usually implemented as analog mirrors.

The switch works as follows. The ingress fiber enters the switch at point A where light is collimated by a microlens. The following lenses image the collimated beam on the diffraction grating at point C. The wavelength dispersed beams fall then onto the MEMS device plane D, where they are reflected with certain tilt angle depending on micromirrors' setting. Note that all reflected beams will be focused on point B, where the angle to space conversion section will image the beam on the output fiber. Each output corresponds to a specific tilt angle of the micromirrors. The strength of the MEMS based implementation of the 1xN switch can be illustrated by examining the performance of the 1x4 switch reported in [55]. This device can switch as many as 128 wavelengths with 50 GHz spacing. The total insertion loss is less than 6 dB. It uses a 100 mm focal length mirror and a 1100 lines/mm grating. The micromirrors can be actuated by ±8° using a voltage of <115 V and the switch can be used as variable attenuator by detuning the tilt angle of the micromirrors. The worst-case switching speed is <2 ms. The device exhibits flat pass-bands. For a single isolated channel, we have 35.6 GHz at -1 dB and 44.5 GHz at -3 dB, which are very wide bands for a device with 50 GHz channel spacing. This broad flat top region enables the switch to be cascaded multiple times and used in 10 Gb/s systems. An additional strength of the MEMS based implementation of the 1xN switch is that the number of output fiber N can be as large as 32 if two-axes tilt mirrors are utilized.

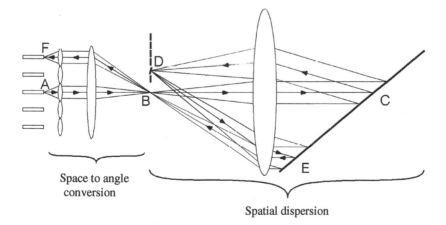

Figure 6-20. Schematic of the optical design of 1x4 wavelength selective switch. N (= 4) output fibers are placed around the sole input fiber at A

6.4.3.4 The $N \times N$ Wavelength Selective Switch

Finally, we consider $N \times N$ matrix switches which can interconnect N input fibers to N output fiber, where each fiber may carry m wavelengths. While these switches can be implemented by interconnections of the $1 \times N$ wavelength selective switches, there are alternative implementations. The $N \times N$ wavelength selective switching functionality can be achieved by using multiplexers/demultiplexers pairs to separate wavelength channels and switch them by a MEMS space switch. This switch can be realized by one large $mN \times mN$ switch or m wavelength-specific $N \times N$ switches. Both architectures offer some advantages. The first, for example, can embrace additional functionalities such as selective wavelength conversion to improve the blocking characteristics of the switch [58]-[59] or partial signal regeneration. The second architecture offers the advantages of scalability and modularity and new $N \times N$ switches can be added as more wavelength are introduced.

Other approaches have been investigated to build wavelength selective $N \times N$ switches, in particular to reduce the complexity and cost associated with the fiber interconnection between the multiplexers/demultiplexers and the switching elements. One of these approaches is based on the use of optical waveguides serving as inputs for beam-steering switches [60]-[61]. Thus a stack of planar waveguides replaces the two-dimensional fiber bundle array as input ports to the switch. The waveguide stack is faced by a microlens array to collimate light coupled off the waveguides, as displayed in Figure 6-21.

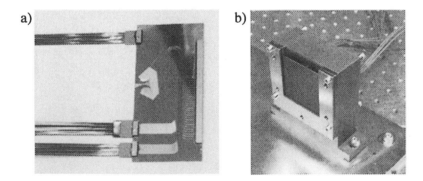

Figure 6-21. Planar waveguide stack as input to beam steering Optical MEMS switches. (a) The Waveguide grating router with fan out. (b) Assembled waveguide stack

Note that other functionalities, such as power monitoring taps, band splitters or interleavers, can be implemented on planar lightwave circuits

(PLCs). The switch reported in [60] is a prototype with 36 fibers each carrying 36 wavelengths. The planar waveguide structure in the device depicted in Figure 6-21 is an arrayed waveguide grating router. It has output channels spaced at 1.25 mm over a 0.76-mm thick substrate. The insertion loss is 20 dB and the pass-band is 30 GHz. The waveguides also provid taps for power monitoring [60].

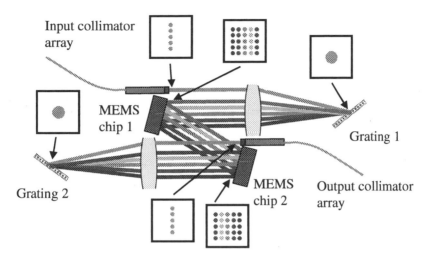

Figure 6-22. Wavelength selective $N \times N$ based on diffraction grating. Top view is shown and the spot topologies are displayed in the squares at different locations in the switch

Another method to build compact $N \times N$ switches is by using shared diffraction grating [33], [62]. A possible implementation is shown in Figure 6-22. The input channels are brought into the switch by a linear array of parallel collimators. Input light is focused on a single spot on grating 1 where it is diffracted. The lens produces a two dimensional array of spots on the MEMS chip 1. The channels are switched in the section between MEMS chip 1 and MEMS chip 2 and end on a common spot on grating 2 which couples the channels into output fibers. Note that the switching section will be bound by the same beam size constraint of Equation (6.35). For example, if we consider a 16x16 switch with 16 wavelengths per fiber, it will require a spot size of approximately 380 μm, and 1 mm spacing between micromirrors. This geometry has to be matched by the WDM channel separation using an appropriate diffraction grating. The relation between spot spacing and channel separation is according to Equation (6.39). For a mirror spacing $b = 1$ mm, a WDM channel separation $d\lambda = 0.8$ nm ≈100 GHz, and using a high dispersion 1100 lines/mm grating ($\alpha = 55.6°$), the required focal length of the lens in Fourier transform configuration is

$$f = \frac{b\lambda_0}{2d\lambda \tan(\alpha)} = 600\text{mm} \tag{6.44}$$

Clearly, this leads to large system size. However, the system can be made four times smaller ($f = 150$mm) if the number of fibers is reduced to 8 and the channel spacing increased to 200 GHz. Note that in this approach the pass-band will be given by the mirror geometry and beam size as discussed in Section 6.4.2. Therefore, minimal gaps between the mirrors in the grating dispersion direction will be required. Also the micromirrors in this arrangement are required to tilt only around one axis. Instead of the two MEMS chips arrangement of Figure 6-22, the switching section can alternatively be implemented with planar 2D MEMS switches [33], [62], where a different switching plane can be used for each wavelength.

Despite the elegance of using a shared diffraction grating to build a wavelength selective NxN MEMS-based switch, only limited information is available about practical realizations of this approach and the optical performance thereof.

REFERENCES

[1] K. Petersen, "Silicon as mechanical Material," Proc. IEEE, Vol. 70, No. 5, pp. 420-57, May 1982.

[2] K. W. Markus, "Commercialization of optical MEMS-volume manufacturing approaches," the 2000 IEEE/LEOS International Conference on Optical MEMS, pp. 7-8, 2000.

[3] L. Y. Lin and E. L. Goldstein, "Opportunities and challenges for MEMS in lightwave communication," IEEE Journal of Selected Topics in Quantum Electronics, Vol. 8, No. 1, pp. 163-72, Jan-Feb 2002.

[4] A. Neukermans and R. Ramaswami, "MEMS technology for optical networking applications," IEEE Communications Magazine, Vol. 39, No. 1, pp. 62-9, Jan. 2001.

[5] A. Keating, "Optical MEMS in switching systems," IEEE/LEOS Annual Meeting 2001, pp. 8-9, San Diego, 2001.

[6] J. A. Walker, K. W. Goossen, and S. C. Arney, "Mechanical anti-reflection switch (MARS) device for fiber-in-the-loop applications," Advanced Applications of Lasers in Materials Processing/Broadband Optical Networks/Smart Pixels/Optical MEMs and Their Applications, IEEE/LEOS 1996 Summer Topical Meetings, pp. 59-60, 1996.

[7] F. Chollet, M. de Labachelerie, and H. Fujital, "Electromechanically actuated evanescent optical switch and polarization independent attenuator," The Eleventh Annual International Workshop on Micro Electro Mechanical Systems, MEMS 98, pp. 476-81, 1998.

[8] H. Toshiyoshi, M. Kobayashi, D. Miyauchi, H. Fujita, J. Podlecki, and Y. Arakawa, "Design and analysis of micromechanical tunable interferometers for WDM free-space optical interconnection," IEEE J. Lightwave Technol., Vol. 17, No. 1, pp. 19-25, Jan. 1999.

[9] C. K. Madsen, J. A. Walker, J. E. Ford, K. W. Goossen, T. N. Nielsen, and G. Lenz, "A tunable dispersion compensating MEMS all-pass filter," IEEE Photon. Technol. Lett., Vol. 12, No. 6, pp. 651-3, June 2000.

[10] V. P. Jaecklin, C. Linder, N. F. de Rooij, J.-M. Moret, and R. Vuilleumier, "Optical microshutters and torsional micromirrors for light modulator arrays," An Investigation of Micro Structures, Sensors, Actuators, Machines and Systems, IEEE MEMS '93, pp. 124-7, Feb 1993.

[11] V. A. Aksyuk, F. Pardo, D. Carr, D. Greywall, H. B.Chan, M. E. Simon, A. Gasparyan, H. Shea, V. Lifton, C. Bolle, S. Arney, R. Frahm, M. Paczkowski, M. Haueis, R. Ryf, D. T. Neilson, J. Kim, C. R. Giles, and D. Bishop, "Beam-steering micromirrors for large optical crossconnects," IEEE J. Lightwave Technol., Vol. 21, No. 3, pp. 634-42, March 2003.

[12] P. De Dobbelaere, S. Gloeckner, S. Patra, L. Fan, D. Reiley, C. King, T. Yeh, J. Gritters, Y. Loke, E. Kruglick, R. Chen, M. Harburn, S. Gutierrez, M. Harisson, P. Marchand, A. Sharma, D. Vacar, J. Hirshkoff, A. Oviedo, J. Ford, A. Belenky, V. Fedoriouk, D. Ferrel, O. Gumuz, P. Keegan, J. Le, S. O'Connor, D. Rines, C. Tran, C. Vacar, J. Ward, A. Husain, H.-J. Schmidtkc, and B. H. Heppner, "Design, Reliability and Qualification of Photonic Crossconnects based on Digital MEMS," Optical Fiber Communication Conference OFC 2002, Paper ThGG70, p. 710, March 2002

[13] O. Solgaard, F. Sandejas, and D. Bloom, "Deformable Grating Optical Modulator," Optics Lett., Vol. 17, No. 9, pp. 688-90, May 1992.

[14] E. Hung and S. Senturia, "Extending The Travel Range Of Analog-Tuned Electrostatic Actuators," Journal of Microelectromechanical Systems, Vol. 8, No. 4, pp. 497-505, Dec 1999.

[15] D.E. Sene, J. W. Grantham, V. M. Bright, and J. H. Comtois, "Development and characterization of micro-mechanical gratings for optical modulation," An Investigation of Micro Structures, Sensors, Actuators, Machines and Systems, IEEE MEMS '96, pp. 222-7, Feb 1996.

[16] G. Perregaux, S. Gonseth, P. Debergh, J.-P. Thiebaud, and H. Vuilliomenet, "Arrays of addressable high-speed optical microshutters," The 14th IEEE International Conference on MEMS 2001, pp. 232-35, Jan 2001.

[17] C. Marxer, C. Thio, M.-A. Gretillat, N. F. de Rooij, R. Battig, O. Anthamatten, B. Valk, and P. Vogel, "Vertical mirrors fabricated by deep reactive ion etching for fiber-optic switching applications," Journal of Microelectromechanical Systems, Vol. 6, No. 3, pp. 277-85, Sept. 1997.

[18] Multi-User MEMS Processes (MUMPS) Introduction and Design Rules, rev. 4, 7/15/96, MCNC MEMS Technology Applications Center, Research Triangle Park, NC 27709.

[19] M. Hoffmann and E. Voges, "Bulk silicon micromachining for MEMS in optical communication systems," Journal of Micromechanics and Microengineering, Vol. 12, No. 4, pp. 349-60, 2002.

[20] H. Y. Wang and Y. L. Bai, "Flip chip technologies and their applications in MEMS packaging," International Journal Of Nonlinear Sciences And Numerical Simulation, Vol. 3, No. 4, pp. 433-6, 2002.

[21] V. A. Aksyuk, F. Pardo, and D. J. Bishop, "Stress-induced curvature engineering in surface-micromachined devices," Proc. SPIE, Vol. 3680, p. 984, March 1999.

[22] W. Liu and J. Talghader, "Thermally Invariant Dielectric Coatings For Micromirrors," Appl. Optics, Vol. 41, No. 16, pp. 3285-93, Jun 2002.

[23] M. S. Rodgers, S. Kota, J. Hetrick, Z. Li, B. D. Jensen, T. W. Krygowski, S. L. Miller, S. M. Barnes, and M. S. Burg, "A New Class of High Force, Low-Voltage, Compliant

Actuation Systems," Solid-State Sensor and Actuator Workshop, Hilton Head Island, South Carolina, June 2000.

[24] M. Horenstein, T. G. Bifano, S. Pappas, J. Perreault, and R. Krishnamoorthy-Mali, "Real time optical correction using electrostatically actuated MEMS devices," Journal of Electrostatics, Vol. 46, pp. 91-101, 1999.

[25] J. M. Dawson, J. Chen, K. S. Brown, P. Famouri, and L. A. Hornak, "Through-wafer optical probe characterization for microelectromechanical systems positional state monitoring and feedback control," Optical Engineering, Vol. 39, No. 12, pp. 3239-46, 2000.

[26] I. Brenner, P. Chu, M. Tsay, C. Pu, M. Chou, J. Dadap, D. Lee, B. Tang, D. Peale, N. Bonadeo, R. Harel, C. Wu, J. Johnson, S. Park, S. Lee, D. Tong, R. Doran, K. Bergman, T. Chau, E. Goldstein, L. Lin, J. Walker, W. Zhong, and R. Gibson, "Nonlinear Servo Control of MEMS Mirrors and Their Performance in a Large Port-Count Optical Switch," Optical Fiber Communication Conference 2003, Vol. 1, pp. 385-6, Atlanta, March 2003.

[27] B. E. A. Saleh and M. C. Teich, "Fundamentals of photonics," John Wiley & Sons, New York, 1991.

[28] N. F. Borelli, Microoptics Technology, Marcel Dekker, New York, 1999.

[29] S. Sinzinger and J. Jahns, Microoptics, John WILEY-VCH, New York, 1999.

[30] H. Kogelnik, "Coupling and conversion coefficients for optical modes," Proc. Sympos. Quasi-Opt., Vol. 14, pp. 333-47, 1964, reprinted in "Microlenses: Coupling light to optical fibers," editors H-D Wu and F.S. Barnes, IEEE Press, New York, 1991.

[31] L. Lin, E Goldstein, and R. Tkach, "Free Space Micromachined Optical Switches with Submillisecond Switching Time for Large-Scale Optical Crossconnects," IEEE Photon. Technol. Lett., Vol. 10, No. 4, pp. 525-7, 1998.

[32] P. De Dobbelaere, K. Falta, S. Gloeckner, and S. Patra, "Digital MEMS for optical switching," IEEE Communications Magazine, Vol. 40, No. 3, pp. 88-95, Mar. 2002.

[33] P. M. Hagelin, U. Krishnamoorthy, J. P. Heritage, and O. Solgaard, "Scalable optical cross-connect switch using micromachined mirrors," IEEE Photon. Technol. Lett., Vol. 12, No. 7, pp. 882-4, July 2000.

[34] D. T. Neilson, V. A. Aksyuk, S. Arney, N. R. Basavanhally, K. S. Bhalla, D. J. Bishop, B. A. Boie, C. A. Bolle, J. V. Gates, A. M. Gottlieb, J. P. Hickey, N. A. Jackman, P. R. Kolodner, S. K. Korotky, B. Mikkelsen, F. Pardo, G. Raybon, R. Ruel, R. E. Scotti, T. W. van Blarcum, L. Zhang, and C. R. Giles, "Fully Provisioned 112×112 Micro-Mechanical Optical Crossconnect With 35.8Tb/s Demonstrated Capacity," Optical Fiber Communication Conference 2000, PD-12, March 2000.

[35] O. Jerphagnon, R. Anderson, A. Chojnacki, R. Helkey, W. Fant, V.Kaman, A.Keating, Bin Liu, C. Pusarla, J. R. Sechrist, D. Xu, Shifu Yuan, and Xuezhe Zheng; "Performance and applications of large port-count and low-loss photonic cross-connect system for optical networks," IEEE/LEOS Annual Meeting 2002, Vol. 1, pp. 299-300, Glasgow Scotland, Sept. 2002.

[36] Y. Uenishi, J. Yamaguchi, T. Yamamoto, N. Takeuchi, A. Shimizu, E. Higurashi, and R. Sawada, "Free-space optical cross connect switch based on a 3D MEMS mirror array," IEEE/LEOS Annual Meeting 2002, Vol. 1, pp. 59-60, Glasgow Scotland, Sept. 2002.

[37] E. Goldstein, L. Lin, and J. A. Walker, "Lightwave micromachines for optical networks," Optics and Photonics News. March 2001.

[38] K. Bergman, N. Bonadeo, I. Brener, and K. Chiang, "Ultra-high capacity MEMS based optical cross-connects," Design, Test, Integration, and Packaging of MEMS/MOEMS 2001, Proc. SPIE Vol. 4408, Cannes (France), pp. 2-5, April 2001.

[39] R. Ryf J. Kim, J. P. Hickey, A. Gnauck, D. Carr, F. Pardo, C. Bolle, R. Frahm, N. Basavanhally, C. Yoh, D. Ramsey, R. Boie, R. George, J. Kraus, C. Lichtenwalner, R. Papazian, J. Gates, H. R. Shea, A. Gasparyan, V. Muratov, J. E. Griffith, J. A. Prybyla, S. Goyal, C. D. White, M. T. Lin, R. Ruel, C. Nijander, S. Arney, D. T. Neilson, D. J. Bishop, P. Kolodner, S. Pau, C. Nuzman, A. Weis, B. Kumar, D. Lieuwen, V. Aksyuk, D. S. Greywall, T. C. Lee, H. T. Soh, W. M. Mansfield, S. Jin, W. Y.Lai, H. A. Huggins, D. L. Barr, R. A. Cirelli, G. R. Bogart, K. Teffeau, R. Vella, H. Mavoori, A. Ramirez, N. A. Ciampa, F. P. Klemens, M. D. Morris, T. Boone, J. Q. Liu, J. M. Rosamilia, and C. R. Giles, "1296-port MEMS Transparent Optical Crossconnect with 2.07Petabit/s Switch Capacity," Optical Fiber Communication Conference 2001, PD-12, March 2001.

[40] R. Ryf, D. T. Neilson, P. R. Kolodner, J. Kim, J. P. Hickey, D. Carr, V. Aksyuk, D. S. Greywall, F. Pardo, C. Bolle, R. Frahm, N. R. Basavanhally, D. A. Ramsey, R. George, J. Kraus, C. Lichtenwalner, R. Papazian, C. Nuzman, A. Weiss, B. Kumar, D. Lieuwen, J. Gates, H. R. Shea, A. Gasparyan, V. A. Lifton, J. A. Prybyla, S. Goyal, R. Ruel, C. Nijander, S. Arney, D. J. Bishop, C. R. Giles, S. Pau, W. M. Mansfield, S. Jin, W. Y. Lai, D. L. Barr, R. A. Cirelli, G. R. Bogart, K. Teffeau, R. Vella, A. Ramirez, F. P. Klemens, J. Q. Liu, J. M. Rosamilia, H. T. Soh, and T. C. Lee, "Multi-service Optical Node Based on Low Loss MEMS Optical Crossconnect Switch," Optical Fiber Communication Conference 2002, ThE3, pp. 410-411, March 2002.

[41] L.-Y. Lin, E. L. Goldstein, and R. W. Tkach, "On the Expandability of Free-Space Micromachined Optical Cross Connects," IEEE J. Lightwave Technol., Vol. 18, No. 4, pp. 482-9, April 2000.

[42] D. G. Cantor, "On Non-Blocking Switching Networks," 22nd international symposium on computer-communications networks and teletraffic, p. 19, New York, April 1972.

[43] G. Shen, Tee Hiang Cheng, S. K. Bose, Chao Lu, and Teck Yoong Chai; "Architectural design for multistage 2-D MEMS optical switches," IEEE J. Lightwave Technol., Vol. 20, No. 2, pp.178-187, Feb. 2002.

[44] L. Wosinska, L. Thylen, and R.P. Holmstrom, "Large-capacity strictly nonblocking optical cross-connects based on microelectrooptomechanical systems (MEOMS) switch matrices: reliability performance," IEEE J. Lightwave Technol., Vol. 19, No. 8, pp. 1065-75, Aug. 2001.

[45] R. R. A. Syms, "Scaling Laws for MEMS Mirror-Rotation Optical Cross Connect Switches," IEEE J. Lightwave Technol., Vol. 20, No. 7, pp. 1084-1094, July 2002.

[46] V. A. Aksyuk, S. Arney, N. R. Basavanhally, D. J. Bishop, C. A. Bolle, C. C. Chang, R. Frahm, A. Gasparyan, J. V. Gates, R. George, C. R. Giles, J. Kim, P. R. Kolodner, T. M. Lee, D. T. Neilson, C. Nijander, C. J. Nuzman, M. Paczkowski, A. R. Papazian, F. Pardo, D. A. Ramsey, R. Ryf, R. E. Scotti, H. Shea, and M. E. Simon, "238x238 Micromechanical Optical Crossconnect," IEEE Photon. Technol. Lett., Vol. 15, No. 4, pp. 587-9, April 2003.

[47] D. T. Neilson, D. S. Greywall, S. Chandrasekhar, L. L. Buhl, H. Tang, L. Ko, N. R. Basavanhally, F. Pardo, D. A. Ramsey, J. D. Weld, Y. L. Low, J. Prybyla, R. Scotti, A. Gasparyan, M. Maueis, S. Arney, S. P. O'Neill, C.-S. Pai, D. H. Malkani, M. M. Meyers, N. Saluzzi, S. H. Oh, O. D. Lopez, G. R. Bogart, F. P. Klemens, M. Luo, J. Q. Liu, K. Teffeau, A. Ramirez, K. S. Werder, J. E. Griffith, C. Frye, M. V. Kunnavakkam, S. T. Stanton, J. A. Liddle, H. T. Soh, T.-C. Lee, O. Nalamasu, and K. C. Nguyen, "High-dynamic Range Channelized MEMS Equalizing Filter," Optical Fiber Communication Conference 2002, pp. 586-8, March 2002.

[48] J. E. Ford, J. A. Walker, K. W. Goossen, and D. T. Neilson, "Broad spectrum micromechanical equalizer," 25th European Conference on Optical Communication ECOC 1999, Tu D3.6, Nice (France), 1999.

[49] E. G. Loewen and E. Popov, Diffraction gratings and applications, Marcel Dekker NY, 1997.

[50] J. E. Ford, J. A. Walker, M. C. Nuss, and D. A. B. Miller, "32 channel WDM graphic equalizer," Advanced Applications of Lasers in Materials Processing/Broadband Optical Networks/Smart Pixels/Optical MEMs and their Applications. IEEE/LEOS 1996 Summer Topical Meetings, pp. 26-7, Aug. 1996.

[51] D. M. Marom and S.-H. Oh, "Filter-shape dependence on attenuation mechanism in channelized dynamic spectral equalizers," The 15th Annual Meeting of Lasers and Electro-Optics Society, LEOS 2002, Vol. 2, pp. 416-7, 2002.

[52] O. Bouevitch, D. Touahri, J. P. Morgan, S. Panteleev, and C. Reimer, "Channel-power equalizer and dynamic gain equalizer based on the optical bench platform," IEEE/LEOS Summer Topical meeting: All-Optical Networking: Existing and Emerging Architecture and Applications/Dynamic Enablers of Next-Generation Optical Communications Systems/Fast Optical Processing in Optical Transmission/VCSEL and Microcavity Lasers, pp. MD1-4-MD1-5, July 2002.

[53] O. Solgaard, D. Lee, Kyoungsik Yu, U. Krishnamoorthy, K. Li, and J. P. Heritage, "Microoptical phased arrays for spatial and spectral switching," IEEE Communications Magazine, Vol. 41, No. 3, pp. 96-102, March 2003.

[54] J. E. Ford, V. A. Aksyuk, D. J. Bishop, and J. A. Walker, "Wavelength Add-Drop Switching Using Tilting Micromirrors," IEEE J. Lightwave Technol., Vol. 17, No. 5, pp. 904-911, May 1999.

[55] D. M. Marom, D. T. Neilson, and D. S. Greywall, "Wavelength selective 1×4 switch for 128 channels at 50 GHz spacing," Optical Fiber Communication Conference 2002, pp. FB7-1-FB7-3, March 2002.

[56] T. Ducellier, J. Bismuth, S. F. Roux, A. Gillet, C. Merchant, M. Miller, M. Mala, Y. Ma, L. Tay, J. Sibille, M. Alavanja, A. Deren, M. Cugalj, D. Ivancevic, V. Dhuler, E. Hill, A. Cowen, B. Shen, and R. Wood, "The NWS 1x4: a High Performance Wavelength Switching Building Block," the 2002 European Conference on Optical Communication (ECOC'2002), Vol. 1, paper 2.3.1, Sept. 2002.

[57] S. Mechels, L. Muller, G. D. Morley, and D. Tillett, "1D MEMS-Based Wavelength Switching Subsystem," IEEE Communications Magazine, Vol. 41, No. 3, pp. 88-94, March 2003.

[58] J. Leuthold, R. Ryf, S. Chandrasekhar, D. T. Neilson, C. H. Joyner, C. R. Giles, "All-optical nonblocking terabit/s crossconnect based on low power all-optical wavelength converter and MEMS switch fabric," Optical Fiber Communication Conference 2001, Vol. 4, PD16, March 2001.

[59] C. Nuzman, J. Leuthold, R. Ryf, S. Chandrasekhar, C. R. Giles, and D. T. Neilson, "Design and implementation of wavelength-flexible network nodes," IEEE J. Lightwave Technol., Vol. 21, No. 3, pp. 648-63, March 2003.

[60] R. Ryf, P. Bernasconi, P. Kolodner, J. Kim, J. P. Hickey, D. Carr, F. Pardo, C. Bolle, R. Frahm, N. Basavanhally, C. Yoh, D. Ramsey, R. George, J. Kraus, C. Lichtenwalner, R. Papazian, J. Gates, H. R. Shea, A. Gasparyan, V. Muratov, J. E. Griffith, J. A. Prybyla, S. Goyal, C. D. White, M. T. Lin, R. Ruel, C. Nijander, S. Amey, D. T. Neilson, D. J. Bishop, S. Pau, C. Nuzman, A. Weis, B. Kumar, D. Lieuwen, V. Aksyuk, D. S. Greywall, T. C. Lee, H. T. Soh, W. M. Mansfield, S. Jin, SW. Y. Lai, H. A. Huggins, D. L. Barr, R. A. Cirelli, G. R. Bogart, K. Teffeau, R. Vella, H. Mavoori, A. Ramirez, N. A. Ciampa, F. P. Klemens, M. D. Morris, T. Boone, J. Q. Liu, J. M. Rosamilia, and C. R. Giles, "Scalable wavelength-selective crossconnect switch based on mems and planar waveguides," the 2001 European Conference on Optical Communication (ECOC'2001), Vol. 6, pp. 76-7, Sept.-Oct. 2001.

[61] T. Kanie, M. Katayama, M. Shiozaki, T. Komiya, K. Saitoh, and M. Nishimura, "A highly dense MEMS optical switch array integrated with planar lightwave circuit," The Fifteenth IEEE International Conference on Micro Electro Mechanical Systems, pp. 560-3, Jan. 2002.

[62] O. Solgaard, J.P. Heritage, and A.R. Bhattarai, "Multi-Wavelength Cross-Connect Optical Switch," US Patent 6374008, Apr. 16th 2002.

Chapter 7

OPTICAL SWITCHING WITH SOAs

Paul R. Prucnal
Ivan Glesk
Paul Toliver
Lei Xu

Semiconductor optical amplifiers (SOAs) have become promising components for optical communication systems. Owing to the quick development of their fabrication technologies, the performance of SOAs has been greatly improved and their cost are reduced. Commercialized SOAs are available now that can provide gain as high as 25 dB (fiber-to-fiber), saturation power of more than 10 dBm and a polarization sensitivity of less than 1 dB over a bandwidth of about 50 nm. Although fiber amplifiers are still preferable for inline amplification in current optical networks, SOAs are finding new applications as a result of their nonlinear properties. Compared with optical fiber, SOAs have the advantages of compact size and large nonlinear coefficients, and offer the possibility of monolithic integration with other optical devices. SOAs can be deployed as building blocks in optical switching, wavelength conversion, signal regeneration, clock recovery, dispersion compensation, etc.

In section 7.1, we discuss SOA structures and the fabrication technologies which are adopted to manufacture high-performance SOAs. Section 7.2 concentrates on static SOA characteristics, gating type SOA space switches and combined wavelength/space switching using SOA-based wavelength conversion. Section 7.3 gives an overview of the recently developed SOA applications for ultrafast switching using nonlinear interferometers. Their working principles will be explained in detail. In section 7.4, we briefly discuss integration of SOA-based optical switches for space, wavelength and ultrafast switches.

7.1   INTRODUCTION TO SOA TECHNOLOGIES

7.1.1   SOA Structure and Suppression of the Facet Reflectivity

The principle underlying the operation of an SOA is very similar to that of a semiconductor laser. The incident light is amplified through stimulated emission. For SOAs with optical gain centered around 1.3 μm or 1.55 μm, they are usually made of $In_{1-x}Ga_xAs_yP_{1-y}$/InP (In: Indium, Ga: Gallium, As: Arsenic, P: Phosphorus). $In_{1-x}Ga_xAs_yP_{1-y}$ is the material for the active region. The subscripts x and y indicate the fraction of In atoms that are replaced by Ga and the fraction of P atoms that are replaced by As, respectively, while InP is the substrate material. A conventional SOA has a double heterostructure, where a thin active semiconductor layer (thickness ~0.1 μm) is sandwiched between p-type and n-type cladding layers of another semiconductor with a larger bandgap. The resulting p-n heterostructure is forward-biased through metallic contacts.

The schematic structure of an SOA with input and output fibers is shown in Figure 7-1. The incident light is coupled into the amplifier through one of the two facets. It then passes through the active region, which is pumped by external current injection. The output light is coupled into the fiber on the other side. Characteristic geometries of the active region are a cross section of ~1 μm (width) $\times$ 0.1 μm (thickness) and a length L of 0.5-2 mm. Compared with a semiconductor laser which usually requires high feedback from the internal facets to form an oscillator, an SOA has very small facet reflectivities to minimize the feedback of light in the cavity.

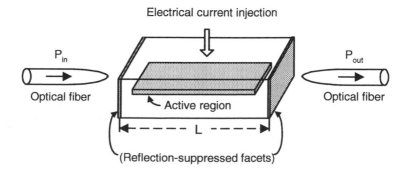

Figure 7-1. Schematic structure of an SOA.

Depending on the residual reflectivity, SOAs can be classified into two types: Fabry-Perot (FP) amplifiers and traveling wave (TW) amplifiers. A

FP amplifier has appreciable reflectivities at the input and output facets, which result in the reflection of light back into the active region, acting as a resonant cavity. Therefore, the SOA gain spectrum is not smooth, but has a series of peaks corresponding to the longitudinal modes of the amplifier chip. The TW amplifier, by contrast, has negligible facet reflectivity and the incident light is amplified in a single pass through the amplifier. The gain spectrum of an ideal TW amplifier has no resonance and is very smooth. TW amplifiers are more suitable for systems applications, due to their broad bandwidth. Therefore, much effort has been devoted to fabricating amplifiers with very low facet reflectivity.

For a FP type SOA, we can estimate the ripple of its gain spectrum by considering the ratio (V) of the maximum and minimum values of the gain (G_{max} and G_{min}) near a cavity resonance [1]:

$$V = \frac{G_{max}}{G_{min}} = \left[\frac{1 + \sqrt{R_1 R_2} G_S}{1 - \sqrt{R_1 R_2} G_S} \right]^2 \tag{7.1}$$

Where R_1 and R_2 are the reflectivities of the SOA facets, and G_S is the single pass gain. For the ideal case $R_1 = R_2 = 0$, V equals 1 and there are no ripples in the gain spectrum. In order to keep $V < 1.26$ (1 dB), the facet reflectivities should satisfy the condition

$$\sqrt{R_1 R_2} G_S < 0.057 \tag{7.2}$$

Hence, 20 dB of amplification requires an average facet reflectivity of less than 5.7 x 10^{-4}. We will discuss the three principal schemes used to achieve such low reflectivities: anti-reflection (AR) dielectric coatings, tilted waveguide and buried facet structures.

1) AR coating. The reflectivity of cleaved facets can be reduced by dielectric coatings. When the coating has a quarter wavelength thickness and optimum index, the reflected waves from the two interfaces created by the coating can totally cancel each other. However, in order to obtain facet reflectivity of 10^{-4} or less, extremely tight control on the refractive index and thickness of dielectric layers is required. AR coatings are inherently narrow band, since the coating thickness can be optimized only for a limited wavelength range.

2) Tilted waveguide structure. Another way to suppress the FP resonance is to slant the active region from the cleavage plane so that the internal light reflected by the cleaved facets does not couple back very well into the waveguide, as is shown in Figure 7-2 (a). With a tilt angle of $\theta = \sim 7°$, reflectivity of 0.2% is demonstrated for a broadband 1.5 μm InGaAsP/InP SOA [2]. This can be further reduced to be less than 10^{-4} by applying $\sim 1\%$ AR coatings on both facets, which is relatively easy to achieve.

3) Buried facet structure. In this scheme, a transparent window region is fabricated between the ends of the active region and the SOA facets, as is shown in Figure 7-2(b). The light beam from the active layer spreads in the window region before arriving at the semiconductor-air interface. The reflected beam from the SOA facet spreads even further on the return trip and does not couple much light into the thin active region. The effective reflectivity of a buried facet decreases with increasing separation between the facet and the end of the active region. However, the coupling efficiency to the fiber decreases as well [3].

In practical applications, these methods are combined to satisfy the requirement on the SOA facet reflectivity.

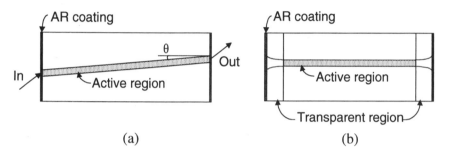

(a)   (b)

Figure 7-2. Schematic of SOA structures with (a) tilted waveguide and (b) buried facet (Top view)

7.1.2 Polarization Sensitivity and its Reduction

In general, the gain in an SOA is dependent on the polarization of the input signal, even though the gain of the semiconductor material is not. In the SOA, the cross-section of the active region facing the coupling optical fibers has a rectangular shape, where the thickness is usually much shorter than the width. This asymmetry results in unequal mode confinement factors for the light passing in parallel to the junction plane and the light perpendicular to it. The light with polarization parallel to the junction plane is the transverse electric (TE) mode. The light with polarization perpendicular to the junction plane is the transverse magnetic (TM) mode, since the magnetic field in this case is parallel to the junction plane. The mode confinement factors represent the fraction of mode energy contained inside the active region. The unequal mode confinement factors for TE and TM modes make the amplifier gain sensitive to the polarization state of the input light. Typically, SOAs can have several dBs of gain difference for the TE and TM modes. Since the polarization of the input light, which is dependent on the local temperature and stress on the fiber, is indeterminate in practical lightwave systems, it is crucial to reduce this polarization dependence for widespread applications of SOAs.

Various strategies have been proposed to overcome this deficiency. For example, two separate SOAs in parallel or series are orthogonally aligned to compensate for the polarization sensitivity [4]. In a configuration using a single amplifier with a double pass, the signal passes the amplifier twice with 90° polarization rotation between the two passes and the gain difference between TE and TM waves is reduced from 4 to 0.2 dB [5]. Although these configurations solve the polarization dependence problem, an SOA which is itself polarization-independent would be highly desirable from a practical point of view. In one scheme, the SOA is designed such that its active region has comparable width and thickness. Hence the mode confinement factors for TE and TM modes will be nearly equal. With a thick active waveguide (0.26 μm thick, 0.4 μm wide and 600 μm long) and window facet structure, a gain difference of 1.3 dB between TE and TM modes and spectral gain ripple of 1.5 dB has been demonstrated [6].

7.1.3   Multiple Quantum Well (MQW) SOAs

The active layer in a quantum well structure SOA has a much smaller thickness than in a conventional bulk structure SOA. In the quantum well SOA the active layer thickness is ~5-10 nm, while in conventional SOAs it is ~100 nm. The small thickness of the quantum well enables us to employ more than one quantum well in the active region. A schematic of a MQW SOA structure is shown in Figure 7-3. An electron in the conduction band of a quantum well is free to move in the y and z directions, but confined in the x direction.

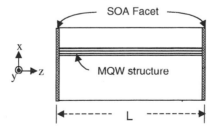

Figure 7-3. Schematic structure of a multiple quantum well (MQW) SOA (Side view).

SOAs incorporating MQW structures have a number of attractive features compared to bulk devices. Because of the small volume of the active region, even moderate injection current is sufficient to create a large carrier density, resulting in a short gain recovery time and broad gain spectrum. The loss coefficient in MQW active regions is significantly smaller than that in conventional SOAs. These properties lead to large saturation output power, and a value of 17 dBm has been demonstrated with a four-quantum-well SOA at 1.5 μm [7]. Saturation output power is defined as the output power for which the amplifier gain is reduced by a factor of 2 (or 3dB) from its unsaturated value. Large saturation power makes MQW SOAs very promising as power amplifiers.

For optical systems using optical amplifiers, the noise properties of amplifiers will largely influence the system performance. The noise performance is commonly characterized by the noise Figure F_n, defined as the ratio of the signal-to-noise ratio (SNR) at the input to the SNR at the output. The theoretical minimum limit of F_n is 3 dB for both SOA and fiber amplifiers. Typical values of the noise Figure for conventional SOAs are in the range 5-7 dB [1]. Quantum well amplifiers are also very promising as low noise amplifiers because F_n is inversely proportional to the difference between the injected carrier density and the carrier density for transparency [9]. Noise Figures of less than 5 dB have been demonstrated with MQW SOAs [8]-[9].

The polarization sensitivity of quantum well structures can be reduced by strained layers. A polarization sensitivity of 0.5 dB and gain of 27.5 dB have been achieved using a tensile-strained-barrier MQW SOA [10].

7.1.4   SOA Applications for Optical Space Switching

SOA based devices and structures have been extensively studied for optical space switching applications. In its linear operation, an SOA is used to amplify the optical signal with the pumping of electrical current. By turning on/off the electrical current, an SOA will be ON or OFF and may perform as a gate. Efficient space switching can be realized with an array of SOA gates.

SOAs have also found important applications in wavelength switching as wavelength converters. In this process optical data carried on one wavelength are "copied" on different wavelength. The process can be performed all-optically so that data will remain all the time in the optical domain. Efficient optical space switches can be constructed from all these devices in combination with an array of optical filters.

Different SOA-based interferometric structures have been developed to achieve ultrafast optical switching devices using SOA nonlinearities. Switching at speed of several hundred Gb/s has been demonstrated despite the relatively slow carrier recovery process in SOAs. We discuss these switching fabrics next, in sections 7.2 and 7.3, in greater detail.

7.2 SPACE AND WAVELENGTH SWITCHING

7.2.1 Static Gain Characteristics of SOAs

The basic equation that describes optical amplification is given by

$$P_{out} = GP_{in},\qquad(7.3)$$

Where P_{in} is the input optical signal power, P_{out} is the output power, and G is the effective amplifier gain as seen from external ports. Note that due to the material properties, the effective gain is dependent on the incoming signal wavelength. In addition, in resonant FP amplifiers where the input and output facet reflectivities R_1 and R_2 are significant factors, the effective amplifier gain has an additional frequency dependency given by [1]

$$G(v) = \frac{(1-R_1)(1-R_2)G_s(v)}{\left(1-\sqrt{R_1 R_2}\,G_s(v)\right)^2 + 4\sqrt{R_1 R_2}\,G_s(v)\sin^2\left[\pi(v-v_m)/\Delta v_L\right]}\qquad(7.4)$$

where v_m are the discrete resonant frequencies of the FP cavity, Δv_L is the spacing of the resonant frequencies, and $G_S(v)$ is the frequency dependent gain of the amplifier for a single pass through the amplification region given by

$$G_s(v) = \exp(g(v))L\qquad(7.5)$$

In this equation, $g(v)$ is the effective gain coefficient of an amplifying medium of length L. Note that for the more common traveling wavelength amplifiers, where facet reflectivities are extremely low ($<10^{-4}$), the effective amplifier gain can be approximated by this single pass gain equation alone. At a given frequency, the effective gain coefficient is given by [1]

$$g = \Gamma g_{mat} - \alpha_{int}\qquad(7.6)$$

and contains both a material gain factor, g_{mat}, as well as an internal loss coefficient, α_{int}. The optical confinement factor, Γ, represents the fraction of the signal that is contained within the cross-sectional area of the amplification region.

Although numerical techniques are generally required in order to fully calculate the material gain properties, a valid approximation at high levels of gain is to assume that it increases linearly as a function of injected carriers, N. The resulting equation is given by [1]

$$g_{mat} = (\sigma_g/V)(N-N_0),\qquad(7.7)$$

Where N_0 is the value of N required at transparency, V is the active gain volume, and σ_g is a parameter referred to as the differential gain of the bulk material.

In order to understand the carrier populations in the active region as a function of electrical drive current and optical power levels, a rate equation approach is commonly employed. Assuming uniform distribution of carriers, a differential equation describing the population N as a function of injection current I and signal power P can be expressed as [1]

$$\frac{dN}{dt} = \frac{I}{q} - \frac{N}{\tau_c} - \frac{\sigma_g(N-N_0)}{\sigma_m}\frac{P}{h\nu}, \tag{7.8}$$

Where q is the electronic charge, τ_c is the carrier lifetime, σ_m is the waveguide mode cross-section, and h is Planck's constant. The right side of this equation has three components; the first term represents the carrier increase due to current injection, the second term takes into account spontaneous emission, and the final term represents stimulated emission.

When a constant level of power P is present, the carrier population approaches a steady state and the material gain can be expressed as [1]

$$g_{mat} = \frac{g_0}{1+P/P_s}, \tag{7.9}$$

Where g_0 is defined as the unsaturated small signal gain and is equal to

$$g_0 = (\Gamma\sigma_g/V)(I\tau/q_c - N_0) \tag{7.10}$$

And P_S is defined as the output power for which the small signal gain is reduced by 3 dB and is given by

$$P_s = h\nu\sigma_m/(\sigma_g\tau_c) \tag{7.11}$$

7.2.2 Noise Characteristics of SOAs and its Impact on System Performance

In addition to the desired amplified signal, SOAs also emit optical noise at their outputs as a result of amplified spontaneous emission (ASE). This noise can be modeled as white Gaussian noise with a single-sided power spectral density of a single polarization equal to [1]

$$S_{sp}(\nu) = (G-1)n_{sp}h\nu, \tag{7.12}$$

where n_{sp} describes the relative carrier population inversion and is given by

$$n_{sp} = N/(N-N_0) \tag{7.13}$$

To understand the impact of optical amplifier noise on system performance, consider the average detected electrical current, $\langle i \rangle$, from a photodetector with responsivity R that is placed at the output of the amplifier

$$\langle i \rangle = RP_{av} = GP_{in} + S_{sp}B_o, \tag{7.14}$$

where B_o is the bandwidth of an optical filter preceding the photodetector. The detector responsivity in the unit of A/W is given by

$$R = \frac{\eta q}{h\nu}, \tag{7.15}$$

where η is defined as the detector quantum efficiency.

As a result of various sources of noise including thermal noise in the receiver electronics, shot noise, spontaneous-spontaneous beat noise, signal-spontaneous beat noise, and shot-spontaneous beat noise, there will be a variance in the detected current, which can be written as [11]

$$\langle i^2 \rangle = \sigma_T^{\,2} + \sigma_s^{\,2} + \sigma_{sp-sp}^{\,2} + \sigma_{sig-sp}^{\,2} + \sigma_{s-sp}^{\,2} \tag{7.16}$$

where the equations describing the five noise terms can be expressed as follows:

$$\sigma_T^{\,2} = (4k_B T/R_L)B_e, \tag{7.17}$$

$$\sigma_S^{\,2} = 2q[R(GP_{in} + S_{sp}B_o) + I_d]B_e, \tag{7.18}$$

$$\sigma_{sp-sp}^{\,2} = 4R^2 S_{sp}^{\,2} B_o B_e, \tag{7.19}$$

$$\sigma_{sig-sp}^{\,2} = 4R^2 GP_{in} S_{sp} B_e, \tag{7.20}$$

$$\sigma_{s-sp}^{\,2} = 4qRS_{sp}B_o B_e \tag{7.21}$$

In these equations, k_B is the Boltzmann constant, T is the absolute temperature, R_L is the photodetector load resistor, B_e is the effective electrical receiver bandwidth, and I_d is the photodetector dark current.

One of the primary metrics for characterizing system performance in lightwave communication systems is the detected electrical signal-to-noise ratio (SNR) given by

$$SNR = \langle i \rangle^2 / \langle i^2 \rangle \tag{7.22}$$

The "noise Figure" of an optical amplifier (SNR_{input}/SNR_{ouput}), where SNR_{input} is assumed to be shot-noise-limited, can be shown to be approximately equal to $2n_{sp}$ at high levels of gain. This implies that for an ideal amplifier ($n_{sp} = 1$), the optimum noise Figure is 3 dB.

One final parameter of key importance is the system bit error rate (BER) of the received data stream. It is commonly computed using

$$BER = 0.5 \times erfc\left(Q/\sqrt{2}\right) \qquad (7.23)$$

where *erfc* stands for the complementary error function, and the variable Q is expressed as

$$Q = \frac{I_1 - I_0}{\sigma_1 + \sigma_0} \qquad (7.24)$$

In this equation, I_1 and I_0 represent the average photocurrents for a "one" bit and a "zero" bit, respectively, while σ_1 and σ_0 represent the corresponding photocurrent standard deviations, respectively.

7.2.3 SOA-Based Space Switches

SOAs can be used as extremely fast optical gating devices by injecting current into the device to turn it on and removing the current to turn it off. In the latter case, the device acts as an efficient absorber. Switching speeds of less than 1 ns are possible, and the extinction ratio, which is defined as the ratio of the output signal power when the device is placed in the on state versus the output signal power when the gate is turned off, can be greater than 45 dB. In addition, by utilizing the SOAs in switch fabric architectures, rapid space switching functionality can be realized for optical cross-connects and for advanced applications such as optical packet switching. For example, by using a 1×N optical splitter followed by an array of SOA gates, a simple 1×N switch with broadcast capabilities can be constructed. Also, by mirroring this structure, an N×1 switch results.

In order to perform more complex functions, such as N×N switching capability, the broadcast-and-select architecture[1] shown in Figure 7-4 is commonly employed. In this architecture, each of the N inputs is connected to a 1×N splitter in order to broadcast the optical signals to N sets of gate arrays, where each set of N amplifier gates selects only one signal to transmit. Although it is possible to construct such a matrix out of discrete components, integration techniques are preferable due to the large number of SOAs required (N^2).

The development of space switches based on SOAs has been an active area of research for over two decades. Although early demonstrations were based on discrete components, recent research has been focused on integrated devices. Researchers have pursued both full monolithic integration, where the passive splitting/combining network and the SOA gates are fabricated on a single common substrate, as well as hybrid integration, in which case the active SOA gate arrays are connected to

---

[1] Switching architectures are discussed in part III

passive waveguide structures through fibers or directly to low-loss planar lightwave circuits (PLC), such as those based on silica.

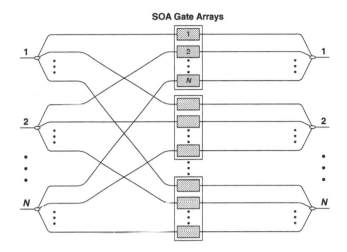

Figure 7-4. Generic SOA space switch based on broadcast- and-select architecture and consisting of N^2 SOA gates.

The first reports of integrated SOA space switches had a relatively small dimension. For example, a monolithically integrated 2x2 switch was demonstrated in 1992 using InGaAsP as the substrate material [12]. The device operated at 1550 nm, had a fiber-to-fiber loss of 12 dB, and was able to achieve an extinction ratio of greater than 40 dB. A number of other research groups also demonstrated integrated 2x2 switches of various architectures, both in the 1310 nm band [13] and 1550 nm band [14], with some devices actually able to achieve a net fiber-to-fiber gain [15]-[17]. Note that some of these devices were only capable of single-polarization operation [12]-[14], while others were polarization insensitive [15]-[17].

One of the first 4×4 fabrics utilized hybrid integration of active SOA arrays placed on top of a silica-based PLC [18]. Operating in the 1310 nm band, the device was capable of a 30 dB extinction ratio, although it had an insertion loss greater than 26 dB. In order to compensate for the large splitting losses incurred in these types of switches, other researchers have used multiple stages of SOAs for amplification and were able to achieve net fiber-to-fiber gains [19]-[20]. As an example, a 3-stage monolithic device containing 24 SOAs [19] achieved 6 dB fiber-to-fiber gain at 1550 nm and had an extinction ratio of 40 dB. However, only one polarization was supported, and three separate current sources of 50 mA, 50 mA, and 100 mA each were required for operation. More recently, a lower current and polarization insensitive version of the device was developed [20].

In order to scale SOA-based space switch fabrics to dimensions larger than 4×4, under the current state of the art, advanced packaging of multiple monolithic devices is required. For example, an 8×8 optical switch based on gain-clamped SOA gates was recently demonstrated using 8×1 switch gates as the basic building block [21]. By generating oscillation at an out of band wavelength, gain-clamped amplifiers are designed to suppress gain interactions that would normally occur between multi-wavelength signals passing through conventional SOA structures. At the input to the switch, 8 signals were broadcast to the array of SOA gates using 1:8 splitters and a passive fiber interconnect. The output signals were then selected by the 8×1 gates. Overall, the 8×8 fabric had an insertion loss of 14-16 dB in the 1550 nm region, a polarization dependent loss of 0.5-0.8 dB, and an extinction ratio of 32-38 dB depending upon the path through the fabric.

7.2.4 SOA-Based Wavelength Selectors

By combining some of the techniques discussed above for SOA-based space switches with wavelength filtering technologies, SOAs can also be utilized for wavelength-selective applications. A generic architecture that is commonly used in this respect is shown in Figure 7-5. It consists of three primary elements: an optical demultiplexer (DEMUX) for separating the incoming wavelengths, an SOA array for gating one or more wavelengths, and, finally, an optical multiplexer (MUX) for recombining the switched wavelengths. By using arrayed waveguide gratings (AWG) [22], which are planar integrated waveguide filter structures, for the DEMUX and MUX wavelength filters, all three components of the wavelength selector can be combined into a single device, either through monolithic or hybrid integration techniques.

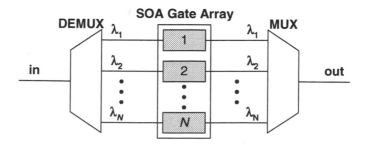

Figure 7-5. Generic wavelength selector architecture consisting of an optical demultiplexer (DEMUX), an SOA gate array, and an optical multiplexer (MUX).

The first integrated wavelength selector was a 4-channel device based on monolithic integration in InP [23]. Although this device was useful in demonstrating the initial concepts, it suffered from a relatively high insertion loss due to waveguides. Using advanced integration techniques, a 4-channel selector monolithically integrated on InP was demonstrated with a minimum 1.5 dB insertion loss in the 1550 nm region [24]. The AWGs were designed for 200 GHz wavelength channel spacing and a 50 dB extinction ratio was achieved. A monolithically integrated 4-channel device with 400 GHz channel spacing was also demonstrated with lossless selection capabilities [25]. Finally, the maximum number of channels demonstrated in a monolithic InP structure has been increased to 16 wavelengths [26]. This device was designed for 100 GHz spacing and was capable of up to 5 dB fiber-to-fiber gain with an average polarization dependent loss of approximately 2.2 dB.

Another approach taken for achieving low loss wavelength-selective devices is through hybrid integration of active SOA gates with PLCs. One of the first devices demonstrated in this respect was a high-speed 4-channel wavelength selector in the 1550 nm band based on a silica PLC platform [27]. The extinction ratio of the device was greater than 35 dB and it was capable of fiber-to-fiber gain in the wavelength range of 1530-1600 nm with a bias current of only 50 mA. In other work, the number of channels in a hybrid wavelength selection device has been increased to 32 channels [28]. An insertion loss of only 2.3 dB was achieved along with an extinction ratio of 46 dB in this work.

7.2.5   Using SOA-Based Wavelength Conversion to Combine Space/Wavelength Selection

The SOA-based space switching gates discussed in Section 7.2.3 did not involve change of wavelength in the switching process. A relatively new area of research on optical switching exploits a combination of all-optical wavelength conversion technologies based on SOAs [29] with wavelength routing to achieve the switching functionality. In this type of architecture, signal wavelengths are generally changed in the switching process. For example, the generic architecture illustrated in Figure 7-6 utilizes wavelength conversion to switch N input signals through an $N \times N$ arrayed waveguide grating (AWG) [30]. An $N \times N$ AWG enables non-blocking routing of individual wavelengths from each of the N input ports to each of the N output ports in a cyclic-frequency manner. A significant advantage of this type of design is that in contrast to the space switches discussed earlier, where the number of SOAs required scales as N^2, the complexity of this architecture increases only linearly with N. One of the first demonstrations of this principle was reported in [31]. The wavelength conversion function was performed using cross-gain modulation in SOAs, and the AWG was used as an optical filter to block the original input signal wavelength. By converting the incoming signal to eight unique wavelengths spaced at 100 GHz in the range of 1548.52 nm to 1554.13 nm, the equivalent functionality of a 1×8 switch was demonstrated.

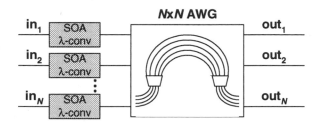

Figure 7-6. Generic architecture of a combined wavelength/space selector switch consisting of an array of wavelength converters followed by an AWG.

7.3 ULTRAFAST SWITCHES USING SOA-BASED NONLINEAR INTERFEROMETER

7.3.1 Gain Dynamics in SOAs

The operation of an SOA depends primarily on the creation of a carrier population inversion that ensures that the stimulated emission is more prevalent than absorption. The population inversion is usually achieved by electric current injection in the p-n junction where the generated electron-hole pairs recombine by means of stimulated emission. In a rate equation approximation, Equation (7.8) can be used to describe the carrier population as a function of injection current and incident optical power.

The propagation of the electromagnetic field E inside the SOA is governed by the wave equation [32]

$$\nabla^2 E - \frac{\left(n_b^2 + \chi\right)}{c^2}\frac{\partial^2 E}{\partial^2 t} = 0 \qquad (7.25)$$

where c is the velocity of light in vacuum, and the dielectric constant includes two part: the background refractive index n_b and the contribution of the charge carriers χ. χ is a function of the carrier density, and can be expressed by

$$\chi(N) = -\frac{\overline{nc}}{\omega_0}(\alpha+i)\sigma_g(\frac{N-N_0}{V}) \qquad (7.26)$$

where $\bar{n}$ is the effective mode index, ω_0 is the photon frequency. The carrier-induced index change is taken into account through a line-width enhancement of the parameter α. For an SOA, α is typically in the range of 3 to 8.

When $\alpha_{int} \ll \Gamma g_{mat}$, Equations (7.6), (7.7) and (7.8) yield [1]

$$\frac{\partial g}{\partial t} = \frac{g_0 - g}{\tau_c} - \frac{g|A|^2}{E_{sat}} \qquad (7.27)$$

where E_{sat} is the saturation energy of the amplifier, g_0 is the small signal gain, and A is the slowly-varying envelope associated with the optical pulse. Equations (7.25) and (7.27) govern optical pulses propagating in SOAs. Separating $A(z, t)$ into its constituent amplitude and phase parts:

$$A = \sqrt{P}\exp(j\phi) \qquad (7.28)$$

a simple relation between the amplitudes and phases of the input and output pulses from the SOA can be obtained [32]

$$P_{out}(\tau) = P_{in}(\tau)\exp[h(\tau)] \qquad (7.29)$$

$$\phi_{out}(\tau) = \phi_{in}(\tau) - \frac{1}{2}\alpha h(\tau) \qquad (7.30)$$

Here the subscript of "in" and "out" represent the input and output pulses, respectively. The time τ is measured in a reference frame moving with the pulse. The function h represents the integrated gain over the transverse direction of the SOA, and is defined by

$$h(\tau) = \int_0^L g(z,\tau)dz \qquad (7.31)$$

From (7.27), we can get that $h(\tau)$ is the solution of the following ordinary differential equation [32]

$$\frac{dh}{d\tau} = \frac{g_0 L - h}{\tau_c} - \frac{P_{in}(\tau)}{E_{sat}}[\exp(h) - 1] \qquad (7.32)$$

The calculated gain change using a short optical pulse at the input is shown in Figure 7-7. E_{sat} was taken to be 1 pJ, and g_0L is 4.5, corresponding to a gain of 90 ($\sim e^{4.5}$). The pulse has a Gaussian shape with pulse width (full width at half maximum) of 10 ps and peak power of 10 mW.

When a short optical pulse is injected into the SOA the carrier density decreases due to enhanced stimulated emission. For high energy input optical pulses, the carrier density decrease can be very large resulting in significant decrease in the gain of the SOA. The process is known as gain saturation. As seen in Figure 7-7, the saturation time can be as short as a few picoseconds. After the short pulse has passed, the gain of the SOA recovers due to the injection of the carriers by the electrical current. The recovery time is decided by the carrier lifetime, which is typically several hundred picoseconds. Figure 7-7 shows the gain change of the SOA with a carrier lifetime of 100 ps and 400 ps. Although the recovery time of this kind of resonant optical nonlinearity is long, the turn on time of the saturation process can be very short. Different kinds of devices using SOAs have been developed to exploit these two processes to build ultra-fast all-optical gates and switches.

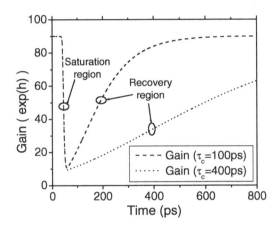

Figure 7-7. Dynamics of a gain change of the SOA with carrier lifetime of 100ps and 400ps after short optical pulse passes the SOA.

7.3.2 All-Optical Switches Based on Nonlinear Optical Loop Mirror

Figure 7-8 shows the schematic of an optical loop mirror, which is also called a Sagnac interferometer. It consists of a 2x2 directional coupler whose two ports (3 and 4) are connected together using a fiber loop. The input light goes to port 1. When the coupling ratio of the coupler K is equal to 0.5, fifty percent of the input light travels clockwise (CW) around the loop and fifty percent travels counterclockwise (CCW). In addition, light coupled across the coupler (between port 1 and port 4) suffers a $\pi/2$ phase lag with respect to light traveling straight through the coupler (between port 1 and port 3). Since the light components traveling in CW and CCW directions traverse the same piece of fiber in the loop, they experience the same amount of phase change ϕ. The transmitted intensity at port 2 (called also the output port) is therefore the sum of a CW field of arbitrary phase ϕ and a CCW field with relative phase (ϕ-π) (Note that the CCW field experience another $\pi/2$ phase lag when going cross the coupler from port 3 to port 2.). The two components will have equal amplitudes, resulting in a zero transmitted intensity at the output port 2 due to the π phase difference. At the port 1, the CW field and the CCW field have the same phase change (ϕ - π/2), resulting a total reflection to the port 1.

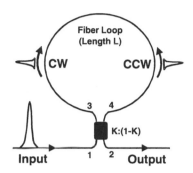

Figure 7-8. Schematic of an optical loop mirror.

Experiments with SOAs inserted into the loop mirror demonstrated that low energy optical pulses could change the gain of the amplifiers sufficiently to produce significant phase shifts in subsequent pulses passing through the amplifier [33]. Efforts were soon underway to form a new class of switching devices using the efficient nonlinearity in SOAs to induce a differential phase change between the two signal pulses counter-propagating in the fiber loop. The first device developed, using a differential phase change induced by the nonlinearity in SOA, was known as a semiconductor laser amplifier in a loop mirror (SLALOM) and was used to investigate "contrast enhancement and optical correlation" [34]. Although the rising edge of the temporal switching window was a few picoseconds, the window's falling edge depended upon the gain recovery time of the SOA which was several hundred picoseconds, which limited the speed of the correlation function.

It was discovered that the temporal width of the switching window could be linearly controlled by adjusting the displacement Δx_{soa} of the SOA from the midpoint of the loop (see Figure 7-9) [35]-[36]. Due to the dynamics of this configuration, the switching window closes earlier than the recovery time of the SOA as the latter is moved closer to the midpoint. Figure 7-9 shows a schematic diagram of such a device, which is known as a Terahertz Optical Asymmetric Demultiplexer (TOAD) [35]-[36].

When a data pulse enters the loop, it is split into CW and CCW traveling components by a 50:50 coupler. The two components pass through the SOA at different times as they counter-propagate around the loop, and recombine interferometrically at the 50:50 coupler. For the TOAD geometry shown in

Figure 7-9, the CCW data pulse component reaches the SOA earlier than the CW data pulse. In the absence of a control pulse, both pulse components experience the same effective medium as they propagate around the loop, and the data is reflected back.

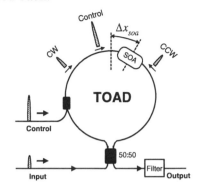

Figure 7-9. Schematic of a terahertz optical asymmetric demultiplexer (TOAD).

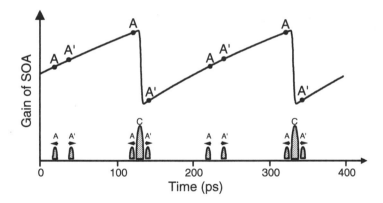

Figure 7-10. Schematic of the timing sequence of the optical power at the SOA and the resulting gain change for optical switching.

If a high energy control pulse is injected into the loop, it depletes the carriers in the SOA and changes its index of refraction. If the control pulse is timed such that it arrives at the SOA after the CCW data pulse and before the CW pulse, a differential phase shift can be achieved between the two counter-propagating data pulses. This differential phase shift can be used to switch the data pulses to the output port. Figure 7-10 shows the schematic of the timing sequence of the CCW data pulse, the control pulse, the CW data pulse component, and the SOA gain change. The SOA gain for CCW and CW pulses is marked by "A" and "A'" in Figure 7-10, respectively. The presence of a control pulse will cause the depletion of the carriers in the SOA and the CCW and CW pulses will experience different gain. Therefore, the two components also experience a large phase difference. When the difference is π, the recombined pulse is transmitted out of the TOAD. Subsequent data pulses experience a gain and a refractive index that are slowly recovering; hence CW and CCW pulses have only a small phase difference and so are reflected on recombination at the coupler. A polarization or wavelength filter can be used at the output to discriminate the switched data signal from the control pulse.

The output signal of the TOAD can be described by the following interferometric equation:

$$P_{out}(t) = \frac{P_{in}(t)}{4}\left\{G_{CW}(t)+G_{CCW}(t)-2\sqrt{G_{CW}(t)G_{CCW}(t)}\cos(\phi_{CW}(t)-\phi_{CCW}(t))\right\}$$

$$(7.33)$$

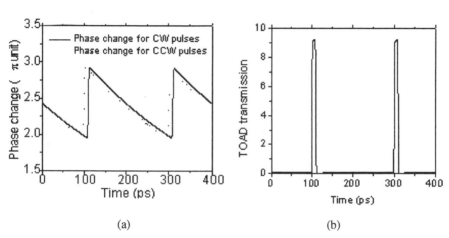

(a)   (b)

Figure 7-11. (a) Calculated phase evolution for CCW and CW pulses. (b) The corresponding TOAD transmission.

The input data signal is represented by $P_{in}(t)$, and it is assumed that the interferometer is balanced so that there is initially no signal at the output. In the presence of control signals, the two components of the input signal (signals CW and CCW) that interfere within the interferometer experience a time-dependent gain, $G_{CW,CCW}(t)$, and phase shift, $\phi_{CW,CCW}(t)$, as they traverse the SOA. From Equation (7.29) and (7.30), we have $G_{CW}(t)= exp(h_{CW}(t))$, $\phi_{CW}(t)= -\frac{1}{2}(\alpha h_{CW}(t))$, $G_{CCW}(t)= exp(h_{CCW}(t))$, and $\phi_{CCW}(t)= -\frac{1}{2}(\alpha h_{CCW}(t))$. Due to the asymmetric position of the SOA in the TOAD loop, the CW pulse will reach the SOA later than the CCW pulse by a time delay of $2\Delta x_{SOA}/c_{fiber}$, where Δx_{SOA} is the offset of the SOA from the center position of the loop and c_{fiber} is the speed of light in fiber. Therefore, we have $h_{CCW}(t)= h(t)$, $h_{CW}(t)= h(t+ 2\Delta x_{SOA}/c_{fiber})$. This differential response results in the input data signal being switched to the output of the device. Figure 7-11(a) shows the phase of both CW and CCW pulses on recombination at the coupler. Due to the SOA offset, the two phase profiles are delayed with respect to each other and a time varying phase difference is obtained with two distinct values of phase differences. Figure 7-11(b) shows the resultant TOAD transmission. The temporal duration of the switching window is determined by the offset of the SOA, Δx_{SOA}. As this offset is reduced, the switching window size decreases. The size of the nominal switching window duration, τ_{win}, is related to the offset position by $\tau_{win} = 2 \, \Delta x_{SOA}/c_{fiber}$.

By precisely controlling the offset position of the SOA, very short or long switching windows with very fast rising and falling edges can be achieved. In case of a very short TOAD switching window, demultiplexing of a single channel from a 250-Gb/s data stream [37] and error-free demultiplexing from a continuous 160-Gb/s data stream [38] have been demonstrated. Also using TOAD with a short switching window an all-optical recognition of the destination address in 250-Gb/s optical packet has been demonstrated [39]. On the other hand, the TOAD with a long switching window can be used as a routing/switching device for applications in very high speed optical networks. It was demonstrated that a combination of two cascaded TOADs, as shown in Figure 7-12 (a), can be used to perform space switching and route optical packets based on their destination addresses in a 250-Gb/s system [40]. In Figure 7-12 (a), TOAD1 has a short window equal to data bit length and acts as an all-optical packet destination address reader and TOAD2 has a long window equal to packet length and acts as an all-optical packet switch/router. An optical buffer was used in this experiment to compensate for the delay encountered while reading and processing the destination address by TOAD1. A diagram of the TOAD configured as an all-optical switch/router is given in Figure 7-12(b). This demonstration also illustrates that a similar approach can be used to perform all-optical high-speed time slots interchange (optical switching in the time domain).

Since the size of the TOAD device only depends upon the SOA length and offset from the center position in the loop, compact TOADs based upon discrete components have been constructed with loop lengths of less than 20 cm.

The TOAD is a good candidate for a variety of applications in optical communication networks because of its ultra-fast operation and because it requires low control pulse energy. To achieve full switching, the differential phase shift between the counter-propagating data pulses should approach π. For a 500 μm long InGaAsP SOA ($n_{SOA} = 3.3$) at 1.55 μm, the index change is on the order of 10^{-3}. It has been estimated that the carrier density change associated with this phase shift is on the order of 10^{17} cm^{-3} [41], which is approximately a 10% change in the carrier density of an appropriately biased SOA in population inversion. Both theoretical calculations [42]-[43] and experimental measurements [44] have verified that a control pulse with 10%

of the SOA saturation energy is sufficient to induce a π phase shift. Depending upon the SOA parameters, bias current, and wavelength, control pulse energies of < 500 fJ are typically needed. This value can fluctuate by nearly 20% without significantly degrading the SNR at the switched output [45].

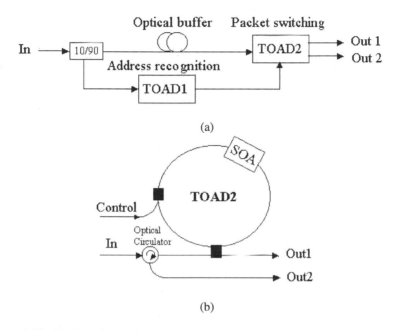

(a)

(b)

Figure 7-12. (a) All-optical switching/routing demonstration. (b) Block diagram of TOAD-based all-optical switch/router.

The TOAD is robust to temperature variations and can be reliably operated without stabilization as data signals propagating in both directions around the loop experience the same effective medium. The architecture can also be adapted to accommodate other short, highly nonlinear materials in place of the SOA to operate the device with different wavelengths or to achieve performance enhancements. The TOAD device and its variations may prove to be the most practical approach to all-optical switching as they can be integrated using a variety of techniques that are discussed below.

7.3.3   SOA-Based Mach-Zehnder All-Optical Switch Geometries

Ultrafast all-optical switching can also be obtained using other interferometric configurations based on a similar operating principle. These architectures improve the integratability and performance of the device, although they may require active stabilization if constructed from discrete components. Two configurations of a Mach-Zehnder interferometer switch are shown in Figure 7-13. The input signal is split 50:50 into two arms of the Mach-Zehnder interferometer structure. After passing through the SOA, the two signals recombine at the output port. For the structure in Figure 7-13(a), the two components can experience destructive interference when the two arms are balanced in the absence of the control signals. Therefore we do not have light output. When control pulses are injected into the interferometer, a differential phase shift is briefly introduced between the two arms of the interferometer causing an input data pulse to be switched to the output port. For the configuration in Figure 7-13(b), the output light can be switched between output port A and B when there is, or is not, control light.

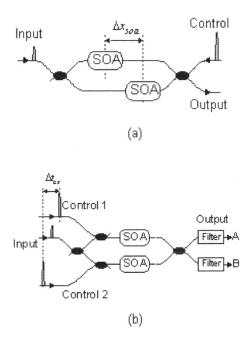

Figure 7-13. All-optical switches based on Mach-Zehnder geometries. (a) Colliding-Pulse Mach-Zehnder (CPMZ) switch; (b) Symmetric Mach-Zehnder (SMZ) switch.

Similar to the TOAD, subsequent data pulses that pass through the switch see the slow recovery of both SOAs and are rejected. Another difference between the two Mach-Zehnder geometries shown is with respect to the propagation direction of control and data signals. In the colliding pulse Mach-Zehnder (CPMZ) shown in Figure 7-13(a), the data and control signals counter-propagate through the interferometer. As a result, a filter is not needed at the output to reject the control, and the control pulses can be coupled into the interferometer without introducing additional coupling losses. The nominal switching window for the CPMZ is determined by the distance between the midpoints of the SOAs such that $\tau_{win}= 2 \, \Delta x_{SOA}/c_{fiber}$. The other architecture known as the symmetric Mach-Zehnder (SMZ), shown in Figure 7-13(b), requires a filter at the output port to block the control signal from the switched data signal since data and control signals co-propagate. Assuming the SOAs are positioned in the same relative location within the interferometer, the nominal switching window for the SMZ is determined by the temporal control pulse separation, Δt_{CS}, of Control 1 and Control 2 prior to entering the interferometer such that $\tau_{win}= \Delta t_{CS}$.

Although the nominal switching window size provides an estimate of the switching window temporal duration, it does not account for the finite length of the SOAs. While the SOA length has little effect on the SMZ geometry, the minimum achievable switching windows for both the TOAD and CPMZ are constrained by the length of the SOAs [46]-[47].

An ultrafast optical space switch can be built with the SMZ structure where the control light can set the switch to be at bar or cross state and the output light can be switched between port A and B. An extinction ratio of 20 dB has been demonstrated by optimizing the SOA bias current and the phases in both arms of the Mach-Zehnder interferometer [48].

7.3.4 UNI All-Optical Switch

Like other interferometric switches, the UNI (Ultra-fast Nonlinear Interferometer) is also based on the carrier depletion of SOAs by high-energy control pulses. The UNI uses one SOA in a single-arm interferometer (SAI) structure as shown in Figure 7-14.

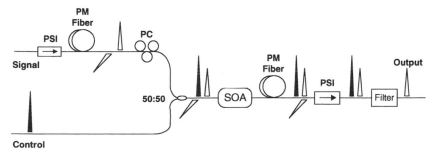

Figure 7-14. Block diagram of UNI. (PC: polarization controller)

The principle of operation of the UNI is as follows. An input signal (data pulse) enters the switch through polarization sensitive optical isolator (PSI). Pulses are split by a polarization-sensitive delay (a length of polarization maintaining (PM) birefringent fiber) into two orthogonal components, which are also separated temporally [49]-[50]. Due to the large difference in refractive indices for the orthogonally polarized modes in the birefringent fiber, these two data components will be delayed from each other by an offset determined by the length of the birefringent fiber and by the degree of birefringence.

If n_x and n_y are the mode indices for the orthogonal fiber modes, then the temporal offset (Δt) is given by

$$\Delta t = \left| n_x - n_y \right| \times \frac{L}{c} \qquad (7.34)$$

where L is the length of the PM fiber and c is the velocity of light in vacuum. Typically birefringent fiber has $\left| n_x - n_y \right| \sim 10^{-4}$. Thus, for an offset of a few picoseconds, a few meters of birefringent fiber are required.

Now, the two data signal components travel through the SOA, into which high-energy controls pulses are injected using the 50:50 coupler. If the control pulses are injected into the SOA such that they arrive at the SOA in between the arrival of the two data pulse components, the two components acquire different phase shifts due to the carrier depletion in the SOA by the high-energy control pulse. The different phase shift can cause polarization state rotation when the two signal components are retimed to overlap in a second polarization sensitive delay (another length of birefringent fiber). Then the polarization-rotated data signals can be subsequently separated by use of a polarizer set at 45^0 to the orthogonal signal polarizations. The control pulse is filtered out at the output of the device. A counter-propagating structure can be used to avoid the wavelength or polarization discriminating element at the output.

Since the UNI is a single arm interferometer device, all the signals travel along the same path and are exposed to identical fluctuations in optical path length. Therefore, the device is stable and no active interferometric bias stabilization is necessary [49]. The switching window duration is determined by the temporal offset produced by the polarization sensitive delay element. Thus, changing the length of the birefringent fiber can vary the switching window size of the UNI. The UNI has been used to demultiplex 40 Gbps and 20 Gbps data streams into 10 Gbps data channels [49].

7.3.5 Gain-Transparent SOA-Based All-Optical Switch

In the SOA-based interferometric switches described so far, optical control pulses at similar wavelength band as the data (Figure 7-15(a)) are used to deplete the carriers in the SOA. The carrier depletion causes a gain change in the SOA and thus a change in the refractive index determined by the Kramers-Kronig relationships [1]. The phase change experienced by the data pulses due to the change in the refractive index is used for interferometric switching. However, this "conventional" operation of the SOA-based interferometric switches has some disadvantages. The gain change that is created simultaneously with the phase change results in reduced extinction ratio for the demultiplexed data channel and an amplitude modulation of the channels that are not demultiplexed. In addition, the data signal quality is degraded by the addition of ASE noise. To a large extent, this can be eliminated by a dual wavelength operation technique [51]. Here, the data signal is chosen at a wavelength that is far off the gain and ASE peaks of the SOA, while the control pulses are chosen at the wavelength close to the peak of the gain spectra of the SOA (Figure 7-15(b)).

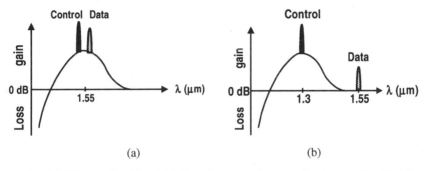

Figure 7-15. (a) "Conventional" and (b) Gain-transparent approaches in operating SOA-based interferometric switch.

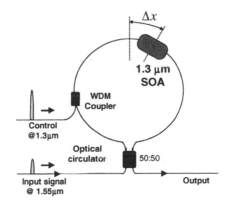

Figure 7-16. Gain transparent (GT) SOA optical switch.

This idea was demonstrated using a TOAD timing principle ($\tau_{win} << \tau_c$) with the SOA gain peak and the control signal wavelength at 1.3 μm and the data signal at 1.55 μm, as shown in the Figure 7-15(b). This type of all-optical demultiplexer was named a Gain Transparent SOA switch (GT-SOA Switch) [51]-[52]. A schematic diagram of this switch is given in Figure 7-16. As the data wavelength is far away from the gain peak of the SOA, the data signal experiences negligible amplitude change. However, the data signal at 1.55 μm still experiences a strong phase change due to the effect of the control pulses at 1.3 μm. This strong phase change has also been used in all-optical wavelength conversion from 1.3 μm to 1.55 μm [53]. Also, since the data wavelength is far away from the ASE maximum of the SOA, a very low amount of noise is added, resulting in a lower noise Figure. Since this dual wavelength technique uses the data wavelength away from the gain and ASE peaks of the SOA, the SOA is essentially transparent to the data signal (i.e., the energy of the data pulses is lesser than the band-gap energy of the material), which explains the name of this effect (gain-transparent, GT).

7.4   TOWARDS INTEGRATED ALL-OPTICAL SWITCHES FOR FUTURE NETWORKS

In this chapter, we have described SOA-based optical switches which are suitable for high capacity WDM optical networks, and can switch optical signals in the space, wavelength, or time domain. While data is typically transported today using WDM, optical time-domain switching continues to out-perform wavelength switching in terms of speed, switching energy, latency, and scalability. Thus, for applications that require high speed, scalable, and rapidly reconfigurable switching, optical time domain switching can offer superior performance when suitably interfaced to WDM data.

One of the main drawbacks of SOA technology continues to be its high cost. Other considerations that will determine their feasibility include size, optical switching energy, sensitivity to polarization and environment, cascadability and scalability. It is not likely that a single SOA with relatively favorable for deployment in comparison to other devices that can be integrated in arrays and manufactured in large volumes. Therefore, integration of SOA-based switches is essential for large-scale commercial deployment. Various research groups have reported integration of the Sagnac, Mach-Zehnder, and Michelson interferometer based all-optical switches using both monolithic and hybrid technologies.

A monolithically integrated Sagnac configuration of the all-optical interferometric switch was used for demultiplexing 20Gb/s streams into composite 10 Gb/s streams and 20 Gb/s to 5 Gb/s [54]. Also, both the CPMZ and SMZ geometries have been demonstrated as integrated high-speed all-optical demultiplexers [55]-[57]. A Michelson interferometer based all-optical switch was integrated and used to demultiplexing 20Gb/s data streams to 5 Gb/s channels [58].

These integration techniques can enable the development of practical SOA-based technologies for space, wavelength and ultrafast switches. These switches, in turn, can enable optical signal processing and lead to development of advanced optical networks where optical signals do not need to be converted into the electrical domain as they traverse the network. This capability will enhance the throughput, flexibility and reconfigurability of the future generations of optical networks.

REFERENCES

[1] G. P. Agrawal, "Fiber-Optic Communication Systems," Wiley, New York, 1997.

[2] C. E. Zah, J. S. Osinski, C. Caneau, S. G. Menocal, L. A. Reith, J. Salzman, F. K. Shokoohi, and T. P. Lee, "Fabrication and performance of 1.5 μm GaInAsP traveling-wave laser amplifiers with angled facets," Electron. Lett., Vol. 23, No. 19, pp. 990-2, 1987.

[3] N. K. Dutta and J. R. Simpson, "Optical amplifiers," Progress in optics, pp. 189-226, 1993.

[4] G. Grosskopf, R. Ludwig, R. G. Waarts, and H. G. Weber, "Optical amplifier configurations with low polarization sensitivity," Electron. Lett., Vol. 23, No. 25, pp. 1387-8, 1987.

[5] N. A. Olsson, "Polarization independent configuration optical amplifier," Electron. Lett., Vol. 24, No. 17, pp. 1075-6, 1988.

[6] I. Cha, M. Kitamura, H. Honmou, and I. Mito, "1.5 μm band travelling-wave semiconductor optical amplifiers with window facet structure," Electron. Lett., Vol. 25, No. 18, pp. 1241-2, 1989.

[7] G. Eisenstein, U. Koren, G. Raybon, T. L. Koch, J. M. Wiesenfeld, M. Wegener, R. S. Tucker, and B. I. Miller, "Large- and small-signal gain characteristics of 1.5 μm multiple quantum well optical amplifiers," Appl. Physics Lett., Vol. 56, No. 13, pp. 1201-3, 1990.

[8] T. Saitoh, Y. Suzuki, and H. Tanaka, "Low noise characteristics of a GaAs-AlGaAs muliple-quantum-well semiconductor laser amplifier," IEEE Photon. Technol. Lett., Vol. 2, No. 11, pp. 794-6, 1990.

[9] K. S. Jepsen, B. Mikkelsen, J. H. Povlsen, M. Yamaguchi, and K. E. Stubkjaer, "Wavelength dependence of noise figure in InGaAs/InGaAsP multiple-quantum-well laser amplifier," IEEE Photon. Technol. Lett., Vol. 4, No. 6, pp. 550-3, 1992

[10] K. Magari, M. Okamoto, and Y. Noguchi, "1.55 μm polarization-insensitive high-gain tensile-strained-barrier MQW optical amplifier," IEEE Photon. Technol. Lett., Vol. 3, No. 11, pp. 998-1000, 1991.

[11] N. A. Olsson, "Lightwave systems with optical amplifiers," J. Lightwave Technol., Vol. 7, No. 7, pp. 1071-82, 1989.

[12] M. Janson, L. Lundgren, A.-C. Morner, M. Rask, B. Stoltz, M. Gustavsson, and L. Thylen, "Monolithically integrated 2x2 InGaAsP/InP laser amplifier gate switch arrays," Electron. Lett., Vol. 28, No. 8, pp. 776-8, 1992.

[13] C. Holtmann, T. Brenner, R. Dall'Ara, P. A. Besse, and H. Melchior, "Monolithically integrated semiconductor optical amplifiers for trans-parent 2x2 switches at 1.3 micrometers," AGEN-Mitteilungen, No. 56/57, pp. 45-8, 1993.

[14] M. G. Young, U. Koren, B. I. Miller, M. Chien, M. A. Newkirk, and J. M. Verdiell, "A compact 2x2 amplifier switched with integrated DBR lasers operating at 1.55 μm," IEEE Photon. Technol. Lett., Vol. 4, No. 9, pp. 1046-8, 1992.

[15] G. Sherlock, J. D. Burton, P. J. Fiddyment, P. C. Sully, A. E. Kelly, and M. J. Robertson, "Integrated 2x2 optical switch with gain," Electron. Lett., Vol. 30, No. 2, pp. 137–8, 1994

[16] K. Hamamoto and K. Komatsu, "Insertion-loss-free 2x2 InGaAsP/InP optical switch fabricated using bandgap energy controlled selective MOVPE," Electron. Lett., Vol. 31, No. 20, pp. 1779–81, 1995.

[17] F. Dorgeuille, B. Mersali, M. Feuillade, S. Sainson, J. Brandon, S. Slempkes, and M. Carre, "Monolithic InGaAsP-InP tapered laser amplifier gate 2x2 switch matrix with gain," Electron. Lett., Vol. 32, No. 7, pp. 686–8, 1996.

[18] Y. Yamada, H. Terui, Y. Ohmori, M. Yamada, A. Himeno, and M. Kobayashi, "Hybrid-integrated 4x4 optical gate matrix switch using silica-based optical waveguides and LD array chips," J. Lightwave Technol., Vol. 10, No. 3, pp. 383–90, 1992

[19] M. Gustavsson, B. Lagerstrom, L. Thylen, M. Janson, L. Lundgren, A.-C. Morner, M. Rask, and B. Stoltz, "Monolithically integrated 4x4 InGaAsP/InP laser amplifier gate switch arrays," Electron. Lett., Vol. 28, No. 24, pp. 2223–5, 1992.

[20] W. van Berlo, M. Janson, L. Lundgren, A.-C. Morner, J. Terlecki, M. Gustavsson, P. Granestrand, and P. Svensson, "Polarization-insensitive, monolithic 4x4 InGaAsP–InP laser amplifier gate switch matrix," IEEE Photon. Technol. Lett., Vol. 7, No. 11, pp. 1291–3, 1995.

[21] F. Dorgeuille, L. Noirie, J-P. Faure, A. Ambrosy, S. Rabaron, F. Boubal, M. Schilling, and C. Artigue, "1.28 Tbit/s throughput 8x8 optical switch based on arrays of gain-clamped semiconductor optical amplifier gates," in Proc. OFC 2000, Baltimore, MD, paper PD18-1, March 2000.

[22] M. K. Smit and C. van Dam, "PHASAR-based WDM-devices: principles, design and applications," IEEE J. Sel. Top. in Quant. Electron., Vol. 2, No. 2, pp. 236–50, 1996

[23] M. Zirngibl, C. H. Joyner, and B. Glance, "Digitally tunable channel dropping filter/equalizer based on waveguide grating router and optical amplifier integration," IEEE Photon. Technol. Lett., Vol. 6, No. 4, pp. 513-5, 1994.

[24] R. Mestric, M. Renaud; F. Pommereau; B. Martin; F. Gaborit; G. Lacoste; C. Janz,; D. Leclerc; and D. Ottenwalder, "Four-channel wavelength selector monolithically integrated on InP," Electron. Lett., Vol. 34, No. 19, pp. 1841–3, 1998.

[25] R. Mestric, M. Renaud, F. Pommereau, B. Martin, F. Gaborit, C. Janz, I. Guillemot, and D. Leclerc, "Loss-less four-channel wavelength selector monolithically integrated on InP," in Proc. OFC '99, San Diego, CA, paper ThB2, Feb. 1999.

[26] R. Mestric, C. Porcheron, B. Martin, B. Pommereau, L. Guillemot, F. Gaborit, C. Fortin, J. Rotte, and M. Renaud, 'Sixteen-channel wavelength selector monolithically integrated on InP," in Proc. OFC'2000, Baltimore, MD, paper TuF6, 2000.

[27] T. Ito, I. Ogawa, N. Yoshimoto, F. Ebisawa, K. Magari, K. Shuto, Y. Kawaguchi, M. Yanagisawa, O. Mitomi, F. Hanawa, Y. Tohmori, Y. Yamada, Y. Yoshikuni, and Y. Hasumi, "Ultra-wide-band high-speed wavelength selector using a hybrid integrated gate module: a 4-channel SS-SOA gate array on PLC platform," in Proc. OFC'97, Dallas, TX, pp. 53-56, Feb. 1997.

[28] F. Ebisawa, I. Ogawa, Y. Akahori, K. Takiguchi, Y. Tamura, T. Hashimoto, A. Sugita, Y. Yamada, Y. Suzaki, N. Yoshimoto, Y. Tohmori, T. Ito, K. Magari, Y. Kawaguchi, A. Himeno, and K. Kato, "High-speed 32-channel OWS using PLC hybrid integration," in Proc. OFC '99, San Diego, CA, paper ThB1, Feb. 1999.

[29] J. M. H. Elmirghani and H. T. Mouftah, "All optical wavelength conversion: technologies and applications in DWDM networks," IEEE Commun. Mag., Vol. 38, No. 3, pp. 86-92, 2000.

[30] A. A. M. Staring, L. H. Spiekman, C. van Dam, E. J. Jansen, J. J. M. Binsma, M. K. Smit, and B. H.Verbeek, " Space-switching 2.5Gbit/s signals using wavelength conversion and phase array routing," Electron. Lett., Vol. 32, No. 4, pp. 377–9, 1996

[31] A. Okada, T. Sakamoto, Y. Sakai, K. Noguchi, and M. Matsuoka, "All-optical packet routing by an out-of-band optical label and wavelength conversion in a full-mesh network based on a cyclic-frequency AWG," in Proc. OFC 2001, Anaheim, CA, paper ThG5, Mar. 2001.

[32] G. P. Agrawal, and N.A. Olsson, "Self-phase modulation and spectral broadening of optical pulses in semiconductor laser amplifiers," IEEE J. Quantum Electron., Vol. 25, No. 11, pp. 2297-306, 1989.

[33] A. W. O'Neill and R.P. Webb, "All-optical loop mirror switch employing an asymmetric amplifier/attenuator combination," Electron. Lett., Vol. 26, No. 24, pp. 2008-9, 1990.

[34] M. Eiselt, "Optical loop mirror with semiconductor laser amplifier," Electron. Lett., Vol. 28, No. 16, pp. 1505-7, 1992.

[35] J. P. Sokoloff, P.R. Prucnal, I. Glesk, and M. Kane, "A Terahertz Optical Asymmetric Demultiplexer(TOAD)," Photonics in Switching, Proc. OSA 16, PD-4, 1993.

[36] J. P. Sokoloff, P.R. Prucnal, I. Glesk, and M. Kane, "A Terahertz Optical Asymmetric Demultiplexer(TOAD)," IEEE Photon. Technol. Lett., Vol. 5, No. 7, pp. 787-90, 1993.

[37] I. Glesk, J.P. Sokoloff, and P.R. Prucnal, "Demonstration of all-optical demultiplexing of TDM data at 250 Gbit/s". Electron. Lett. Vol. 30, No. 4, pp. 339-41, 1994.

[38] K. Suzuki, K. Iwatsuki, S. Nishi, and M. Saruwatari, "Error-free demultiplexing of 160Gbit/s pulse signal using optical loop mirror including semiconductor laser amplifier," Electron. Lett., Vol. 30, No. 18, 1501-3, 1994.

[39] I. Glesk, J. P. Sokoloff, and P. R. Prucnal, "All-optical address recognition and self-routing in a 250Gbit/s packet-switched network," Electron. Lett, Vol. 30, No.16, pp. 1322-3, 1994.

[40] I. Glesk, K.I. Kang, and P.R. Prucnal, "Demonstration of ultrafast all-optical packet routing," Electron. Lett., Vol. 33, No. 9, pp. 794–5, 1997.

[41] D. Cotter, R.J. Manning, K.J. Blow, A.D. Ellis, A.E. Kelly, D. Nesset, J.D. Phillips, A.J. Poustie, and D.C. Rogers, "Nonlinear optics for high speed digital information processing," Science, Vol. 286, No.5444, pp. 1523-8, 1999.

[42] R. J. Manning, A.D. Ellis, A.J. Poustie, and J. Blow, "Semiconductor laser amplifiers for ultrafast all-optical signal processing," J. Opt. Soc. Am. B, Vol. 14, No. 11, pp. 3204-16, 1997.

[43] K. I. Kang, I. Glesk and P.R. Prucnal, "Ultrafast optical time demultiplexers using semiconductor optical amplifiers," Intl. J. High Speed Electron. Sys., Vol. 7, pp. 125-151, 1996.

[44] K. I. Kang, T.G. Chang, I. Glesk, and P.R. Prucnal, "Nonlinear-index-of-refraction measurement in a resonant region by the use of a fiber Mach-Zehnder interferometer," Appl. Opt., Vol. 35, No. 9, pp. 1485-8, 1996.

[45] D. Y. Zhou, K.I. Kang, I. Glesk, and P.R. Prucnal, "An analysis of the signal-to-noise ratio and design parameters of a terahertz optical asymmetric demultiplexer," J. Lightwave Technol., Vol. 17, No. 2, pp. 298-307, 1999.

[46] K. I. Kang, T.G. Chang, I. Glesk and P.R. Prucnal, "Comparison of Sagnac and Mach-Zehnder ultrafast all-optical interferometric switches based on a semiconductor resonant optical nonlinearity," Appl. Opt., Vol. 35, No. 3, pp. 417-26, 1996.

[47] P. Toliver, R.J. Runser, I. Glesk, and P.R. Prucnal, "Comparison of three nonlinear interferometric optical switch geometries," Opt. Commun., Vol. 175, No. 4-6, pp. 365-373, 2000.

[48] J. Leuthold, J. Eckner, Ch. Holtmann, R. Hess, and H. Melchior, "All-optical 2×2 switches with 20dB extinction ratios," Electon. Lett. Vol. 32, No. 24, pp. 2235-6, 1996.

[49] N. S. Patel, K.A. Rauschenbach, and K.L. Hall, "40-Gbps demultiplexing using an ultrafast nonlinear interferometer (UNI)," IEEE Photonics Technol. Lett., Vol. 8, No. 12, pp. 1695-7, 1996.

[50] K. Tajima, S. Nakamura, and Y. Sugimoto, "Ultrafast polarization-discriminating Mach-Zehnder all-optical switch," Appl. Phys. Lett., Vol. 67, No. 25, pp. 3709-11, 1995.

[51] S. Diez, R. Ludwig, and H. G. Weber, "Gain-transparent SOA-switch for high-bitrate OTDM add/drop multiplexing," IEEE Photon. Technol. Lett., Vol. 11, No.1, pp. 60-2, 1999.

[52] S. Diez, R. Ludwig, and H.G. Weber, "All-optical switch for TDM and WDM/TDM systems demonstrated in a 640Gbit/s demultiplexing experiment," Electron. Lett., Vol. 34, No. 8, pp. 803-5, 1998.

[53] J. P. R. Lacey, G.J. Pendock, and R.S. Tucker, "All-optical 1300nm to 1550nm wavelength conversion using cross-phase modulation in a semiconductor optical amplifier," IEEE Photon. Technol. Lett., Vol. 8, No. 7, pp. 885-7, 1996.

[54] E. Jahn, N. Agrawal, W. Pieper, H.-J. Ehrke, D. Franke, W. Furst, and C.M. Weinert, "Monolithically integrated nonlinear Sagnac interferometer and its application as a 20 Gbit/s all-optical demultiplexer," Electron. Lett., Vol. 32, No. 9, pp. 782-4, 1996.

[55] R. Hess, M. Caraccia-Gross, W. Vogt, E. Gamper, P.A. Besse, M. Duelk, E. Gini, H. Melchior, B. Mikkelsen, M. Vaa, K.S. Jepsen, K.E. Stubkjaer, and S. Bouchoule, "All-optical demultiplexing of 80 to 10 Gb/s signals with monolithic integrated high-performance Mach-Zehnder interferometer," IEEE Photonics Technol. Lett., Vol. 10, No. 1, pp. 165-7, 1998.

[56] D. Wolfson, A. Kloch, T. Fjelde, C. Janz, B. Dagens and M. Renaud, "40-Gb/s all-optical wavelength conversion, regeneration, and demultiplexing in an SOA-based all-active Mach-Zehnder interferometer," IEEE Photonics Technol. Lett. Vol. 12, No.3, pp. 332-4, 2000

[57] P. V. Studenkov, M.R. Gokhale, J. Wei, W. Lin, I. Glesk, P.R. Prucnal, and S.R. Forrest, "Monolithic integration of an all-optical Mach-Zehnder demultiplexer using an asymmetric twin-waveguide structure," IEEE Photonics Technol. Lett., Vol. 13, No. 6 pp. 600-2, 2001.

[58] B. Mikkelsen, M. Vaa, N. Storkfelt, T. Durhuus, C. Joergensen, R.J.S. Pedersen, S.L. Danielsen, K.E. Stubkjaer, M. Gustavsson, and W. van Berlo, "Monolithic integrated Michelson interferometer with SOAs for high-speed all-optical signal processing" (paper TuD4), Proc. of OFC, TuD4, 1995.

Chapter 8

OTHER OPTICAL SWITCHING TECHNOLOGIES

Tarek S. El-Bawab

The second part of this book so far has covered six main optical switching technologies in detail. These technologies share a number of common features. First, they have been around for some time, in whatever form and implementation. During this time, they received considerable research attention from both the industrial and academic communities. This resulted in large presence in engineering and scientific literature. The work carried out with these technologies include theoretical studies, experimental demonstrations, and, in many cases, large-scale system/network field trials. Second, based on this heritage, all of these technologies were strong deployment candidates for optical-switching based systems in the 1990s, and some were even incorporated in vendors' products. Interest in each of these technologies however has obviously varied over time and there are some today that are stronger deployment candidates than others. Third, all these technologies are pursued by numerous research and development groups in many parts of the world. They are not proprietary technologies of any particular optical-component vendors, although certain implementations are obviously patented.

We now conclude part II by an overview of other optical switching technologies which may share some of the above features, but not all of them. In this chapter, very different technologies are covered, with little in common among them. Therefore, an attempt is made to organize the material in such a manner that is as logical as possible. The goal here is to continue to build upon the structure we have developed so far throughout the previous chapters. We shall cover at least three groups of technologies that are scattered among the five main categories identified in chapter 1. Sections 8.1 and 8.2 discuss Opto-Mechanical (OM) switching technologies of the general category to which Optical-MEMS switches (chapter 6) belong. These are largely transparent technologies which can operate over broad wavelength range. Section 8.3 discusses electrically controlled holographic switching approaches. According to the classification adopted in chapter 1, these are electro-optic in nature. They are more wavelength dependent than the ones discussed in the first two sections. Finally, section 8.4 concludes the chapter by a brief overview of a selection of other technologies. In particular, we consider a data-format dependent technology which works best with specific data-pulse characteristics. Though it represents a research topic that is far from maturity and commercial deployment, this technology has the potential advantage of very fast switching speeds.

The purpose of this chapter is to complement the knowledge gained about optical switching technologies throughout the previous six chapters. The discussion hereafter will not be as detailed as it was in previous chapters. Yet, our goal is to provide a high-level overview with sufficient depth for the reader to appreciate as many of the existing approaches to build optical switches as possible. Many references are provided at the end of this chapter to help the reader explore more details as needed.

8.1   OPTO-MECHANICAL SWITCHING: THE MOVING-FIBER AND MOVING-OPTICAL-COMPONENT TYPES

This technology is based on the use of mechanical positioning systems where movement of optical fibers or optical components against each other permits switching of light from input fibers/ports to output fibers/ports of the system [1]-[30]. In principle, motion can be invoked manually, but an actuation mechanism is usually deployed. Electromagnetic actuation is often used. Stepper-motor driven translation, or rotation and electrostatic actuation are also used. Thermal actuation has been used too. Switches based on this technology are classified into two main types [31], namely the fiber moving type, known as the *fiber switch*, and the optical-component moving type where the moving component can be prism, lens, mirror, collimator or other optical component(s). This latter type is sometimes referred to as the *moving-beam* type [24]. Switches based on motion of fiber connectors can be considered as fiber switches.

There are many different approaches to building these Opto-Mechanical (OM) switches. We merely survey a few examples, configurations and characteristics to give the reader some flavor of what this technology is all about.

8.1.1   The Fiber Switch

Simple fiber switches can be built based on the generic principle depicted in Figure 8-1. This is a 1xN (N= 1, 2, 3, ...) switch with movable input fiber. For the special case where N= 1, the switch takes the simplest form of a 1x1 switch, and is also referred to as an ON/OFF (connect/disconnect) switch[1]. For N > 1, the switch has the capability of directing an input optical signal to one of N different outputs. Nx1 switches are also possible.

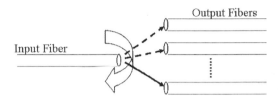

Figure 8-1. Generic principle of a 1xN opto-mechanical (OM) fiber switch

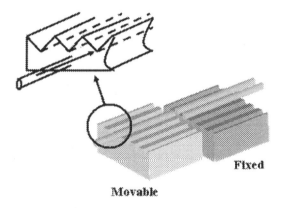

Figure 8-2. Fiber switch realization using V-grooves

Several implementations have been reported in literature to demonstrate this principle. The mechanical force acting on fiber is produced in various manners. It can be generated by a magnet acting on ferromagnetic coating upon the fiber itself or on an ultra light magnetic pipe which circumvents the fiber. An electromagnet acting on a movable fiber-holding ferrule, a fiber-holding block or on an otherwise movable carrier of fiber arrays, may be used as well. Movable and/or fixed fiber arrays can be set over *V-grooves* as outlined in Figure 8-2. These grooves are manufactured with very high precision. Groove depths and pitches of few micrometers (μm) are often

---

[1] ON/OFF switches can also be realized with multiple inputs/outputs

demonstrated. Arrayed fiber connectors and ribbons are used with V-grooves, and without them, as inputs/outputs to the switching device. Positioning pins are used for precise alignment of ferrules, plugs, and other fiber-holding blocks. In some implementations, compression springs are introduced to retain one switching positions as default. Permanent magnets may be used to secure *latching* operation[2] of the switch.

A 1x2 fiber switch with electromagnetic actuation was reported in [17]. The switch has a surface area of 3 cm^2, insertion loss of 2.2 dB, crosstalk of -60 dB and 12-milliseconds (msec) switching time.

A polarization-maintaining 1x2 fiber switch is described in [19]. It has 0.5-dB insertion loss, 42-dB return loss and 2-msec switching time. This switch sustained ten-thousand switching cycles (operations) with marginal change in characteristics.

Another 1x2 switch based on electromagnetic actuation is demonstrated in [24]. This switch has a volume of 3.3 cm^3 and typical characteristics of 0.36-dB insertion loss, 49-dB return loss, and a crosstalk of -70 dB. The switching time of the device is 2 msec and its switching power is 9 mW. The device exhibits latching operation and is tested under various environmental and mechanical conditions. A loss change of only 0.1 dB is reported after 10^5 switching cycles.

Work in [27] has the objective of developing a high-performance electromagnetic-actuation based OM switch that is suited for mass production. Two linked 1x2 switches are fabricated in a package the volume of which is less than 3.7 cm^3. Insertion loss ranges from 0.27 to 0.81 dB, and the cause of this relatively large variation range is not accounted for. However, return loss is 50 dB, switching time is 3.5 msec and high durability is demonstrated after more than 10^7 switching cycles.

While electrical actuation is the most common, thermal actuation of fiber switches has also been deployed ([18], [21], [22], [25] and [26]). A 1x2 switch has been reported in this respect. In this switch, a V-groove fiber clamp is used to accurately align, and fix, the input fiber against one of two outputs. The application of heat on one arm of the input-fiber-bearing U-shaped cantilever results in asymmetric longitudinal expansion of this arm causing angular movement of the fiber end. This fiber bending, in turn, results in switching of light between the two outputs. Other thermal design(s) were shown to be possible [26]. Thermal energy is also used to actuate the fiber clamp itself. The switch operation requires thermal power of about 1 Watt, but it is argued that this energy is only needed during switching time (the switch is latching), which is less than 400 msec. The

[2] A switch is said to be latching if it requires no power to maintain a configuration. Power is only required for transition from one switching state to another, but not to maintain a state. A latching switch may also be referred to as self sustaining.

device has a surface area of less than 1.3 cm². Insertion loss can be as low as 0.7 dB and crosstalk is less than − 60 dB. No degradation in performance was observed after 15000 switching cycles without hermetic packaging. Larger switches (1x4 and 2x2) were demonstrated along with an array of twelve 1x2 switching devices using the same principle.

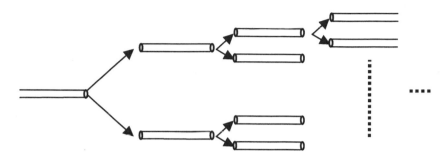

Figure 8-3. Construction of 1xN OM switch using multiple 1x2 units

1xN switches with relatively large N values can be realized by cascading multiple stages of 1x2 switches [6], [24], as shown in Figure 8-3. On the other hand, a one-stage 1x100 switch for remote fiber testing systems was demonstrated in [23]. The switch is relatively compact (15x11x3.5 cm³). It has average insertion loss of 0.26 dB at 1.3-μm wavelength and 0.17 dB at 1.5-μm wavelength. Return loss is better (larger) than 50 dB. Switching time between adjacent ports is less than 0.8 second whereas the worst-case switching time (among ports that are furthest to each other) is about 3 seconds.

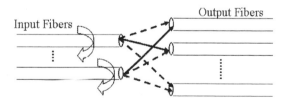

Figure 8- 4. Generic principle of NxM fiber switch

Figure 8-4 demonstrates the fiber-switch concept in the more general and combinational case of multiple input and output fibers/ports, an NxM matrix switch. Implementations are possible with N< M, N> M or N= M. Movable V-grooves, ferrules, and connectors can be used to build the switching system. A 2x2 example based on V-groove arrays with external loop fibers is shown in Figure 8-5. This concept was also demonstrated with ferrules [11].

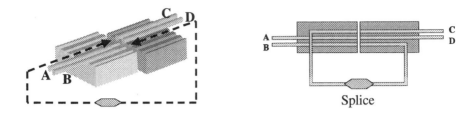

Figure 8-5. A 2x2 fiber switch using V-groove arrays (left side). The interconnection diagram is shown in the right side in the cross-state (A connected to D and B connected to C). By moving one of the two arrays, the bar-state (A-C and B-D) is obtained

A 100x100 non-blocking crossbar switch matrix was implemented based on a combination of movable ferrules and sleeve connectors along with a robot hand to control their motion [13]. The insertion loss in this system was about 1 dB, but switching time was up to 1.3 minutes, which is not suitable for most network applications. The overall system volume was about 0.2 m³.

Large single-stage NxM switches are possible to build from 1xN units, but large numbers of units would be involved (2N for an NxN switch) and large numbers of fiber interconnections would also be required. Therefore, multi-stage switching systems are proposed. A 3-stage 512x512 OM non-blocking *Clos* architecture[3] was demonstrated. This system is based on connectors' motion and utilizes a pair of robot hands to carry out all switching operations, one at a time [14]. The matrix, which was partially implemented, requires nearly 2000 fiber interconnections, but fiber tangling is avoided by using an intelligent zigzag transfer motion. Using single-mode fiber, the mean insertion loss of the switch was 0.97 dB for 200 measurements and the mean return loss was 40.7 dB for 20 measurements. No degradation in loss performance was observed after 500 switching cycles. However, the possibility of abrupt degradation by wear debris is acknowledged. The mean switching time was estimated by simulation to be about 57 seconds. Actual switching time strongly depends on the control algorithm and how it may strike a balance between fast switching and avoiding fiber tangling. The overall system volume was as large as 0.9 m³.

Fiber switch designs may use index-matching fluids wherein the switch is immersed in order to confine optical signals to their intended paths and to reduce reflections. These fluids may also serve to reduce loss [19], [23] and provide some lubrication to ease mechanical motion [7]. However, index matching fluids may not be suitable in thermally actuated switches [25]-[26]

---

[3] Switch fabric architectures are discussed in chapter 9

since heating can alter the characteristics of the fluid. Reflections can be reduced in this case by antireflection coating on the end faces of fiber.

8.1.2 Moving-Component Switches

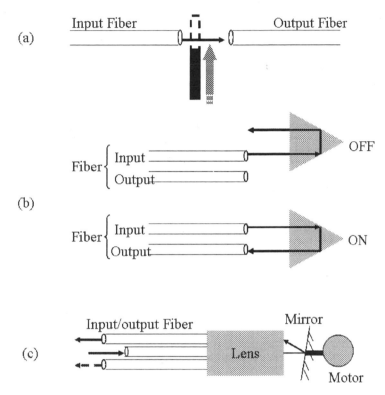

Figure 8-6. Examples of OM switch configurations of the moving optical component type

Figure 8-6 depicts three generic examples of how OM switches of the moving component type may be designed [31]. Figure 8-6(a) is a 1x1 ON/OFF switch where light is passed to output fiber/port (ON-state) or blocked from reaching it (OFF-state) depending on the position of a shutter or a movable opaque object. As with ON/OFF fiber switches, this design principle is applicable to arrays of input/output fibers. The system in Figure 8-6(b) provides the same functionality using a moving prism and Figure 8-6(c) is a 1xN switch based on a lens and moving mirror. A Gradient Index Lens (GRIN) can be used and the mirror can rotate to reflect light from the input onto one of N outputs [4]. For simplicity, the figure assumes the case where N= 2, but larger values of N are indeed possible.

A one-stage multi-mode 1xN switch based on a rotating reflector to scan an image of the input fiber onto a circular array of output fibers is described in [4]. An advanced 1x160 version of this fiber-scanning switch, with two-axis reflector, is also proposed in [20].

In [5], a 4x4 switch was based on movable rhombic glass-block elements. Switching is accomplished by moving the glass block with respect to the lightpath, whereby cross and bar states are configured. In [15], a switching system based on a precision robot mechanism that manipulates (moves) fluid droplets against the signal path is reported.

OM switches based on other movable optical components and more sophisticated systems are proposed as well. A large port-count, 576x576, switching system based on collimators' orientation is demonstrated in [28]. The principle is shown in Figure 8-7 in a simplified manner. It is argued that the ratio between the total number of switch elements and the number of switch ports becomes larger when the number of ports increases for all switch architectures except the beam-steering ones. The problem for beam steering, on the other hand, is to have high isolation among ports and low insertion losses. The switch in [28] is of the beam steering type where the pointing direction of the collimators is controlled by servo motors. The system is based on a unit module of 4 collimators and 8 motors. The lateral spacing between collimators is 19 mm and the distance between input and output collimators is 158 cm. The beam diameter is as large as 3 mm. The frame housing the whole fabric has a volume of about 0.35 m^3. Switching time was claimed to be 100 msec. A measurement setup, two collimators each of which can be actuated for ± 8° of motion with a resolution of 0.0016°, yielded insertion loss of 0.7 dB. The average return loss was 70 dB. Another switch based on a similar operation principle, but using Risley beam deflectors with two wedge-shaped prisms, is also reported [30].

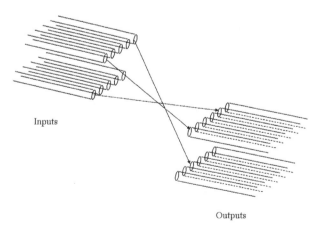

Inputs

Outputs

Figure 8-7. Concept of the beam steering cross-connect discussed in [28]. Input and output fiber ends are coupled to collimators that are capable of steering the light beams. Collimators' motion is driven by servo motors

Optical switching by the technique known as *Frustrated Total Internal Reflection* (FTIR) represent another form of OM switching of the moving component type [32]. As depicted in Figure 8-8, the switch is mainly composed of a prism and a switch plate. The switch plate is made of glass with its back surface (the reflecting surface) positioned at an angle with respect to the front surface (the one that comes in contact with the prism). A Piezoelectric Bimorph Transducer is used to actuate the switch plate. Input and output light are coupled from fiber into the prism (and vice versa) using GRINs.

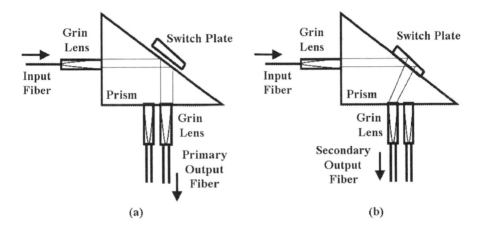

Figure 8-8. Design and principle of operation of a 1x2 FTIR switch

The operation of this switch is simply based on Snell's law. Figure 8-8(a) shows the switch condition in the absence of actuation. This is defined as the open position or the OFF state. In this case, the switch plate is not engaged and the input beam, which is incident with a greater angle than the critical angle, is totally reflected off the prism hypotenuse into a primary output fiber. When actuation is present, Figure 8-8(b), the switch plate makes intimate contact with the prism surface leading to the closed position, or ON state, and to frustrating the previous reflection pattern. The light beam passes into the switch plate and reflection occurs from its back surface onto a secondary output of the switch.

The switch plate only alternates between being off the light evanescent field (OFF state) or in it (ON state). This digital (binary) mode of operation limits the need for sophisticated alignment. In the ON state, electrostatic force retains the intimate prism/plate contact against mechanical shocks and vibrations. The insertion loss of the switch is 1 dB, its return loss is 54 dB, the crosstalk is -55 dB, and the switching time is 2 μsec. It is claimed that multiple switching units of this type can be cascaded to produce larger switches, up to 256x256. This technology is relatively old but re-emerged in the 1990s after progress was made in the techniques to reduce residual contact reflections.

8.1.3 Potential of Opto-Mechanical Switches of the Moving Fiber/Component Types

Opto-mechanical (OM) switches of the moving fiber and moving component types have several advantages. They can be manufactured in large quantities using standard mass production techniques. They have low insertion loss (especially the fiber switch) and high return loss. Their loss performance depends on design, switch size, and fiber type. These switches can operate over a large wavelength range. They are bit rate independent and polarization independent. They can exhibit latching operation, which reduces power consumption and enhances the stability of the switch. They are also stable against many environmental variations. The main drawbacks of these switches are their large switching times and their limited scalability. Fine fiber alignment is also important and can be a limiting factor, but this requirement varies. In the moving fiber case, repeatable fine alignment is critically important while insertion loss is fairly low. Moving component switches on the other hand are easier to align because fibers are fixed, but they have relatively higher loss. Finally, mechanical systems are known to show wear and fatigue if activated for millions of cycles. Although many experimental results have shown impressive performance in this regard, there are certain configurations that are better than others and some may be constrained to applications with a limited rate and/or number of switching cycles.

OM switches are deployed in measuring instruments and are successful in this field. Optical instruments are vital for laboratories and field work (in telecommunications and in other areas). This switching technology can also be used in applications where its performance advantages are desirable while its drawbacks are not influential. As such, they can be deployed in applications where high switching speeds are not required. For example, they can be used to build controllable main distribution frames (MDF) [15]. These frames facilitate easy access, static cross-connection, management, and maintenance of fiber lines. OM switching is also useful in many sensor applications.

The advents of integrated optics, Planar Lightwave Circuits (PLC) and silicon micromachining led to introducing a miniature generation of these OM switches [33]-[38]. The use of micromachining improves the scalability of these switches and makes them more suitable for mass production. It is argued that classical fiber switches require large switching energy due to the mass associated with fiber and its carrier system and that this can be avoided by replacing fiber with the much lighter integrated optical waveguide. Also, the use of planar waveguides avoids the need for expensive three-dimensional high-precision assembly and alignment, thereby reducing cost. Waveguide-based OM switches may be fabricated using bulk or surface micromachining and can be integrated into PLCs or MEMS-based systems.

An example of this miniature approach is discussed in [33] where self-aligning 1x12 and 1x24 optical switches are demonstrated. These switches are based on the translation of silica waveguides past each other. Insertion loss of these switches was as low as 0.2 dB. Another example is reported in [35], where an input fiber is coupled to the fixed end of a movable waveguide. The motion of the waveguide is triggered by applying control voltage between a metallic probe and metal coating on the waveguide itself. Light passes through the guide to its movable end where it can be coupled to an output fiber. The output fiber is positioned such that output power is maximized at zero control voltage. The estimated insertion loss of the device is about 5 dB.

In [34], a two 1x2 micromechanical switching system was demonstrated. In this system, the lightpath is switched by moving two waveguide-bearing cantilevers whose tips are connected by an electromagnetic actuation mechanism. The system volume is about 0.62 cm^3 and it exhibits insertion loss of 3.1 dB, crosstalk of –40 dB and the switching time is 40 msec. It is shown that the increase in loss due to waveguide bending and due to the switching gap is negligible compared to propagation loss of the waveguide. Of the 3.1 dB loss obtained, 2.4 dB is attributed to displacements resulting from fabrication inaccuracy and it is suggested that it is possible to attain a total loss of 1 dB with this switching device.

8.2 OPTO-MECHANICAL SWITCHING: THE MOVING BUBBLE TYPE

This class of switching technologies is based on air-bubble controlled planar-silica-waveguide crosspoints. This method of optical switching was first proposed in early 1990s [39]. However, the two approaches to this technology which are discussed below have not become widely known before the second half of the 1990s. Both approaches are proprietary, with the first proposed in Japan and the second in US.

8.2.1 Thermo-Capillarity Optical Switching

Earlier research leading to thermo-capillarity switching is reported in [15], [40]. This optical switching technology is based on a thermally-induced capillarity effect in a groove, or a slit[4], which is etched at the point of intersection of two planar silica waveguides. The technology is therefore called *thermo-capillary* or *thermo-capillarity* optical switching. Switches based thereon are also referred to as OLIVE (Oil Latching Interfacial-tension Variation Effect) switches [41]-[50].

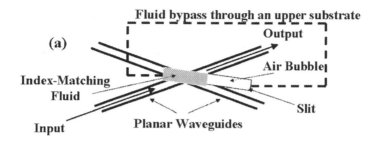

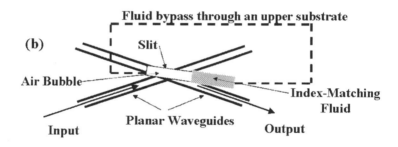

Figure 8-9. A thermo-capillarity crosspoint: (a) the OFF state, (b) the ON State

The principle of operation and the structure of a 1x2 switching device based on this technology are depicted in Figure 8-9. Two waveguides intersect at a point where the slit is etched across the chip and partially filled with index-matching oil. The remainder of the slit volume is filled with an air bubble. When the bubble is pushed apart from the exact point of intersection, incoming light from one waveguide passes through the slit without reflection, as shown in Figure 8-9(a). When the bubble is placed at this point, the lightpath is switched to the other waveguide by *total internal reflection* (TIR) at the silica-air interface on the side wall of the slit, as

---

[4] In the literature of the moving-bubble based technologies, the groove within which the bubble slides is either referred to as a *slit* or as a micro *trench*, depending on which of the two approaches (section 8.2.1 or 8.2.2) is discussed. The term slit is used in the version with a 10-μm wide groove whereas the term trench is used in the other version where the width of the groove is 15-25 μm. The two terms are of course interchangeable, but we maintain this dual terminology to account for groove-width differences, which can be as large as 150%.

shown in Figure 8-9(b). The slit where the bubble slides is sealed by an upper chip (not shown in the figure) and a bypass is etched through this top chip to connect the ends of the slit and to permit circulation of the fluid with bubble movement.

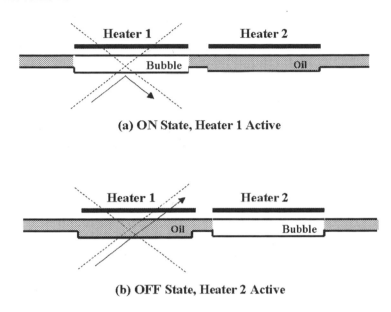

(a) ON State, Heater 1 Active

(b) OFF State, Heater 2 Active

Figure 8-10. Design of the slit in the thermo-capillarity switch

A pair of heaters is used to control the motion of the bubble. Heaters produce thermal gradient along the slit and the bubble is driven to higher temperature (fluid is driven to lower temperature) by virtue of surface tension of the air-fluid interface. Figure 8-10 illustrates the design of the slit and heaters' arrangement. The waveguides intersect at the left side of the figure. In the ON state, heater 1 is active, the bubble is driven to waveguides' side, and input signals on one waveguide are reflected into the crossing waveguide. If heater 2 is activated instead, the gradient in the surface tension/energy drives the bubble away from waveguides thereby configuring the OFF state. In this state, the lightpath propagates straight through the slit without reflection. The switch is self-latching. In order to explain how this works, we consider, say, the transformation from ON state to OFF state. When the left end of the bubble reaches the narrow center of the slit, the bubble is trapped to the right side and heating is no longer needed to maintain this position.

According to the above description, the switch is *bubble driven*. An alternative approach exists where switching is accomplished by controlling the position of the fluid column and this is referred to as the *fluid-column driven* type. Both types are equivalent in principle although the design varies in terms of slit structure and fluid volume.

Thermo-capillarity switching was demonstrated in the form of 1x3 prototypes. The driving power of each heater was 0.1 Watt, and the heating time was about 0.1 second. For the bubble driven type, the through insertion loss was 0.5 dB whereas the reflection insertion loss was 4.3 dB. The crosstalk was -33 dB and -21 dB in the former and latter cases, respectively. In the fluid-column driven type, insertion loss was 2.0 dB and 4.5 dB for the through and reflection cases, respectively. The corresponding crosstalk figures were -28 dB and -14.9. This switch exhibited one year of operation under diverse thermal conditions (5-100°C) without performance degradation.

In order to fabricate larger matrix switches, a technique for encapsulating the index matching fluid into multiple crosspoints simultaneously, and homogeneously, is necessary [46]. Since the slit is typically 10 μm wide and 40-50 μm deep and the length of the fluid droplet is about 100 μm ($\pm$ 10-20%), the volume of fluid to be injected in each crosspoint is in the pico-liter range. A combination of surface tension, the dominant force at this scale, and external pressure is used to control the fluid injection process. For given slit dimensions and fluid surface tension, the capillary force involved can be known. Thus, the fluid volume can be determined by controlling external pressure. Using this technique, a 1x8 switch was fabricated over a 12x12 mm^2 surface area.

A 2x2 thermo-capillarity switch is also reported in [48] and larger *NxN* matrix switches are reported in [47]. A special *NxN* fluid-injection system is developed for filling the slits. A main injection path in the upper chip is shared among every row of crosspoints. The upper chip also hosts small tributary paths to connect each crosspoint to its main path. After excess fluid is removed from all paths, they are sealed with sealing agents. In order to facilitate flexible high-density wiring of $2N^2$ heaters and to maintain a small switch size, multi-layered wiring across the chip and wire sharing are used. The wiring is configured in rows and columns and a heater is activated by applying voltage between a certain row/column pair. The trade-off is that not all optical paths can be switched at the same time. A 16x16 prototype is demonstrated using this design. The waveguide chip size was 23x23 mm^2. At 1.55 μm wavelength, and with single mode input/output, insertion loss was 4 dB and 10 dB for the shortest and longest optical paths, respectively.

Initial measurements of switching time in thermo-capillarity prototypes were not impressive (50-100 msec). Two measures were taken to improve switching time. First, a refractive index-matching fluid with lower viscosity was developed. It is claimed that this was accomplished without compromising the optical characteristics of the fluid and its chemical stability. Second, changes in device design and improvements in fabrication precision made it possible to shorten the length of the fluid-column and reduce its traveling distance along the slit. A 2x2 prototype switch (16x16 mm^2) was used to demonstrate the improvement in switching time by these measures. At room temperature, 6-msec switching time was achieved for 10^6 cycles using 0.15 Watt driving power. At 1.55-μm wavelength, the through-path loss was 0.11 dB and reflection loss was 1.3 dB. Crosstalk was below –60 dB. Detailed analysis of the switching time is carried out in [49]-[50].

Thermo-capillarity optical switching is characterized by low crosstalk, small size, suitability for large-scale integration, and is bi-stable self-latching. Low insertion loss figures are reported in some experiments. It is also claimed that the technology is wavelength tolerant, polarization independent, reliable, and has potential for long-term stability. Design improvements made its switching speed comparable to that of most other optical switching technologies. Reports in literature suggest that this technology was deployed in back-up switching systems in the access network in Japan.

8.2.2   Bubble-Jet Optical Switching

Figure 8-11 describes the principle of operation of the *Bubble-Jet* optical switching technology [51]-[56], which is similar to thermo-capillarity switching in many aspects. The switch utilizing the bubble-jet technology is also known as the Champagne switch [52]-[53]. The switch is based on silica PLC where two arrays of waveguides cross each other to form a matrix of crosspoints. At waveguide intersections, vertical trenches are etched and filled with index-matching fluid. Figure 8-11(a) depicts the default state where no actuation is applied. In this case, input light is passed through the trench(s) into the next collinear waveguide segment on the chip. To switch an input lightpath to a certain output port, thermal actuation takes place at the crosspoint where the input and output paths intersect, as shown in Figure 8-11(b). This actuation results in the formation of a small bubble, which displaces the fluid in the trench. Incident light undergoes TIR into the crossing waveguide as a result of refractive index discontinuity at the sidewall of the trench.

The matrix of trenches on the PLC chip is matched to a corresponding matrix of resistors on a silicon actuator chip (envision the PLC chip inverted top-to-bottom and placed above the actuator chip). The resistors act as heaters and the thermal actuation process is based on the commercially available thermal inkjet technology, used in printers. Vertical holes penetrate the actuator chip on the side of each crosspoint to supply the index-matching fluid to the trenches. The fluid wicks by capillary action through these holes into the trenches from a reservoir underneath the actuator chip. These vertical holes through the actuator chip also serve to accommodate expansion and contraction of the bubble. The PLC chip is hermetically sealed to the actuator chip and the latter is sealed to the fluid reservoir. Formation of the thermal actuators on a separate substrate allows for independent optimization of the etching process for smooth trench sidewalls and optimization of the PLC for minimum insertion loss.

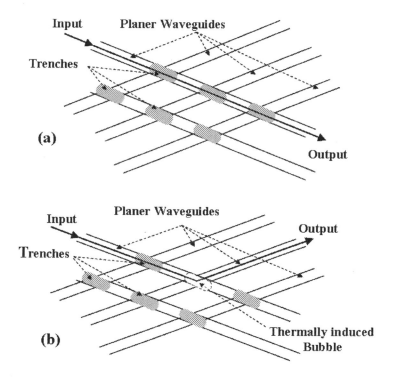

Figure 8-11. The Bubble-Jet PLC switching chip. The figure shows the switching process at one crosspoint: (a) Switching device inactive, (b) Switching device active

While insertion loss through waveguides can be low (0.1 dB/cm estimated), work had to be done to reduce trench loss. Theoretical analysis and experimental measurements suggested that insertion loss in the fluid-filled trench is 0.2 dB and 0.14 dB for 25-µm-wide and 15-µm-wide trenches, respectively. Loss due to a single reflection point was theoretically estimated to be about 0.1 dB. In a 4x4 device (5 mm wide and 8 mm long), loss through four filled trenches was 0.9 dB while reflection insertion loss was 2.9 dB. For 32x32 prototypes, maximum connector-to-connector through-path loss was typically 3.9 dB, while switched path loss was in the 3 to 9 dB range. Some techniques are reported to improve insertion-loss performance in the 32x32 switching system [56]. These include alternative waveguide design and improving the quality of trench walls. About 2.5 dB improvement is said to be possible by incorporating these two techniques. Analysis of loss performance of the Bubble-Jet switch is provided in [53]. The measured switching time of this technology is under 1 msec, crosstalk is −70 dB and polarization dependent loss is 0.1 dB.

Based on several design considerations, the 32x32 fabric was selected as a main building block for optical cross-connect (OXC) systems based on Bubble-Jet technology. It is suggested that a 512x512 Clos architecture can be constructed from 32x32 and 16x32 units. It is shown that add/drop functionalities can be implemented easily using this technology.

The physics associated with the bubble behavior and the requirements of reliable packaging represent some engineering challenges. Reliability of the Bubble-Jet switch was examined in terms of the possibility to change states and maintain them (even after an extended period of inactivity) and in terms of long-term survivability of the materials used (including reservoir and sealing material) and their compatibility with the fluid. Preliminary tests were satisfactory after more than 10^4 switching cycles, and 10^6 actuator firings. Designing for minimum bubble-formation errors is an important target and high-quality hermetic sealing is necessary to secure a targeted lifetime of 20 years.

8.3   HOLOGRAPHIC OPTICAL SWITCHING

Holography is a method for forming optical images. Therefore, a classical comparison with photography may help the reader who is not familiar with holography to recognize its fundamental attributes. We use the analogy made in [57]. Basic photography provides a method of recording the two-dimensional irradiance distribution of an image. Each scene consists of a large number of reflecting and/or radiating points of light. The waves from each of these elementary points, all together, contribute to a complex wave which may be called the *object wave*. Conventional photography collapses this wave into an image of the radiating object. Instead of recording this image of the object, holography provides the means to record the object wave itself. This wave is recorded in such a way that a subsequent illumination of the record serves to reconstruct the original object wave.

Assume a single monochromatic beam of light which has originated from a small optical source. Let this coherent beam be split into two components, one of which is directed toward an object while the other is directed to a photographic recording medium. Let the latter beam component be the reference wave. The former beam is scattered, or diffracted, by the object. The scattered wave constitutes the object wave, which is now allowed to fall on the recording medium. Since the object and reference waves are mutually coherent, they will form a stable interference pattern when they meet at the recording medium. This pattern is a complex system of fringes, i.e., spatial variations of irradiance which are recorded by the recording medium in detail. This record is called a *hologram*, from which the term *holography* was derived. Now, the hologram consists of a complex distribution of clear and opaque areas corresponding to the recorded interference fringes. If the hologram is illuminated under certain conditions, a duplicate of the original reference wave can be obtained from the light transmitted through the hologram [57].

A hologram can serve as a controllable, or programmable, diffraction grating for switching purposes. Several approaches to Holographic optical switching have been reported in literature [58]-[65]. Usually, switches of this kind are capable of multi-dimensional steering of light using electrical control/activation of a holographic system or a photorefractive grating. There are several demonstrations of this method and different technologies are involved. In the following, we look at some arbitrary examples.

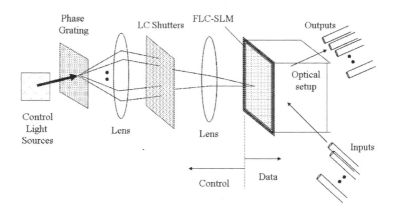

Figure 8-12. Simplified diagram of the switch in [58]: input light is diffracted by the hologram and the optical setup and directed to certain output

Optical switches are demonstrated based on the use of *ferroelectric-liquid-crystal* (FLC) *spatial light modulators* (SLMs) as hologram recording media[5]. Figure 8-12 shows a system where a FLC/SLM is placed at the interface between an optical switch and its optical control system [60]. The experiment therefore aims also at demonstrating the feasibility of an all-

---

[5] Note that liquid crystals are used here in a different manner than that of chapter 5 where switching was based on polarization control or guided-wave refractive-index control. Liquid crystals have numerous applications in optical switching [64] and in optical communications in general [66].

optical switch. The switch has a control light source per every input port. In order to interconnect any input/output pair, a designated control light beam is generated and split by a phase grating into several beams. Two beams are selected out of the split and steered to fall onto the FLC/SLM while a voltage is applied on it. A hologram that is specific to the interference pattern of these two control beams is written on the FLC/SLM and is retained after both the control beams and the voltage are turned off. On the data side, an optical setup guides the lightpath to incident on the FLC/SLM where it is reflected by the hologram toward the desired output position. Every output position corresponds to certain selection of control beams and a related hologram. Switching between multiple input/output pairs is accomplished via time sharing. A 4x1204 system was demonstrated at 670-nm wavelength, but the insertion loss of the system was fairly high (28-36 dB).

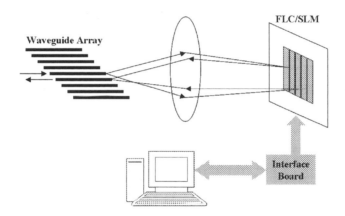

Figure 8-13. Holographic 1x8 optical switch, part of the Roses project, after [63]

Another FLC/SLM-based 1x8 holographic switch was developed for sensor polling and demonstrated in [63]. Unlike the previous example, the control mechanism in this switch is electronic. The switch is shown in Figure 8-13. A waveguide array comprises the input/outputs lineup and a Fourier lens is used to collimate the input beam onto the reflective FLC/SLM. The latter acts as a diffraction-grating beam-steering element where the diffraction angle is dictated by the hologram downloaded from a personal computer. By diffraction angle control, light is directed to the designated output. Measures are taken to discard the undesirable components of reflected light. Tight opto-mechanics are used to precisely align and control the relative position of all parts and no dynamic alignment is required. Insertion loss and crosstalk figures are 16.9 and -19.1 dB, respectively [63]. Preliminary demonstration of a two-FLC/SLM 3x3 switching system based on the same principle with trans-missive SLM was also reported. The insertion loss in this system was 19.5 dB and the crosstalk was -35.5 dB [63]. Both the 1x8 and 3x3 demonstrations operate at the wavelength of 1550 nm. Reflective SLM however are more favorable than transmissive SLM because of the difficulty to address individual pixels on the latter.

A holographic optical switch based on the use of the voltage-controlled photo-refractive effect in Potassium Lithium Tantalate Niobate (KLTN) crystals, doped with Copper and Vanadium, is reported in [61]. The illumination of a KLTN crystal with the interference pattern of two mutually coherent beams results in photo-ionizing the charge carriers. If an electrical field is applied across the crystal, a space charge field is induced which modulates the refractive index of the crystal. This results in an index grating (a hologram) that can be exploited to provide optical switching functionality. The refractive index profile is correlated with the illumination pattern. In a 2x2 cross-bar switch based on this technology, lightpaths propagate through the crystal unaffected in the absence of an electric field (the bar state). When the field is present, the hologram is activated leading to diffracting input beams to the other outputs (the cross state). The crosstalk performance of the switching device is sensitive to polarization and environmental instabilities. This performance is improved, at the expense of extra loss, by placing more crystals, instead of just one, in the signal path.

A 4x4 KLTN-based switch was demonstrated where no significant change in diffraction efficiency was observed after 50 hours of continuous illumination (1.3-μm wavelength, 2-mW beam, and dark ended room). The loss was down to 4 dB for small systems and crosstalk was about -30 dB. Based on studies in holographic aging, 10 years of lifetime in the dark was estimated for this switch.

By writing and angularly multiplexing more holograms into each crystal, larger switching systems can be built. Potentially, additional holograms can be added per each KLTN crystal for additional operating wavelengths leading to multi-wavelength switching fabrics and larger switching capacities. Two problems exist however. First, the larger the number of stored holograms gets, the lower the average diffraction efficiency and the higher insertion loss will be. Second, the number of holograms per crystal is limited in practice. Meanwhile, the pass-band of the crystal is defined by the grating stored in it. Typically, this pass-band has a Gaussian shape in case where apodized beams are used to record the hologram. Therefore, pass-band can be a design issue in practice.

Hence, tradeoffs are involved in designing KLTN-based holographic switches. Also, high-voltage power supplies are required for switching voltages (several-hundred volts) and switching efficiency is lowered if the voltage is reduced. Scalability and long-term reliability remain to be verified. These issues constitute open areas for research.

One of the advantages of holographic switching is that it involves no moving parts. The technology has the potential for integrating a number of optical functionalities on the same device, which can improve performance, reduce cost, and enable flexible switching devices [65]. These functionalities include space switching, wavelength selection, wavelength drift compensation, channel monitoring, variable optical attenuation, dynamic reconfiguration, and parallelism (multiple interconnections may be handled simultaneously, which is useful also for multicasting).

Switching time of Holographic switches varies from moderate, in the FLC/SLM based type, to very low, 5-10 nanosecond in the KLTN-crystal based type. Their loss characteristics are not impressive compared to most other optical switching technologies.

8.4 OTHER TECHNOLOGIES

Part II of this book has provided a fairly extensive coverage of optical switching technologies. However, the subject is diverse and there are other approaches that we may have not covered. Some of these may constitute different technologies, but most of them are variations of technologies we have discussed. In the following, we give a few examples.

In section 8.2.1, we discussed thermo-capillarity switching. The reader who is interested in this approach may need to know that electro-capillarity switching was also explored [67].

In chapter 7, we discussed the semiconductor optical amplifier (SOA) as the primary type of optical amplifiers that is used in optical switching applications. In principle, other amplifiers can be used for this purpose too and some work has been reported about using EDFAs in this regard (see [68] for example).

In the same chapter, some switching methods which utilize nonlinear effects have been discussed. One should note that optical switches based on nonlinear fiber directional couplers have been proposed as well. We explore this kind of switches in little more detail.

The Nonlinear Directional Coupler (NLDC) was introduced in the beginning of the 1980s and was first referred to as the Nonlinear Coherent Coupler (NLCC) [69]. There was growing ambition at that time to use optical devices and technologies in data processing and digital logic applications, and NLDC was one of the candidates in this respect. The device utilizes coherent interactions among two optical waveguides that are placed in close proximity. Due to the evanescent field overlap, as discussed in chapters 2-4, these waveguides exchange power periodically. Nonlinearity can be intensified and exploited to modify this exchange of power leading to power-based optical switching functionality [69]-[73]. This phenomenon has been explored for other devices as well [70], [73].

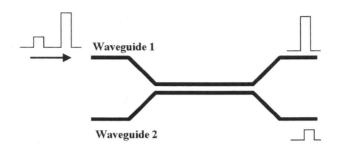

Figure 8-14. Simplified principle of operation of optical switching by the nonlinear directional coupler

Figure 8-14 depicts the principle of this switching technology in a simplified manner. Optical pulses are launched through the input of waveguide 1. Let these pulses have variable optical power (intensity). With proper coupler design, the device acts like a linear coupler for low-power pulses and transfers most of their power to waveguide 2. In the presence of Kerr nonlinearity, high-power pulses are passed through to the output of waveguide 1. The device thereby acts as a pulse router or selector. Signal encoding techniques along with control mechanisms can be utilized to transform this functionality into optical switching. However, the extinction ratio can be poor and the switching process depends on the pulse phase besides its power. The phase varies along the pulse width. If pulses are *solitons*[6] (or soliton likes) high performance switching can be obtained.

NLDC based switching utilizes the instantaneous changes in refractive index due to changing light intensity, which is attributed to the *Kerr effect*. Sub-picosecond (psec) switching time is envisioned by this technology and very high bit-rate operation is possible. Research in this kind of ultra-fast optical processing and switching technologies is important for future optical communication visions, such as ultra-high-bit rate systems (multi Tera-bit/second and above) and optical packet switching. Today, intensity-based nonlinear optical switching is only in the research stage. Interested readers can explore more about this technology by consulting the literature cited in references [69]-[73].

---

[6] Solitons are pulses that propagate without changing their shape [74].

REFERENCES

[1] P. Hale and R. Kompfner, "Mechanical Optical-Fiber Switch," Electron. Lett., Vol. 12, p. 338, 1976.

[2] Y. Ohmori and H. Ogiwara, "Optical Fiber Switch Driven by PZT Bimorph," Appl. Opt., Vol. 17, No. 22 , pp. 3531-2 , 1978

[3] H. Yamamoto and H. Ogiwara, "Moving Optical-Fiber Switch Experiment," Appl. Opt., Vol. 17, No. 22, pp. 3675-8, 1978.

[4] W. Tomlinson, R. Wagner, A. Strnad, and F. Dunn, "Multiposition Optical-Fiber Switch," Electron. Lett., Vol. 15, No. 6, pp. 192-4, 15th March 1979.

[5] Y. Fujii, J. Minowa, T. Aoyama, and K. Doi, "Low-Loss 4x4 Optical Matrix Switch for Fiber-Optic Communications," Electron. Lett., Vol. 15, p. 427, 1979.

[6] W. C. Young and L. Curtis, "Cascaded multipole switches for single-mode and multimode optical fibres," Electron. Lett., Vol. 17, No. 16, pp. 571-3, Aug. 6 1981.

[7] W. C. Young and L. Curtis, "Single-Mode Fiber Switch with Simultaneous Loop-Back Feature", Paper ThC3-1, conference paper, unconfirmed source.

[8] T. Satake, S. Nagasawa, and N. Kashima, "Single-Mode 1x50 Fiber Swich for 10-Fiber Ribbon", Trans. of the Institute of Electronics, Information and Communication Engineers (IEICE of Japan), Vol. E70, No.7, pp. 623-4, July 1987.

[9] S. Nagasawa, H. Furukawa, T. Satake, and N. Kashima, "A New Type of Optical Switch with a Plastic-Molded Ferrule," Trans. of the Institute of Electronics, Information and Communication Engineers (IEICE of Japan), Vol. E 70, No.8, pp. 696-8, Aug. 1987.

[10] K. Hogari and T. Matsumoto, "Electrostatically Driven Fiber-Optic Micromechanical On/Off Switch and its Applications to Subscriber Transmission Systems," IEEE J. Lightwave Technol., Vol. 8, No. 5, pp. 722-7, 1990.

[11] S. Nagasawa, H. Kobayashi, and F. Ashiya, "A Low-Loss Multifiber Mechanical Switch for 1.55 µm Zero-Dispersion Fibers," Trans. of the Institute of Electronics, Information and Communication Engineers (IEICE of Japan), Vol. E 73, p. 1147, 1990.

[12] G. Pesavento, "Optomechanical MxN Fiberoptic Matrix Switch," in Proc. Opt. Tech. for Signal Processing Syst., SPIE, Vol. 1474, pp. 57-61, Orlando, FL, April 1-3, 1991.

[13] T. Katagiri and M. Tachikura, "Cassette-type non-blocking 100x100 Optomechanical Matrix Switch," IEICE Trans. Commun., Vol. E75-B, No. 12, pp. 1373-5, 1992.

[14] M. Tachikura, T. Katagiri, and H. Kobayashi, "Strictly nonblocking 512x512 optical fabric switch based on three-stage Clos network," IEEE Photonics Technol. Lett., Vol. 6, No. 6, pp. 764-766, June 1994.

[15] T. Kanai, A. Nagayama, S-I Inagaki, and K. Sasakura, "Automated Optical Main-Distribution-Frame System," IEEE J. Lightwave Technol., Vol. 12, No. 11, pp. 1986-92, 1994.

[16] S. Nagaoka, "Compact and High-Performance Single-Mode Fiber Switches," in Proc. IOOC, Paper WA1-4, pp. 14-5, Hong Kong, 1996.

[17] K. Sato, M. Horino, T. Akashi, N. Komatsu, and D. Kobayashi, "Mechanical Optical Switch of a Plane Type with Electromagnetic Actuators," Proc of Pacific Rim Conference on Lasers and Electro-Optics 1997, CLEO/Pacific Rim '97, pp. 201-202, July 1997.

[18] M. Hoffmann, P. Kopka, T. Grobß, and E. Voges, "All-Silicon Bistable Micromechanical Fiber Switches," Electron. Lett., Vol. 34, No. 2, pp. 207-8, 22nd January 1998.

[19] S. Nagaoka, "Compact Latching Type PANDA Fiber Switch," IEEE Photonics Technol. Lett., Vol. 10, No. 2, pp. 233-4, February 1998.

[20] J. E. Ford and D. J. DiGiovanni, "1x*N* fiber bundle scanning switch," IEEE Photonic Technol. Lett., Vol. 10, No. 7, pp. 967-9, 1998.

[21] M. Hoffmann, P. Kopka, and E. Voges, "Bistable Micromechanical Fiber-Optic Switches on Silicon," Broadband Optical Networks and Technologies: An Emerging Reality/Optical MEMS/Smart Pixels/Organic Optics and Optoelectronics, the 1998 IEEE/LEOS Summer Topical Meetings, pp. II/31-2, 20-24 July 1998.

[22] P. Kopka, M. Hoffmann, and E. Voges, "Micromechanical Fiber Switch Arrays on Silicon," the 24th European Conference on Optical Communications (ECOC), Vol. 1, pp. 39-40, September 1998.

[23] H. Furukawa, Y. Nomura, and H. Yokosuka, "2x100 Direct Fiber Transfer Opto-Mechanical Switch," part of: D. Faulkner and A. Harmer (Eds.), "Technology and Infrastructure", IOS Press, 1998 © AKM.

[24] S. Nagaoka, "Compact Latching-Type Single-Mode-Fiber Switches Fabricated by a Fiber-Micromachining Technique and their Practical Applications," IEEE J. Select. Topics in Quantum Electron., Vol. 5, No. 1, pp. 36-45, January/February 1999.

[25] M. Hoffmann, P. Kopka, and E. Voges, "All-Silicon Bistable Micromechanical Fiber Switch Based on Advanced Bulk Micromachining," IEEE Journal of Select. Topics in Quantum Electron., Vol. 5, No. 1, pp. 46-51, January/February 1999.

[26] M. Hoffmann, P. Kopka, and E. Voges, "Lensless Latching-Type Fiber Switches Using Silicon Micromachined Actuators," the 2000 Optical Fiber Communications Conference (OFC'2000), Vol. 3, pp. 250-2, 7-10 March 2000.

[27] M. Horino, K. Sato, Y. Hayashi, M. Mita, and T. Nishiyama, "Plane-Type Fiberoptic Switches," Electronic and Communications in Japan, Part 2, Vol. 85, No. 1, pp. 41-9, 2002 (Translated from Denshi Joho Tsushin Gakkai Ronbunshi, Vol. J83-C-1, No. 8, pp. 681-8, August 2000).

[28] B. Lee and R. Capik, "Demonstration of a very low-loss, 576x576 servo-controlled, beam-steering optical switch fabric," Proc. European Conference on Optical Communication, Vol. 4, pp. 95-97, Sept. 2000.

[29] F. Gonte, Y-A Peter, H. Herzig, and R. Dandliker, "Massive Free-Space 1x*N* fiber switch using an adaptive membrane mirror," Proceedings of SPIE, Vol. 4493, pp. 64-70, February 2002.

[30] T. Matsui, F. Oohira, M. Hosogi, and T. Yamamoto, "Free-Space optical switch modules using Risley optical beam deflectors" Proc. of the 2004 IEEE/LEOS international conference on optical MEMS and their applications (Optical MEMS 2004), P-12, pp. 102-3, Takamatsu, Japan, Aug. 22-26, 2004.

[31] N. Kashima, "Passive Optical Components for Optical Fiber Transmission," Ch. 13, Artech House, 1995.

[32] R. H. Laughlin and T. G. Hazelton, "Frustrated total internal reflection an alternative for optical cross-connect architectures," IEEE Lasers and Electro-Optics Society Annual Meeting, LEOS'98, Vol. 2, pp. 171-2, 1-4 Dec. 1998.

[33] H. Presby and C. Narayanan, "Mechanical silica optical switch," Electron. Lett., Vol. 34, No. 5, pp. 484-5, 5th march 1998.

[34] M. Horino, K. Sato, T. Akashi, E. Furukawa, and D. Kobayashi, "Micromechanical Optical Switch Using Planar Lightwave Circuits," Electronics and Communications in Japan, Part 2, Vol. 83, No. 5, pp. 13-20, 2000 (Translated from Denshi Joho Tsushin Gakkai Ronbunshi, Vol. J82-C-1, No. 6, pp. 335-41, June 1999).

[35] T. Eng, S. Sin, S. Kan, and G. Wong, "Micromechanical optical switching with voltage control using SOI Movable Integrated Optical Waveguides," IEEE Photo Tech Lett, Vol. 7, No. 11, pp. 1297-9, 1995.

[36] E. Ollier, P. Labeye, and F. Revol, "Micro-opt Mechanical Switch Integrated on Silicon," Electron. Lett., Vol. 31, No. 23, pp. 2003-5, 1995.

[37] E. Ollier, P. Labeye, and F. Revol, "A Micro-opt-mechanical Switch Integrated on Silicon for Optical Fiber Network," IEEE LEOS 1996 Summer Topical Meetings, pp. 71-2, 5-9 Aug. 1996.

[38] E. Ollier and P. Mottier, "Integrated Electrostatic Micro-Switch for Optical Fiber Networks Driven by Low Voltage," Electron. Lett., Vol. 32, No. 21, pp. 2007-9, 1996.

[39] J. L. Jackel, J. J. Johnson, and W. J. Tomlinson, "Bistable Optical Switching Using Electrochemically Generated Bubbles," Optics Lett., Vol. 15, No. 24, pp. 1470-2, December 15, 1990.

[40] Y. Hanaoka, F. Shimokawa, and Y. Nishida, "Low-loss intersecting grooved waveguides with low Δ for a self-holding optical matrix switch," IEEE Trans. on Components, Packaging, and Manufacturing Technology, Part B: Advanced Packaging, Vol. 18, No. 2, pp. 241-4, May 1995.

[41] M. Makihara, F. Shimokawa, and Y. Nishida, "Self-Holding Waveguide Switch Controlled by Micromechanics," IEICE Trans. Electron., Vol. E80-C, No. 2, pp. 274-9, February 1997.

[42] M. Sato, M. Makihara, F. Shimokawa, and Y. Nishida, "Self-latching waveguide optical switch based on thermo-capillarity," 11th International Conference on Integrated Optics and Optical Fibre Communications and 23rd European Conference on Optical Communications, Vol. 2, pp. 73-6, 22-25 Sept. 1997.

[43] M. Sato, F. Shimokawa, S. Inagaki, and Y. Nishida, "Waveguide optical switch for 8:1 standby system of optical line terminals," the 1998 Optical Fiber Communication Conference and Exhibit (OFC'98), Technical Digest, pp. 194-5, 22-27 Feb. 1998.

[44] T. Ohsaki and K. Kaneko, "Optical cross-connect switches for optical access network," Proc. 24th European Conference on Optical Communication (ECOC '98), Vol. 1, pp. 247-8, Madrid, Spain, 1998.

[45] M. Makihara, N. Sato, F. Shimokawa, and Y. Nishida, "Micromechanical optical switches based on thermocapillary integrated in waveguide substrate," IEEE J. Lightwave Technol., Vol. 17, No. 1, pp. 14-8, Jan. 1999.

[46] H. Tago, M. Sato, and F. Shimokawa, "Multi-element thermo-capillary optical switch and sub-nanoliter oil injection for its fabrication," Twelfth IEEE International Conference on Micro Electro Mechanical Systems, MEMS '99, pp. 418-23, 17-21 Jan. 1999.

[47] M. Makihara, F. Shimokawa, and K. Kaneko, "Strictly non-blocking N×N thermocapillarity optical matrix switch using silica-based waveguide," Optical Fiber Communication Conference (OFC 2000), Vol. 1, pp. 207-9, 7-10 March 2000.

[48] T. Sakata, H. Togo, and F. Shimokawa, "Reflection-type 2x2 optical waveguide switch using the Goos-Hänchen shift effect," Appl. Phys. Lett., Vol. 76, No. 20, pp. 2841-3, May 2000.

[49] T. Sakata, H. Togo, M. Makihara, F. Shimokawa, and K. Kaneko, "Improvement of switching time in a thermocapillarity optical switch," Optical Fiber Communication Conference and Exhibit (OFC 2001), Vol. 3, pp. WX3-1 - WX3-3, 2001.

[50] T. Sakata, H. Togo, M. Makihara, F. Shimokawa, and K. Kaneko, "Improvement of switching time in a thermocapillarity optical switch," IEEE J. Lightwave Technol., Vol. 19, No. 7, pp. 1023-27, July 2001.

[51] J. Fouquet, S. Venkatesh, M. Troll, D. Chen, H. Wong, and P. Barth, "A compact, scalable cross-connect switch using total internal reflection due to thermally-generated bubbles," the 1998 IEEE Lasers and Electro-Optics Society Annual Meeting (LEOS '98), Vol. 2, pp. 169-70, Dec. 1998.

[52] J. Fouquet, S. Venkatesh, M. Troll, D. Chen, S. Schiaffino, and P. Barth "Compact, Scalable Fiber Optic Cross-Connect Switches," 1999 Digest of the LEOS Summer Topical Meetings, Nanostructures and Quantum Dots/WDM Components/VCSELs and Microcavaties/RF Photonics for CATV and HFC Systems, pp. II59-60, 26-30 July 1999.

[53] J. Fouquet "Compact Optical Cross-Connect Switch based on Total Internal Reflection in a Fluid-Containing Planar Lightwave Circuit," Optical Fiber Communication Conference (OFC'2000), Vol. 1, pp. 204-6, 7-10 March 2000.

[54] S. Venkatesh, R.Haven, D. Chen, H. Reynolds, G. Harkins, S. Close, M. Troll, J. Fouquet, D. Schroeder, and P. McGuire, "Recent advances in bubble-actuated cross-connect switches," The 4th Pacific Rim Conference on Lasers and Electro-Optics, CLEO/Pacific Rim 2001,Vol. 1, pp. I414-5, 15-19 July 2001.

[55] D. Chen, S. Close, J. Fouquet, R. Haven, R. Reynolds, S. Schiaffino, D. Schroeder, M. Troll, and S. Venkatesh, "An Optical Cross-Connect Switch Based on Macro-Bubbles," Micro-Electro-Mechanical Systems (MEMS) ASME Conference, 2000.

[56] S. Venkatesh, J. Fouquet, R. Haven, M. DePue, D. Seekola, H. Okano, and H. Uetsuka, "Performance Improvements in Bubble-Actuated Photonic Cross-Connect Switches," The 15th Annual Meeting of the IEEE Lasers and Electro-Optics Society (LEOS 2002), Vol. 1, pp. 39-40, 10-14 Nov. 2002.

[57] H. M. Smith, "Principles of Holography," Wiley-Interscience, 1969

[58] H. Yamazaki and M. Yamaguchi, "Experiments on Multi-Channel Holographic Optical Switch with the Use of a Liquid-Crystal Display," Opt. Lett., Vol 17, No. 17, pp. 1228-30, September 1992

[59] D. O'Brien, R. Mears, T. Wilkinson, and W. Crossland, "Dynamic Holographic Interconnects that Use Ferroelectric Liquid Crystal Spatial Light Modulators," Appl. Opt., Vol. 33, No. 14, pp. 2795-803, May 1994.

[60] H. Yamazaki, T. Matsunaga, S. Fukushima, and T. Kurokawa, "Large-scale holographic switch with a ferroelectric liquid-crystal spatial light modulator," Proceedings of the IEEE Lasers and Electro-Optics Society 10th Annual Meeting, (LEOS'97), Vol. 1, pp. 128-9, 10-13 Nov. 1997.

[61] B. Pesach, G. Bartal, E. Refaeli, A. Agranat, J. Krupnik, and D. Sadot, "Free-Space Optical Cross-Connect Switch by Use of Electroholography," Appl. Opt., Vol. 39, No. 5, 10 February 2000.

[62] N. Wolffer, B. Vinouze, R. Lever, and P. Gravey, "8x8 Holographic Liquid Crystal Switch," Proc. 26th European Conference on Optical Communication (ECOC 2000), Vol. 3, pp. 275-6, Berlin, Germany, September 2000

[63] W. Crossland, I. Manolis, M. Redmond, K. Tan, T. Wilkinson, M. Holmes, T. Parker, H. Chu, J. Croucher, V. Handerek, S. Warr, B. Robertson, I. Bonas, R. Franklin, C. Stace, H. White, R. Woolley, and G. Henshall, "Holographic optical switching: the "ROSES" demonstrator," IEEE J. Lightwave Technol., Vol. 18, No. 12, pp. 1845-54, Dec 2000.

[64] A. d'Alessandro and R. Asquini, "Liquid crystal devices for photonic switching applications: state of the art and future developments," Mol. Cryst. Liq. Cryst., Vol. 398, pp. 207-21, 2003.

[65] B. Fracasso, J. de la Tocnaye, M. Razzak, and C. Uche, "Design and performance of a versatile holographic liquid-crystal wavelength-selective optical switch," IEEE J. Lightwave Technol., Vol. 21, No. 10, pp. 2405-11, Oct. 2003.

[66] K. Chang (Editor), "Handbook of Microwave and Optical Components," Vol. 4, Ch. 8 (U. Efron, Liquid Crystals: Materials, Devices, and Applications), pp. 357-470, Wiley-Interscience, 2001

[67] M. Sato, "Electrocapillarity Optical Switch," IEICE Trans. Commun., Vol. E77-B, No. 2, pp. 197-203, February 1994.

[68] M. N. Islam, L. Rahman, and J. R. Simpson, "Special erbium fiber amplifiers for short pulse switching, lasers, and propagation," IEEE J. Lightwave Technol., Vol. 12, No. 11, pp. 1952-62, Nov. 1994.

[69] S. M. Jensen, "The Nonlinear Coherent Coupler," IEEE J. Quantum Electron., Vol. QE-18, No. 10, pp. 1580-3, October 1982.

[70] D. R. Rowland, "All-Optical Devices Using Nonlinear Fiber Couplers," IEEE J. Lightwave Technol., Vol. 9, No. 9, pp. 1074-82, September 1991.

[71] M. J. Potasek and T. Campbell, "All-Optical Switch Using Multi-Terabit/sec pulses," Summaries of papers presented at the Conference on Lasers and Electro-Optics, 2001, CLEO'01, Technical Digest, pp. 180-1, 6-11 May 2001.

[72] M. J. Potasek, "Multiterabit-per-second All-Optical Switching in a Nonlinear Directional Coupler," IEEE J. Select. Topics in Quantum Electron., Vol. 8, No. 3, pp. 714-21, May/June 2002.

[73] M. J. Potasek, "All-Optical Switching for High-Bandwidth Optical Networks," Optical Network Magazine, Vol. 3, No. 6, pp. 30-43, November/December 2002.

[74] G. P. Agrawal, "Fiber-Optic Communication Systems," Wiley-Interscience, Third Edition, 2002.

PART III
OPTICAL SWITCHING FABRICS, SYSTEMS, AND NETWORKS

Chapter 9

PLANAR OPTICAL SPACE SWITCH ARCHITECTURES

David K. Hunter

Optical space switches are important subsystems in optical cross-connects (OXCs) and reconfigurable optical add/drop multiplexers (R-OADM), while also having applications in more long-term communications scenarios, such as optical packet switching. This chapter develops the theory of operation of planar optical space switches, and illustrates the inter-relationship that exists between device characteristics, choice of architecture and space switch performance. The chapter also describes how switching elements from a variety of technologies can be combined into larger switch fabrics in different ways.

Recent technology developments such as 3D MEMS, covered in Part II of this book, have radically changed the performance tradeoff landscape for very large, relatively slow OXCs typically having a reconfiguration time in the order of milliseconds. However, the original vision of interconnected waveguide-based switches still has relevance, especially for sub-nanosecond multiplexing and switching, including the emerging field of fast optical packet switching. Appropriate design can yield fabric performance significantly better than the individual switching elements can separately achieve.

An optical space switch has multiple inputs and outputs, and can establish an optical connection between any idle input and any idle output, without involving conversion of the signal into electronic form. Similar systems, intended for purely electronic implementation, have been studied for over 50 years, and much of that work is amenable to optical implementation. Furthermore, the particular characteristics of optical switches (such as loss, crosstalk and problems of integration) have suggested a number of new architectures which are also discussed here.

Besides describing the optical space switches and their implementation, the theory of their blocking performance is developed. (For further reading on theoretical results in switching, including additional topics such as Banyan networks and Baseline networks, see Reference [1]). Also, formulae are developed for optical loss, crosstalk and noise, which are important aspects of optical space switch performance. This survey is limited to architectures that have actually been built or fabricated, or are strong candidates for implementation in the future.

9.1   SWITCH MODELS

Before discussing the space switch architectures themselves, the models used to analyze the optical performance of the architectures are introduced.

An $n \times m$ switch has n inputs and m outputs. 1×2 and 2×2 switching devices, which are components of many optical space switches, are often implemented using Lithium Niobate. These are based on interferometric operation under the electro-optic effect, with a switching speed in the order of nanoseconds. Whether based on directional couplers or Mach Zehnder interferometers, as discussed in chapter 2, these devices generally have a periodic response to drive voltage, with optimal crosstalk only at a single operation point. Device and control tolerancing in conjunction with multiple wavelengths require functionality over an operation band, reducing the effective crosstalk performance, while strategies to provide polarization independence further limit performance. Devices with saturating responses can significantly reduce these constraints, though the devices are generally larger, with increased electrode capacitance resulting in slower switching speeds. Emerging electro-optic polymer technology may offer new design tradeoffs unavailable with inorganic materials.

Current injection in transparent semiconductor materials (InP, GaAs and Si) provides an alternative way to switch. Switch fabric performance is dependant upon the basic waveguide variability, with InP being most variable and silicon being least. Switching speeds can again be of the order of nanoseconds, depending on the electrical properties. Fiber-to-waveguide losses are generally high in the absence of effective mode tapers, due to the high NA (numerical aperture) of semiconductor waveguides.

Silica technology is another option, dependent on the thermo-optic effect for its operation, with typical switching speeds of a few milliseconds. The low waveguide losses and larger substrate sizes enable complex switch fabrics to be realized. This technology was discussed in chapter 4.

By contrast to the space switches described above, semiconductor optical amplifiers (SOAs) can provide a gating function with extinction ratios beyond 50 dB, and with a gain to compensate losses in the switch fabric, as discussed in chapter 7.

The first model analyzed here is very much simplified, and is designed to permit a comparison of switch architectures with multiple 2 × 2 devices on one substrate.

The crosstalk in dB of a single 2 × 2 device is represented by X, and its loss (exclusive of any fiber-to-substrate coupling) is L dB. X is always negative. The loss at each fiber/substrate or substrate/fiber interface is W dB. SXR is the signal-to-crosstalk ratio of the whole switching network, while the total insertion loss is A. (In reality, crosstalk terms at the same wavelength as the main signal would produce interferometric noise, which significantly compromises system operation, but the details of this is outside the scope of this chapter. The formulae for signal-to-crosstalk ratio (SXR) developed in this chapter describe the ratio of the desired signal to these interfering terms. The phase noise in the interfering terms is converted to intensity noise when they, and the signal, are converted to electrical form by a square law detector.)

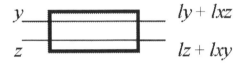

Figure 9-1. Crosstalk and loss model for a directional coupler switch. Multiplication by l indicates that loss has occurred due to the switch propagation losses while multiplication by x indicates that a fraction of a signal has gone the "wrong" way in the switch and is now part of a crosstalk term. This model applies primarily to interferometric switches such as directional couplers, but not to SOAs

Table 9-1 shows a typical set of values for representative material systems; the set of values for silica is used when comparing architectures in Section 9-11. It is assumed that X and L do not vary for different devices in the same optical space switching system. The model used, excluding waveguide-fiber coupling losses, is shown in Figure 9-1 for the so-called "bar state", (where the signal entering the top (bottom) input exits the top (bottom) output), with[1] $x = 10^{X/10}$ and $l = 10^{-L/10}$. The input/output relationship in the "cross-state" (where the signals cross over in the switch) is similar, except that the outputs are changed over.

---

[1] These equations are a consequence of the definition of decibels, for example, $X = 10\log_{10} x$.

	X (crosstalk)	L (device loss)	W (coupling loss)
Lithium Niobate (EO)	–25dB	0.25dB	0.5dB
Gallium Arsenide	–20dB	0.5dB	3dB
Silica (saturating) [2]	–40dB	0.36dB	0.4dB

Table 9-1 Values representing crosstalk and loss for different waveguide-based directional-coupler technologies; used for comparing architectures in section 9-11.

If a signal enters the space switch and passes through k switches, each of which carries one other signal of identical intensity, the resulting SXR is approximately[2]:

$$SXR \approx -X - 10 \log_{10} k \quad \text{dB} \tag{9.1}$$

The attenuation calculations in this chapter only take account of the loss due to the 2×2 switches and fiber/substrate interfaces. No attempt is made to consider additional waveguide losses, or losses and crosstalk in waveguide bends and crossovers, as the complexity introduced would make the calculations intractable. The equations for attenuation should therefore only be taken as a rough guide.

Alternatively, a space switch may be constructed from Semiconductor Optical Amplifiers (SOAs). This may be modeled as a chain of SOAs, since any crosstalk introduced into the fabric is regarded as negligible. The signal-to-noise ratio (SNR) is [3]:

$$SNR = \frac{\left(G_{sig} P_{in} \right)^2}{4 G_{sp} n_{sp} h v_c B_e \left(G_{sig} P_{in} + G_{sp} n_{sp} v_c B_o \right)} \tag{9.2}$$

The following notation was used:

P_{in} = signal input power

n_{sp} = excess spontaneous emission factor

h = Planck's constant

v_c = SOA's central frequency

---

[2] This equation can be derived as follows. sxr is the signal-to-crosstalk expressed as a ratio (i.e. not in decibels) and S is the signal intensity. Then the intensity of a single crosstalk term will be xS and therefore $sxr = S/kxS = 1/kx$. By taking logarithms of both sides and multiplying by 10, one obtains Equation 9-1. When deriving this equation, loss is assumed to be equal for all paths through the network, making the loss for signal and crosstalk terms cancel out. Loss can therefore be assumed to be any value without affecting the result, and may hence be set to zero.

B_e = receiver electrical noise bandwidth

B_o = bandwidth of an optical filter placed on front of the receiver

The signal gain along the chain is:

$$G_{sig} = \left(L_s L_{in} L_{out} G\right)^M \qquad (9.3)$$

L_s = system loss in each stage

L_{in} = SOA input coupling loss

L_{out} = SOA output coupling loss

G = gain of one SOA (See equation 7.4, chapter 7).

The spontaneous emission gain is:

$$G_{sp} = \left(\frac{G_{sig}^{1/M}}{L_s L_{in}} - L_{out}\right) \frac{1 - G_{sig}}{1 - G_{sig}^{1/M}} \qquad (9.4)$$

In the architectural comparison of Section 9-11, to achieve a 10^{-9} bit error rate (BER), SNR < 144 is required, and the total power leaving the end-face of the last SOA must be less than the nominal saturation power. For further details of this model, see Reference [3].

Having discussed two simple models that may be used to analyze the performance of optical switch fabrics, we now turn our attention to the construction of these fabrics, and the various possible varieties of non-blocking operation.

9.2   CROSSBAR SWITCHES

Originally, crossbar switches were electromechanical devices made up of orthogonal crossbars, with crosspoint switching devices at the junctions. This formed a rectangular array of crosspoints (Figure 9-2), each of which could connect one of the N input lines to one of the M output lines. This is referred to as an $N \times M$ switch, and is drawn as a box with N inputs and M outputs.

Crossbar switches may be implemented optically (Figure 9-3). The diagram shows a 4 × 4 matrix constructed from 16 2 × 2 switches. In general, an $N \times N$ matrix uses N^2 switches, one for each crosspoint. (In the literature on optical switching, the term crosspoint is often used to mean a 2 × 2 switch, regardless of the architecture). The control algorithm is very simple; when no calls are set up, all the switches are set to the cross-state. To connect an idle input to an idle output, the switch that is connected to both is put into the bar-state. To disconnect the call, the same switch is put back into the cross-state. For example, to connect input 2 to output 3, switch J would be put into the bar-state.

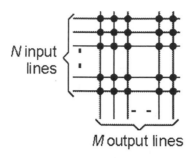

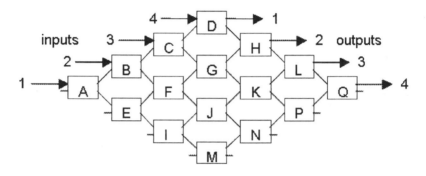

Figure 9-2. An N × M crossbar switch.

Figure 9-3. A 4 × 4 optical crossbar switch.

From Figure 9-3, one can see that the minimum path length through the switch matrix is one switch (input 4 to output 1), and the maximum is $2N - 1$ switches (input 1 to output 4). So the minimum value of insertion loss A is $L + 2W$ and the maximum is $L(2N - 1) + 2W$. While this does mean that the attenuation through the matrix is not constant, the variation is not so marked as it appears from these rough calculations. This is because the attenuation is equalized to some extent by the varying lengths of waveguide required to reach the matrix itself from the edge of the substrate.

In Figure 9-3, the worst possible signal-to-crosstalk ratio occurs when a signal enters on input 1 and leaves on output 1. This is because the maximum number of other signals (i.e. 3) can pass through the switches B, C, and D, which are used by the signal in question. Neglecting loss in the switches:

$$SXR \approx -X - 10\log_{10} 3 \qquad (9.5)$$

and for general N,

$$SXR \approx -X - 10\log_{10}(N - 1) \qquad (9.6)$$

A more accurate formula, which considers the switch attenuation, is [4]:

$$SXR = -X - NL - 10\log_{10}\left(\frac{1-10^{\frac{-(N-1)L}{10}}}{1-10^{\frac{-L}{10}}}\right) \tag{9.7}$$

As $L \to 0$, it reduces to the previous equation; for a comparison of L and W, refer to Table 9-1.

There were several early implementations of the crossbar architecture using the electro-optic effect in lithium niobate, for example an 8×8 device [5]. Directional coupler switch devices are very much longer than they are wide. In addition, there are limits on the minimum possible waveguide bend radius that yields an acceptable loss. For these two reasons, the size of crossbar that can be implemented on one substrate is limited.

An OXC based on "delivery and coupling" switches has been proposed [6]. This switch is essentially a thermo-optic crossbar switch which has the facility to combine many inputs on different wavelengths and send them to one output. The feasibility of 8×16 switch boards for providing 320 Gb/s throughput has been confirmed [7]. An architecture derived from the crossbar, made from 2×2 devices, has been proposed, having an equal insertion loss for all paths [8]. A 16×16 switch of this type, with integrated drive electronics, has been fabricated on one silica substrate. Each 2×2 switch also operated on the thermo-optic effect. The worst case extinction ratio and loss were 50 dB and 6.7 dB respectively [9].

Routing of 10 Gb/s packets through a compact 4×4 InGaAsP/InP vertical coupler SOA crossbar fabric has been demonstrated [10], [11]. The guard band between packets was as short as 2 ns and the adjacent packet suppression ratio was −50 dB.

Crossbars may also be implemented in free space with 2D MEMS (for example, a 16×16 device has been fabricated [12].) The switching speed is a few milliseconds, satisfactory for OXC applications, however the minimum and maximum insertion loss may vary by as much as 5 dB due to variable collimation geometry [12].

The 32×32 "Champagne" switch [13], discussed in chapter 8, is also a crossbar switch. It is based upon total internal reflection from the sidewalls of trenches etched in the crosspoints of a silica Planar Lightwave Circuit (PLC) matrix. Normally, liquid is present at a crosspoint, and the light is transmitted straight through. When a small bubble displaces the liquid, light is reflected at the crosspoint.

9.3   MATRIX VECTOR MULTIPLIERS

In *matrix vector multipliers* (MVM) [3], or broadcast and select switches, each input is connected to every output by means of couplers and an SOA gate that can make or break the connection. There are two variants (Figure

9-4 and Figure 9-5) – the lumped-gain version, which has only one stage of amplification besides the gates themselves, and the distributed gain version, which has several stages of gain in addition to the gates. Besides any SOAs used purely for amplification, there are N^2 SOAs functioning as on-off gates, where N is the number of inputs and outputs. At most N of these on-off gates are powered at any one time.

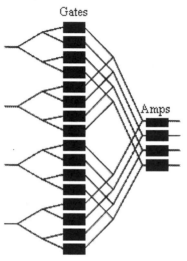

Figure 9-4. An example of a 4 × 4 matrix vector multiplier (MVM) switch with lumped gain. The black rectangles represent SOA gates, which can either be "gates" or amplifiers ("amps").

4 × 4 monolithic matrix vector multiplier switches have been fabricated in InGaAsP/InP [14]. This particular example has three stages – a stage of amplification at both the inputs and outputs (4 + 4 SOAs), and a stage of 16 SOAs to implement switching itself.

9.4   TREE-BASED ARCHITECTURES

Tree architectures [15] are constructed from tree-structured splitters and combiners, hence their name. They exhibit excellent crosstalk performance, but at the expense of using more devices than many other architectures. On the other hand, for N inputs and outputs, a crossbar, as described in Section 9-2, needs $2N - 1$ stages of switch devices, while a tree switch needs only $2\log_2 N$ stages. For a directional coupler implementation, this allows longer devices (having a longer interaction length) and hence lower operating voltages.

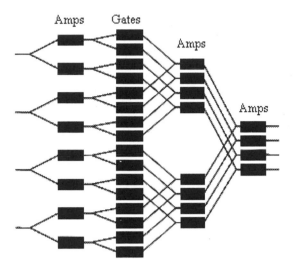

Figure 9-5. An example of a 4 × 4 matrix vector multiplier (MVM) switch with distributed gain. The black rectangles represent SOA gates, which can either be "gates" or amplifiers ("amps").

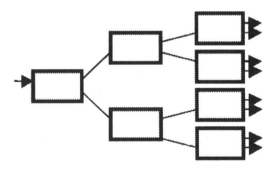

Figure 9-6. A 1 × 8 active splitter.

An "active" 1 × N splitter (i.e. a switched demultiplexer) may be constructed from 2 × 2 switches; a 1 × 8 example is shown in Figure 9-6. One input on each 2 × 2 switch is left unused. An active combiner (i.e. a switched multiplexer) is simply an active splitter in reverse. Only $\log_2 N$ control signals are required to drive such an active splitter or combiner. This is because, in each stage, only one of the switches is used at once and hence they can all share one control signal.

Corresponding passive splitters and combiners are made from passive fiber couplers, or they may be fabricated with waveguide splitters. The symbols representing splitters and combiners are shown in Figure 9-7. Using a passive splitting or combination stage reduces control complexity but increases loss by a factor of N. In a tree architecture, passive splitting and active combining offers a broadcast function.

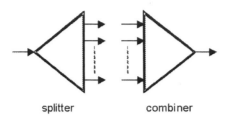

splitter   combiner

Figure 9-7. Splitter and combiner symbols.

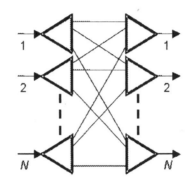

Figure 9-8. The tree architecture.

The *tree architecture* itself is shown in Figure 9-8. If active splitters and active combiners are employed, the architecture contains $2N(N-1)$ switches; nearly twice as many as the crossbar switch. The SXR is greatly improved over the crossbar switch, since for a signal to find its way spuriously from an input to a wrong output, it must go to a wrong switch device output, and experience an attenuation of X, at least twice. In fact, the worst-case signal-to-crosstalk ratio is:

$$SXR \approx -2X - 10\log_{10}(\log_2 N) \qquad (9.8)$$

If the active splitters and combiners are fabricated on separate substrates, the loss is:

$$A \approx 2L\log_2 N + 4W \qquad (9.9)$$

If everything is fabricated on one substrate, this becomes:

$$A \approx 2L\log_2 N + 2W \qquad (9.10)$$

Clearly, providing a broadcast function will give poorer crosstalk performance.

Polarization independent 4×4 implementations of this architecture on single substrates have been reported in lithium niobate, facilitated by the electro-optic effect [16], [17]. A 16×16 lithium niobate switch fabric of this type, using multiple substrates with passive splitters and active combiners, was fabricated to act as the center stage of a time-division switching system [18].

A 16×16 polymeric device with a tree-based architecture has been fabricated on one substrate, employing the thermo-optic effect [19]. It consists of 480 1×2 switches, with 704 S-bends and 227 waveguide intersections. A 1×128 multiplexer/demultiplexer on a single silicon substrate has been fabricated, employing the thermo-optic effect [20]. 256 such substrates could be interconnected by fiber to form a 128×128 switch.

9.5   SWITCH FABRICS

In the context of optical space switches, a *switching element* is either a 2×2 device or one of the architectures described above; these are connected together to form larger switches called *switch fabrics* (or just *fabrics*). Switch fabrics are organized into *stages*, each consisting of a column of switching elements. In every stage, except for the last one, each switching element output is connected to a switching element input on the next stage. The set of connections between adjacent stages is known as an *interconnect*. The inputs of the first stage switching elements form the input terminals of the switch fabric, and the outputs of the final stage switching elements form the output terminals. All the switch fabrics that are considered here have the same number of inputs as outputs. An example of a 3-stage switch fabric is shown in Figure 9-9; stages 1 and 3 are made from 2×2 switching elements and stage 2 is made from 3×3 switching elements. It will be assumed throughout the remainder of this chapter that signals travel from left to right through a switch fabric.

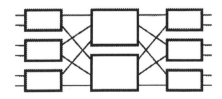

Figure 9-9. An example of a 3-stage switch fabric.

A *call* in a switch fabric is a connection between an input terminal and an output terminal. The switch fabric has a controller associated with it that must be able to accommodate new calls as they arrive, and disconnect old calls once they finish. An *assignment* is a set of calls that are in progress, where each input or output terminal can carry at most one call. An input or output terminal is *free* if it does not carry a call. A *maximal assignment* is an assignment where no input (or output) terminals are free.

A switch fabric is said to be *blocking* if there are one or more assignments that it cannot realize. This is equivalent to saying that it is not always possible to set up a call between a pair of free input and output terminals. Switch fabrics that are not blocking are said to be *nonblocking*; they may be categorized into three types.

In a *strict-sense nonblocking* fabric, there will always be at least one free route through the fabric for a new call, where the call may be set up without rerouting existing calls. Any free route may be used without blocking ever taking place. *Wide-sense nonblocking* fabrics are similar, but some rule must be used to decide what route to choose when activating a new call, otherwise blocking may occur later. In a *rearrangeably nonblocking* fabric, new calls can always be accommodated, but it may be necessary to reroute existing calls through the fabric to do this. The resulting break in service is, of course, unacceptable in an OXC.

This chapter is concerned with strict-sense and rearrangeably nonblocking fabrics. Comparatively little is known about wide-sense nonblocking fabrics, and they will not be considered here to any extent.

9.6   CLOS NETWORKS

Many switch fabrics are based on a type of 3-stage fabric that was first studied by Clos, and are generally referred to as *Clos networks* [21]. The first stage consists of r $n \times m$ switching elements, the second stage m $r \times r$ switching elements, and the third stage r $m \times n$ switching elements (Figure 9-10). Each switch in stages 1 and 2 has exactly one connection to each switching element in the next stage.

If $m \geq 2n - 1$, a Clos network is strict-sense nonblocking [21]. To prove this, it is sufficient to show that, irrespective of how the existing calls are set up, there is always a free path through the network for a new call. Suppose a new call is to be set up between an input on a first stage switching element A and an output on an output switching element B. The worst case is shown in Figure 9-11; $n - 1$ calls will be already carried through switch A, so $n - 1$ center stage switches are used up. Also, $n - 1$ calls pass through B, each of them using up a further center stage switch. Since it is the worst case, none of the center stage switches used by calls in A are used by calls in B. Thus, $2n - 2$ center stage switches are used in the center stage. To provide a free path from A to B, a further center stage switch must be provided, so there must be at least $2n - 2 + 1 = 2n - 1$ center stage switches.

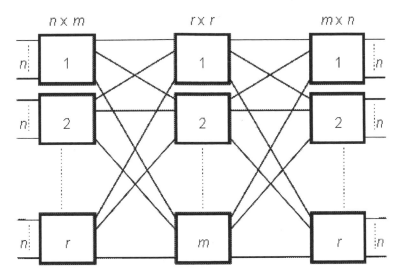

Figure 9-10. A general 3-stage Clos network.

In the days of early electromechanical systems, the attraction in using this switch architecture, rather than one large crossbar switch, was in the reduction in the number of crosspoints, since this was then a good indication of cost. At the present day, for electronic implementations, this cost reduction benefit has become less pronounced, particularly with the advent of VLSI (Very Large Scale Integration). However, in guided wave optical space switches, the number of switching devices is generally still a good rough estimator of cost.

For large systems, a further reduction in crosspoint count can be made by replacing each center stage switch by another 3-stage Clos network, or subnetwork. The resulting network would have five stages. By repeatedly applying this procedure, fabrics with 7, 9, and 11... stages may be produced. As the number of inputs and outputs becomes larger, the number of stages required to obtain an optimal crosspoint saving increases.

In a $2t+1$ stage network, with $n = {}^{t+1}\!\sqrt{N}$ throughout, the number of crosspoints $C(2t+1)$ is given by [21]:

$$C(1) = N^2$$
$$C(3) = 6N^{3/2} - 3N$$
$$C(5) = 16N^{4/3} - 14N + 3N^{2/3}$$
$$C(7) = 36N^{5/4} - 46N + 20N^{3/4} - 3N^{1/2} \qquad (9.11)$$
$$C(9) = 76N^{6/5} - 130N + 86N^{4/5} - 26N^{3/5} + 3N^{2/5}$$
$$C(2t+1) = \frac{n^2(2n-1)}{n-1}\left[(5n-3)(2n-1)^{t-1} - 2n^t\right]$$

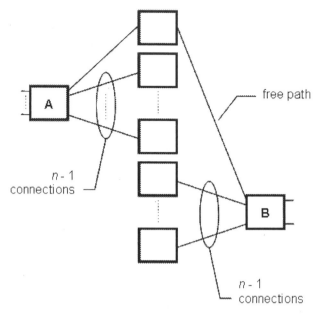

Figure 9-11. The worst-case number of center stage switching elements required in a 3-stage
Clos network: $m = 2n - 1$.

It can be shown that this is not the minimum number of crosspoints; for example, for a three-stage network with large N, the optimal value of n is approximately [21]:

$$n \cong \sqrt{\frac{N}{2}} \tag{9.12}$$

yielding approximately the following number of crosspoints

$$C(3) \cong 4\sqrt{2}N^{3/2} - 4N \tag{9.13}$$

It can be shown that for a 3-stage Clos network to be wide-sense nonblocking, $m \geq \lfloor 2n - n/r \rfloor$ [22], where $\lfloor x \rfloor$ is the largest integer less than or equal to x. This result does not specify how a center stage switching element is chosen for each new call i.e. it is the best result possible. If $r = 2$, it holds if a center stage switch which is already in use must, if possible, be assigned to each new call [23]. For large r or small n the saving over strict sense nonblocking is small or non-existent e.g. if $n = 2$ (2×3 switches are used), there is no saving.

To set up a path in such a strict-sense nonblocking fabric, the controller must search iteratively and systematically through possible routes through the network until a free route is found.

A design for a 128 × 128 optical switching system has been proposed, implementing a 5-stage strict-sense nonblocking Clos network [24]. The system uses lithium niobate electro-optic switching elements, which are based on a variant of the tree architecture, with SOAs to compensate for system losses. The system was not built in its entirety, but experiments were carried out to demonstrate its viability. Several important studies have taken place on the viability of various OXC architectures derived from Clos networks [25], [26].

Moreover, Clos networks have been proposed as a means for interconnecting many small 2D MEMS crossbars to make a large switch fabric [12].

9.7   THE SLEPIAN-DUGUID THEOREM

The Slepian-Duguid theorem [23] shows that a Clos network is rearrangeably nonblocking if $m \geq n$. To prove this, it is sufficient to consider a maximal assignment with $m = n$. The theorem does not provide an algorithm for connecting an assignment of inputs onto outputs; it merely states that any such assignment is possible.

This result relies on a combinatorial theorem due to Hall [27], which is often called "Hall's marriage theorem" for reasons that will become clear. It can be stated informally like this. Suppose there are a certain number of boys and the same number of girls, and it is necessary to pair each boy off with a girl that he knows in order to be married. Hall proved that a sufficient and necessary condition for this is that if any group of boys (of size say k) is selected, the number of girls they know between them must be at least k. It is clear that this condition is necessary, but the fact that it is sufficient for the pairing to take place makes the theorem both elegant and surprising.

As an example with $k = 3$, suppose there is a group of 20 boys and 20 girls, and three boys are picked: Brian, Jim and Geoff. Suppose that Brian knows Elaine, Aileen and Jennifer, Jim knows only Elaine and Jennifer and Geoff knows Janice and Elaine. Between them, the three boys know four girls (Elaine, Aileen, Jennifer and Janice), at least as many as the number of boys being considered (three). If all other possible subsets of the 20 boys in the group (there are $2^{20} - 1$ subsets in total) also fulfill such a condition, then Hall's marriage theorem says they may each be paired off with a girl they know. If this were the case, then apart from the other 17 boys, Geoff might marry Elaine, Brian could marry Aileen and Jim could marry Jennifer.

Now consider the Clos network (Figure 9-10). Each of the r boys can be said to "own" a unique input stage switching element and each girl can be said to own a unique output stage switch. A boy is said to know a girl if a call passing through his switching element also passes through the girl's switching element. Each collection of k boys must know at least k girls, because between them the k boys' own switching elements which handle kn calls (remember a maximal assignment is being considered with n calls in each input-stage switching element). These kn calls cannot be distributed over fewer than k output switching elements.

Hence each boy can be paired off with one girl that he knows, and the r calls that this pairing represents (one call for each boy-girl pair) can be routed through the top center stage switching element. This is because one call comes from each input stage switching element and one call goes to each output stage switching element. Now remove this top center-stage switching element from the switch fabric, removing all the calls passing through it, and reduce all the input and output-stage switching elements to $n-1 \times n-1$. This reduction in size happens because the inputs and outputs that were occupied by the calls passing through the top center-stage switch are no longer required. Again pair off each boy with a girl that he knows, and continue repeatedly as described above until the entire switch fabric is routed.

9.8 BENEŠ NETWORKS

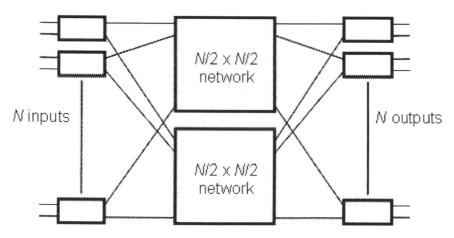

Figure 9-12. Recursive definition of an $N \times N$ Beneš network.

The Slepian-Duguid theorem can be viewed as showing how smaller fabrics, forming the center stage of a Clos network, can be used to construct a Clos network which is itself a larger fabric. *Beneš networks* [23], [28] are created by repeatedly applying the Slepian-Duguid Theorem in this way. This discussion will be confined to such fabrics built from 2×2 switches, since such devices are readily available. Such fabrics are defined in Figure 9-12, this corresponding to a Clos network with $m = n = 2$, which is rearrangeably nonblocking.

We consider N in Figure 9-12 to be an integral power of 2. If $N = 4$, each center stage switch is a single 2×2 directional coupler switch. If $N > 4$, each center stage switch can be replaced by a $N/2 \times N/2$ Beneš network. By repeating this substitution, a rearrangeably nonblocking Beneš network of arbitrary size can be produced. A Beneš network with 16 inputs/outputs is shown in Figure 9-13.

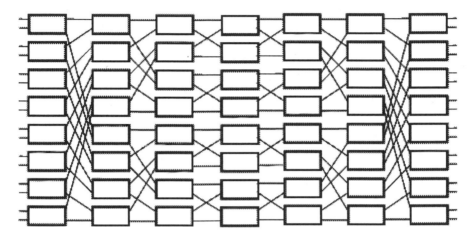

Figure 9-13. A 16×16 Beneš network.

These switch fabrics are organized into $2\log_2 N - 1$ stages of $N/2$ switching elements; the total number of switching elements required is $N\log_2 N - N/2$. For a more general Beneš network made from $b \times b$ switching elements, there are $2\log_b N - 1$ stages of N/b switching elements, with $(N/b)(2\log_b N - 1)$ switching elements in total.

The theoretical attenuation for this type of network is:

$$A \approx (2\log_2 N - 1)L + 2W \qquad (9.14)$$

The SXR is:

$$SXR \approx -X - 10\log_{10}(2\log_2 N - 1) \qquad (9.15)$$

since each signal passes through $2\log_2 N - 1$, 2×2 switches.

Because a Beneš network is rearrangeable, the switch setting may be dramatically changed even if only one new call is set up, which can be as complicated as re-connecting all input-output pairs. The *looping algorithm* is the simplest algorithm to do this [29]. It will be assumed that a maximal assignment is involved, although the algorithm is easily modified if this is not the case.

Let the inputs be denoted by $u_1, \ldots, u_N$ and the outputs by $v_1, \ldots, v_N$. π is a mapping of the inputs to the outputs, representing a maximal assignment, and π^{-1} is its inverse, mapping outputs onto inputs. For any input u_i, the other input sharing the same first-stage switch is denoted by "co u_i", and similarly, "co v_i" is the output sharing the same output switch as v_i. Initially, all the inputs and outputs are unconnected; the algorithm connects them up as follows, where the variables S and T represent various inputs and outputs respectively while the algorithm runs:

1) Select any unconnected u_i and set $S = u_i$. If no such input exists then the algorithm terminates since all the inputs are now connected to an output.
2) Connect S to $\pi(S)$ through the top subnetwork in the center stage.
3) Set $T = \text{co } \pi(S)$.
4) Connect T to $\pi^{-1}(T)$ via the lower subnetwork in the center stage.
5) Set $S = \text{co } \pi^{-1}(T)$.
6) If S has not been connected to an output, go back to step 2. Otherwise, a "loop" has been completed, so go to step 1 to start a new loop.

The algorithm works by traversing the network between the inputs and the outputs, until it gets back to its starting point, thus forming a loop. An example of a loop, with arrows showing how it was created, is shown in Figure 9-14. Table 9-2 and Table 9-3 show the complete assignment that is to be realized. For each input $u_1, \ldots, u_8$, Table 9-2 shows the corresponding output (one of $v_1, \ldots, v_8$) to which it is to be connected. Table 9-3 relates each output to its corresponding input in a similar way. The loop in Figure 9-14 contains four calls, but a loop may vary in size from two to N calls, and there can be from one to $N/2$ loops, depending on the maximal assignment being realized.

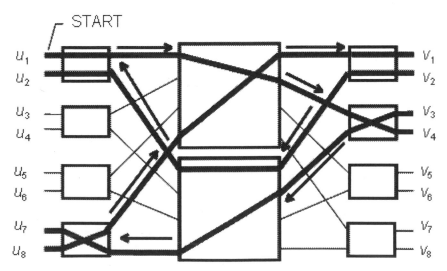

Figure 9-14. One loop in controlling a Beneš network.

u_i	$\pi\left(u_i\right)$
u_1	v_4
u_2	v_2
u_3	v_6
u_4	v_8
u_5	v_5
u_6	v_7
u_7	v_3
u_8	v_1

Table 9-2 Mapping of inputs onto outputs in an 8×8 Beneš network.

v_i	$\pi^{-1}\left(v_i\right)$
v_1	u_8
v_2	u_2
v_3	u_7
v_4	u_1
v_5	u_5
v_6	u_3
v_7	u_6
v_8	u_4

Table 9-3 Mapping of outputs onto inputs in an 8×8 Beneš network.

Calculating the switch settings and subnetwork subassignments in this way takes $O(N)$ time since N calls must be set up sequentially. By applying the algorithm repeatedly for each stage in the Beneš network recursion, routing of a complete fabric (including each subnetwork and their respective subnetworks etc.) can be carried out in $O(N \log N)$ time. The execution time for the algorithm is therefore roughly proportional to $N \log N$. Hence, one of the major disadvantages of a Beneš network compared to other types is the complex control required to accommodate even one new call. It is possible to ameliorate this situation to some extent by employing parallel processors [30], although the algorithm does not adapt to parallel implementation very efficiently. The looping algorithm is discussed further in chapter 10.

Although a 16×16 polarization dependent Beneš network has been fabricated [31] operating by means of the electro-optic effect in lithium niobate, better crosstalk performance can be obtained by dilating the switch architecture using two conjoined chips, as discussed in the next section.

OXC architectures have been proposed, based on acousto-optic tunable filters and wavelength converters that employ difference frequency generation [32], [33]. This is logically equivalent to a Beneš network. A method for constructing OXCs from Fiber Bragg Gratings (FBGs) has been proposed [34] using a Beneš architecture; each FBG is tunable by means of compression applied by a stepper motor.

9.9 DILATED BENEŠ NETWORKS

For switch fabrics built from devices having poor crosstalk performance, the SXR calculated by Equation (9.15) may not be adequate. An obvious way of improving the SXR is simply to use better devices; another way is to dilate the switch fabric. Dilation trades off improved SXR for a greater number of switches [35]. The object of dilation is to modify the switch fabric so that each device only carries one signal at once, drastically reducing crosstalk. Suppose the switch fabric has k stages, numbered left to right from 1 to k; in an 8×8 Beneš network (Figure 9-15), $k = 5$. There are two steps to dilation. First, a four-switch structure is substituted for each switch (Figure 9-16), carrying out the same function as a single switch, but only permitting one signal to pass through any of its four switch devices at once. The resulting switch fabric has $2k$ stages (Figure 9-17).

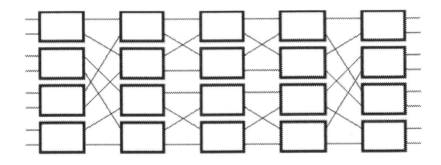

Figure 9-15. An 8 x 8 Beneš Network

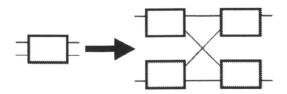

Figure 9-16. First stage in dilating a Beneš network.

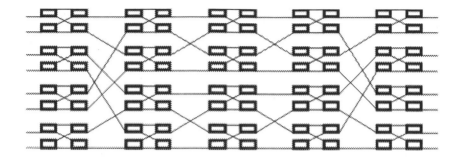

Figure 9- 17. An 8 × 8 Beneš network after the first stage of dilation.

In the second step, each switch device in an even-numbered stage (except the rightmost stage, stage $2k$) is combined with the switch that it is connected to in the following stage, to form one switch. The required transformation is shown in Figure 9-18; the finished switch fabric has $k + 1$ stages (6 stages in Figure 9-19), and has roughly twice as many switch devices as the original switch fabric. The settings for each switch in a dilated fabric can be derived from those for an undilated fabric by using a small number of logic gates for each switching device.

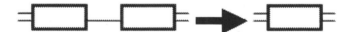

Figure 9-18. The second stage in dilating a Beneš network.

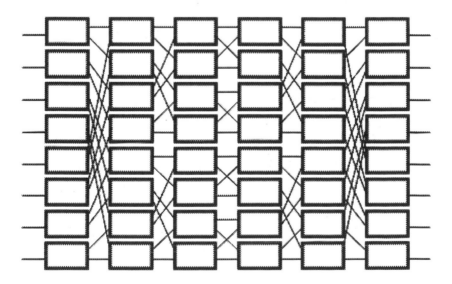

Figure 9-19. An 8 × 8 dilated Beneš network.

The theoretical attenuation for this type of switch fabric is:

$$A \approx 2L\log_2 N + 2W \tag{9.16}$$

In reference [35], the crosstalk is described as "negligible"; it is, however approximately [36]:

$$SXR \approx -2X - 20\log_{10}\left(2\log_2 N - 1\right) + 3 \tag{9.17}$$

This figure is over twice as great as the SXR of a conventional Beneš network.

Dilated Beneš networks as large as 16 × 16 have been constructed using the electro-optic effect in lithium niobate [37]-[38], fabricated on two substrates which were butt-coupled together.

9.10 STACKED SWITCH FABRICS

To improve the performance of a switch fabric, for example, by upgrading it from rearrangeably nonblocking to strict-sense nonblocking operation, multiple copies may be stacked in parallel (Figure 9-20). The simplest example of this is the *Cantor network* [39]. Like Clos networks, these are strictly nonblocking. They have the advantage of being constructed from 2 × 2 switching elements, rather than 2 × 3 or 3 × 5 etc. as for strict-sense nonblocking Clos networks. They are constructed from two half networks each named $L(i)$, where i is an integer related to the size of the network. $L(i)$ may be defined as follows. $L(i)$ is a network of switching elements which has $a(i)$ inputs and $b(i)$ outputs. These quantities depend on i (and the size of the network $L(i)$), and expressions to derive them will shortly be obtained.

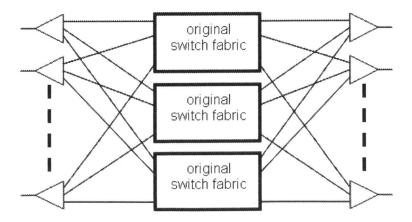

Figure 9-20. A stacked switch fabric, with three copies of the original fabric in this case.

When calculating the number of outputs that can be connected to a free input without disturbing existing calls (i.e. the number of outputs that can be reached from a free input), the worst case involves having all other inputs busy. Hence let $c(i)$ be the number of outputs of $L(i)$ that the free input could be connected to without disturbing the $a(i) - 1$ existing calls. $L(1)$ is defined as a $1 \times M$ active splitter (e,g. Figure 9-6), so $a(1) = 1$, $b(1) = M$ and $c(1) = M$. Networks $L(i)$, where $i > 1$, are defined in terms of $L(i - 1)$ in Figure 9-21. From the diagram, $a(i) = 2a(i - 1)$ and $b(i) = 2b(i - 1)$. Suppose, without loss of generality, that the free input in question is on the lower $L(i - 1)$. Then if all the $a(i - 1)$ inputs on the upper $L(i - 1)$ are disconnected, $2c(i - 1)$ outputs on $L(i)$ can be reached from the free input. If these $a(i - 1)$ inputs are connected up again, in the worst case, $a(i - 1)$ outputs of $L(i)$ which could have been reached otherwise will now become unreachable. Therefore $c(i) = 2c(i - 1) - a(i - 1)$. General solutions are: $a(i) = 2^{i-1}$, $b(i) = M\,2^{i-1}$, $c(i) = M\,2^{i-1} - (i-1)2^{i-2}$. We now have expressions for the size of $L(i)$ and the number of outputs that can be reached from a free input. These results are now used to derive a condition for the Cantor network to be strict-sense nonblocking.

Now suppose that there is another network $L(j)$ and it is reflected about the vertical axis so that it has $b(j)$ inputs, $a(j)$ outputs, and at least $c(j)$ inputs reachable from a free output. Call this network $L^{-1}(j)$. The outputs of the original $L(j)$ are connected to the inputs of $L^{-1}(j)$ so that the two outputs on one final stage switching element in $L(j)$ connect to the same switching element on the input stage of $L^{-1}(j)$. Each of these $b(j)/2$ pairs of switching elements can be replaced by a single 2×2 switching element; these switching elements form the center stage of the switch fabric produced from $L(j)$ and $L^{-1}(j)$. This switch fabric is the completed Cantor network.

At least $c(j - 1)$ of these 2×2 switching elements are reachable from a free input, and $c(j - 1)$ from a free output. To ensure that the switch fabric is strict-sense nonblocking, there must be at least one center stage switching element in common, reachable from both the free input and the free output. A sufficient condition for this to happen is that $2c(j - 1) > b(j)/2$ since there are $b(j)/2$ 2×2 switching elements in the center stage. This reduces to $M > j - 2$. The number of inputs and outputs is $N = a(j) = 2^{j-1}$ so the condition becomes

$$M > \log_2 N - 1 \qquad (9.18)$$

i.e. $\log_2 N$ is the smallest value of M that yields a strictly nonblocking Cantor network. This type of network requires N $1 \times M$ active splitters and N $M \times 1$ active combiners. There are $2\log_2 N - 1$ stages of 2×2 switch devices, each consisting of $MN/2$ such devices i.e. $MN\log_2 N - MN/2$ devices in total.

Figure 9-22 is an 8 × 8 Cantor network. There are many ways in which the switches could be connected up to form such a network; here, the network can be thought of as three 8 × 8 Beneš networks, 8 demultiplexers and 8 multiplexers. This generalizes, and an $N \times N$ Cantor network can be constructed from $\log_2 N$ Beneš networks, N demultiplexers and N multiplexers [40].

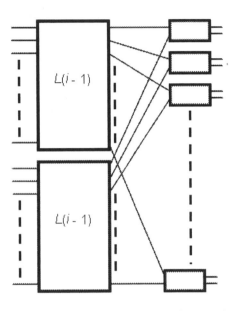

Figure 9-21. Forming a network $L(i)$ from networks $L(i-1)$, when creating a Cantor network.

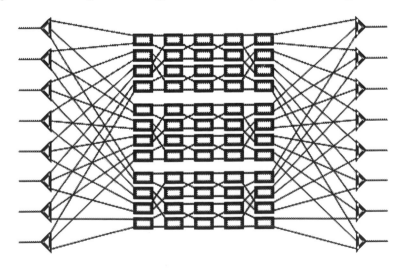

Figure 9-22. An 8 × 8 Cantor network.

A type of stacked network, known as the *Extended Generalized Shuffle* (EGS) network, has been fabricated from 448 electro-optic lithium niobate directional coupler switching elements in 23 packaged modules [41]. The system was operated continuously and without maintenance for 20 months. EGS networks represent a very general class of switch fabrics, which includes Clos and Cantor networks as special cases.

9.11 PERFORMANCE COMPARISON

To conclude, we briefly compare some of the architectures that were introduced above in the context of switch fabrics where switching elements may be cascaded. First, consider those based on SOA gates. In Reference [3], the cascadability of MVM switches was determined analytically, based upon considerations of saturation and signal-to-noise ratio. These findings are summarized here.

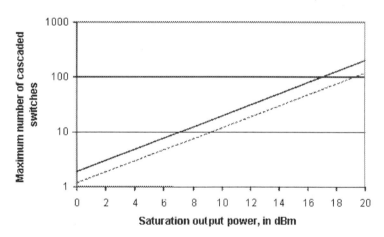

Figure 9-23. Maximum number of 8×8 switches that can be cascaded versus saturation output power for distributed-gain MVMs (solid line) and lumped-gain MVMs (dashed line). $G_{sig} = 1$, $B_e = 1$ GHz, $\lambda = 1.3$ µm, $n_{sp} = 5$, $p_{eff} = 2$, $L_{in} = L_{out} = 0.5$

Using the model described in Section 9-1, it was found that 200 distributed-gain MVM switches could be cascaded [3], assuming 100 mW SOA saturation power (Figure 9-23). For the purposes of the study saturation powers of 1-100 mW were considered; 100 mW is achievable with quantum well devices [3]. For large switch sizes, the distributed-gain MVM cascades considerably better than the lumped-gain MVM switch, which has gain in only two stages. These figures only take saturation effects into consideration, and in practice, noise would also limit cascadability, with a power penalty of 1 dB for 5-7 cascaded SOAs.

Secondly, Figures 9-24 to 9-26 show the switch device count, SXR and loss for architectures based on thermo-optic silica-on-silicon 2×2 switching elements (Table 9-1). The crossbar switch does not scale well in terms of crosstalk or loss, and both the tree and crossbar architectures use a large number of devices, particularly when making a large switch. Hence it is not surprising that these architectures are intended for use as switching elements, since they do not scale to the size generally required for switch fabrics. The tree and dilated Beneš architectures yield the best crosstalk performance. On the other hand, Beneš networks are themselves types of switch fabric so are not designed for use as switching elements. This would in any case be difficult since they are rearrangeably nonblocking and not strict sense nonblocking.

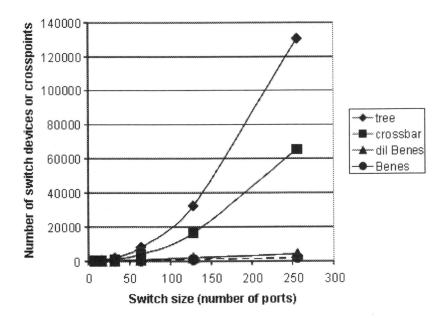

Figure 9-24. Number of switch devices or crosspoints versus number of ports N ($N{\times}N$ architectures are assumed).

This chapter has enumerated the main architectural principles behind the design of planar optical space switches. Architectures such as the tree, MVM and crossbar can be used to construct small switching "elements", which are then combined to form larger switch "fabrics". An indication has been given of optical switches found in research laboratories, and their fabrication technologies, such as lithium niobate directional couplers (and variants), semiconductor optical amplifiers, MEMS, and thermo-optic switches in silica and polymers. While the choice of technology affects performance, it should be clear that the choice of architecture has a significant influence also.

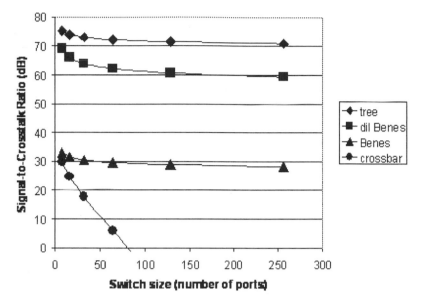

Figure 9-25. Signal-to-crosstalk ratio versus switch size.

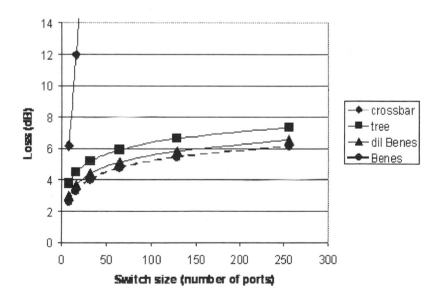

Figure 9- 26. Loss versus switch size.

Acknowledgments

It is a pleasure to thank the following colleagues for reading previous drafts of this chapter, and making many useful and insightful suggestions for improvements:
- Peter Duthie, Madeleine Glick and Sven Östring, all formerly of Marconi Labs, Cambridge, UK,
- Kevin Williams of the Engineering Department, University of Cambridge, UK, and
- Ian Henning of the Department of Electronic Systems Engineering, University of Essex, Colchester, UK.

REFERENCES

[1]   D. K. Hunter, "Switching Systems," Encyclopedia of Information Technology," Vol. 42, supplement 27, A. Kent, J. G. Williams, C. M. Hall (Eds.), Marcel Dekker, New York/Basel, pp. 335-70, 2000.

[2]   T. Goh, M. Yasu, K. Hattori, A. Himeno, M. Okuno, and Y. Ohmori, "Low Loss and High Extinction Ratio Strictly Nonblocking 16×16 Thermooptic Matrix Switch on 6-in Wafer Using Silica-Based Planar Lightwave Circuit Technology," IEEE/OSA J. Lightwave Technol., Vol. 19, No. 3, pp. 371-9, March 2001.

[3]   R. F. Kalman, L. G. Kazovsky, and J. W. Goodman, "Space Division Switches Based on Semiconductor Optical Amplifiers," IEEE Photon. Technol. Lett., Vol. 4, No. 9, pp. 1048-51, September 1992.

[4]   H. S. Hinton, "A Nonblocking Optical Interconnection Network Using Directional Couplers," IEEE GLOBECOM 1984, pp. 26.5.1-26.5.5, 1984.

[5]   P. Granestrand, B. Stoltz, L. Thylén, K. Bergvall, W. Döldissen, H. Heinrich, and D. Hoffman, "Strictly Nonblocking 8×8 Integrated Optical Switch Matrix," Electronics Letters, Vol. 22, No. 15, pp. 816-8, 17th July 1986.

[6]   S. Okamoto, A. Watanabe, and K. Sato, "A New Optical Path Crossconnect System Architecture Using Delivery and Coupling Matrix Switch", Transactions of IEICE Japan, October 1994, Vol. E77-B, No. 10, pp. 1272-4.

[7]   A. Watanabe, S. Okamoto, M. Koga, K.-I. Sato, and M. Okuno, "Packaging of 8×16 Delivery and Coupling Switch for a 320 Gb/s Throughput Optical Path Cross Connect System," IEEE/OSA J. Lightwave Technol., Vol. 14, No. 6, pp. 1410-22, June 1996.

[8]   T. Shimoe, K. Hajikano, K. Murakami, "A Path-Independent-Insertion-Loss Optical Space Switching Network", International Switching Symposium 1987, Phoenix, Vol. 4, paper C12.2, pp. 999-1003, March 1987.

[9]   T. Shibata, M. Okuno, T. Goh, M. Yasu, M. Itoh, M. Ishii, Y. Hibino, A. Sugita, and A. Himeno, "Silica-based 16×16 Optical Matrix Switch Module with Integrated Driving Circuits," OFC 2001, Anaheim, CA, March 2001.

[10] S. Yu, M. Owen, R. Varrazza, R. V. Penty, and I. H. White, "Demonstration of High-Speed Optical Packet Routing using Vertical Coupler Crosspoint Space Switch Array," Electron. Lett., Vol. 36, No. 6, pp. 556-8, 16th March 2000.

[11] S. Yu, R. Varrazza, M. Owen, R. V. Penty, I. H. White, D. Rogers, S. Perrin, and C. C. Button, "Ultra-low Crosstalk, Compact Integrated Optical Crosspoint Space Switch Arrays Employing Active InGaAsP/InP Vertical Waveguide Couplers," CLEO 1999, post-deadline paper CPD24, Baltimore, March 1999.

[12] T.-W. Yeow, K. L. E. Law, and A. Goldenberg, "MEMS Optical Switches," IEEE Commun. Mag., Vol. 39, No. 11, pp. 158-63, November 2001.

[13] J. E. Fouquet, "Compact Optical Cross-Connect Switch Based on Total Internal Reflection in a Fluid-Containing Planar Lightwave Circuit," OFC 2000, Baltimore, Maryland, paper TuM1, March 2000.

[14] E. Almström, C. P. Larsen, L. Gillner, W. H. van Berlo, M. Gustavsson, and E. Berglind, "Experimental and Analytical Evaluation of Packaged 4 × 4 InGaAsP/InP Semiconductor Optical Amplifier Gate Switch Matrices for Optical Networks," IEEE/OSA J. Lightwave Technol., Vol. 14, No. 6, pp. 996-1004, June 1996.

[15] R. A. Spanke, "Architectures for Large Nonblocking Optical Space Switches," IEEE J. Quantum Electron., Vol. 22, No. 6, pp. 964-7, June 1986.

[16] P. Granestrand, B. Lagerström, P. Svensson, and L. Thylén, "Tree-Structured Polarisation Independent 4 × 4 Switch Matrix in LiNbO₃," Electron. Lett., Vol. 24, No. 19, pp. 1198-200, 15th September 1988.

[17] P. Granestrand, B. Lagerström, P. Svensson, L. Thylén, B. Stoltz, K. Bergvall, J.-E. Falk, and H. Olofsson, "Integrated Optics 4 × 4 Switch Matrix with Digital Optical Switches," Electron. Lett., Vol. 26, No. 1, pp. 4-5, 4th January 1990.

[18] I. M. Burnett and A. C. O'Donnell, "Synchronous Optical Switching in a Time Multiplexed Network," EFOC/LAN 1990, Eighth Annual European Fibre Optic Communications and Local Area Networks Conference, Munich, 27-29 June 1990.

[19] F. L. W. Rabbering, J. F. P. van Nunen, and L. Eldada, "Polymeric 16 × 16 Digital Optical Switch Matrix," ECOC 2001, Amsterdam, postdeadline paper, October 2001.

[20] T. Watanabe, T. Goh, M. Okuno, S. Sohma, T. Shibata, M. Itoh, M. Kobayashi, M. Ishii, A. Sugita, and Y. Hibino, "Silica-based PLC 1 × 128 Thermo-Optic Switch," ECOC 2001, Amsterdam, October 2001.

[21] C. Clos, "A Study of Non-Blocking Switch fabrics," Bell System Technical Journal, Vol. 32, pp. 406-24, March 1953.

[22] J. C. Boyd, J. M. Hunter, E. McKenzie, and D. G. Smith, Final Report, UK Post Office Contract 534325, March 1976.

[23] V. E. Beneš, "Mathematical Theory of Connecting Networks and Telephone Traffic," Academic Press, 1965.

[24] C. Burke, M. Fujiwara, M. Yamaguchi, H. Nishimoto, and H. Honmou, "128 Line Photonic Switching System using LiNbO₃ Switch Matrices and Semiconductor Traveling Wave Amplifiers," IEEE/OSA J. Lightwave Technol., Vol. 10, No. 5, pp. 610-5, May 1992.

[25] S. Okamoto, A. Watanabe, and K.-I. Sato, "Optical Path Crossconnect Node Architectures for Photonic Transport Network," IEEE/OSA J. Lightwave Technol., Vol. 14, No. 6, pp. 1410-22, June 1996.

[26] H.-H. Witte, "Comparison of a Novel Photonic Frequency-based Switching Network with Similar Architecture," IEICE Trans. Commun., Vol. E77-B, No. 2, pp. 157-64, February 1994.

[27] P. Hall, "On Representatives of Subsets," J. of the London Mathematical Society, Vol. 10, pp. 26-30, 1935.

[28] F. K. Hwang, "On Beneš Rearrangeable Networks," Bell System Technical Journal, Vol. 50, No. 1, pp. 201-7, January 1971.

[29] D. C. Opferman and N. T. Tsao-Wu, "On a Class of Rearrangeable Switching Networks," Part I, Control Algorithm," Bell System Technical Journal, Vol. 50, No. 5, pp. 1579-600, May-June 1971.

[30] G. F. Lev, N. Pippenger, and L. G. Valiant, "A Fast Parallel Algorithm for Routing in Permutation Networks," IEEE Trans. Computers, Vol. 30, No. 2, pp. 93-100, February 1981.

[31] P. J. Duthie, M. J. Wale and I. Bennion, "Size, Transparency and Control in Optical Space Switch Fabrics, A 16×16 Single Chip Array in Lithium Niobate and its Applications," Technical Digest of the 1990 International Topical Meeting on Photonic Switching, Kobe, Japan, 12-14 April 1990.

[32] N. Antoniades, S. J. B. Yoo, K. Bala, G. Ellinas, and T. E. Stern, "An Architecture for a Wavelength-Interchanging Cross-Connect Utilizing Parametric Wavelength Converters," IEEE/OSA J. Lightwave Technol., Vol. 17, No. 7, pp. 1113-25, July 1999.

[33] N. Antoniades, K. Bala, S. J. B. Yoo, and G. Ellinas, "A Paramteric Wavelength Interchanging Cross-Connect (WIXC) Architecture," IEEE Photon. Technol. Lett., Vol. 8, No. 10, pp. 1382-4, October 1996.

[34] Y.-K. Chen and C.-C. Lee, "Fiber Bragg Grating-Based Large Nonblocking Multiwavelength Cross-Connects," IEEE/OSA J. Lightwave Technol., Vol. 16, No. 10, pp. 1746-56, October 1998.

[35] K. Padmanabhan and A. N. Netravali, "Dilated Networks for Photonic Switching," IEEE Transactions on Communications, Vol. 35, No. 12, pp. 1357-65, December 1987.

[36] D. K. Hunter and D. G. Smith, "New Architectures for Optical TDM Switching," IEEE/OSA J. Lightwave Technol., Vol. 11, No. 3, pp. 495-511, March 1993.

[37] J. E. Watson, M. A. Milbrodt, K. Bahadori, M. F. Bautartas, C. T. Kemmerer, D. T. Moser, A. W. Schelling, T. O. Murphy, J. J. Veselka, and D. A. Herr, "A Low-Voltage 8×8 Ti:LiNbO$_3$ Switch with a Dilated- Beneš Architecture," IEEE/OSA J. Lightwave Technol., Vol. 8, No. 5, pp. 794-801, May 1990.

[38] T. O. Murphy, C. T. Kemmerer, and D. T. Moser, "A 16×16 Ti:LiNbO$_3$ Dilated Beneš Photonic Switching Module," 1991 Topical Meeting on Photonic Switching, Salt Lake City, UT, Postdeadline Paper PD3, March 1991

[39] D. G. Cantor, "On Construction of Nonblocking Switch Networks," Symposium on Computer Communications Networks and Teletraffic, Polytechnic Institute of Brooklyn, 4-6 April 1972.

[40] G. M. Masson, G. C. Gingher, and S. Nakamura, "A Sampler of Circuit Switch Fabrics," IEEE Computer, pp. 32-48, June 1979.

[41] E. J. Murphy, T. O. Murphy, A. F. Ambrose, R. W. Irvin, B. H. Lee, P. Peng, G. W. Richards, and A. Yorinks, "16 × 16 Strictly Nonblocking Guided-Wave Optical Switching System," IEEE/OSA J. Lightwave Technol., Vol. 14, No. 3, pp. 352-8, March 1996.

Chapter 10

SWITCH FABRIC CONTROL

Wojciech Kabacinski
Grzegorz Danilewicz
Mariusz Glabowski

10.1 CONTROL ALGORITHMS

Different optical space switch architectures have been discussed in chapter 9. In general, a switch fabric may have N input terminals and N output terminals, and is usually constructed of smaller switching elements (switches) arranged into one or more stages. When a new call arrives at the fabric, a connecting path has to be established between one of its input terminals and one of its output terminals. Meanwhile, when a call is finished, its connecting path is no longer needed and may be released for use by other incoming calls.

Optical transport networks are expected to provide circuit-switched lightpaths between its users. These lightpaths are established by optical cross-connects (OXCs) which are linked by optical fiber cables, as depicted in Figure 10-1(a). Each OXC contains input modules, output modules, a switch fabric, and a control system, as shown in Figure 10-1(b). Signals are received by input modules where control/signaling information are passed to the control system. The latter instructs the switch fabric to be configured in such a way to interconnect inputs to outputs as desired. It contains a routing table with information about how to route new calls. Connecting paths in successive OXC fabrics, along with fiber transmission links among them, form the end-to-end lightpath across the network. Signals leaving the switching fabric are passed to output modules where control information pertinent to next hop may be added. Input/output modules perform other functions and this is discussed in chapter 11.

Usually the control system sends commands to a controller (possibly a separate microprocessor or a program implemented in the system) to setup (release) connecting paths. Depending on the switch fabric architecture, there may be one or more paths between an arbitrary pair of input/output terminals, or sometimes there may not be a path available for a new call, because of blocking.

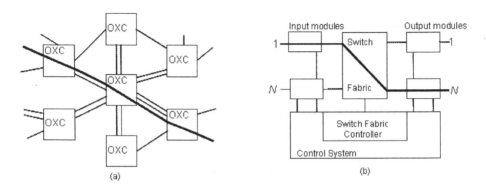

Figure 10-1. (a) An arbitrary optical network (b) Generic architecture of an OXC. The switch fabric can be based on any of the architectures discussed in chapter 9, and others

When a new call is to be set up, the controller has to find a connecting path in the switch fabric, check whether it is available (i.e. switching elements are not occupied by other calls), issue respective control signals to change states of switches, and update the current state of the switch fabric. These tasks are mainly performed by control algorithms. The state of the switch fabric is usually stored in a table containing information about all connecting paths already set up. When any of these paths is released, information about it is deleted from the table and the switch configuration is reset accordingly. Apart from switches, there may also be a need to control other elements, such as optical amplifiers.

The control algorithm of a switching fabric may vary depending on its capability to establish different types of connections, such as unicast, multicast or broadcast. A *unicast connection* is a connection between one input terminal and one output terminal. A *multicast connection* is a connection between one input terminal and a set of output terminals. When this set contains all output terminals of the switch fabric, the connection is called a *broadcast connection*. Numerous control algorithms have been developed for different connection types and various switch fabric architectures. Algorithms designed for one fabric topology usually cannot be used in other topologies. However, they can be used in similar fabrics made of different technologies, electronic or optical. Special requirements or concerns about certain technology (such as those of crosstalk reduction) may call for sophisticated algorithms, which are not necessary with other technologies [1].

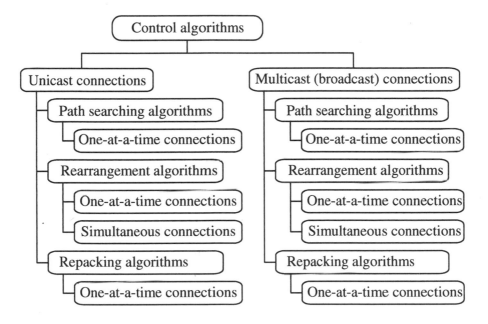

Figure 10-2. Classification of control algorithms in multi-path switch fabrics

Control algorithms in multi-path switch fabrics can be divided into three major groups (Figure 10-2). *Path searching algorithms* are used for finding a connecting path through a switch fabric for one call at a time. Depending on the fabric architecture, a given algorithm may, and may not, always be successful. Another group of control algorithms is called *rearrangement algorithms*. These algorithms can be used when a path searching algorithm fails. Their task is to find connecting paths which can be re-routed to unblock a path for a new call. Some of these algorithms are designed to find a connecting path for one call at a time. Others are designed to find connecting paths for all new calls simultaneously. In *call repacking algorithms* some connecting paths may also be re-routed to accommodate new calls. However, in contrast to rearrangement algorithms, re-routing is executed after one of the existing calls is terminated. The role of repacking algorithms is to "pack" existing calls more efficiently, and thereby prevent switch fabrics from being in a blocking state when a new call arrives. The goal of this chapter is not to cover in detail all aspects of switch-fabric control algorithms. We merely try to help readers in the areas of optical switching and networking to appreciate the role, scope and issues involved in these control algorithms. In the remaining part of this chapter, we consider only the first two classes of algorithms and focus on unicast connections. More information on call repacking algorithms can be found in [2]-[3]. Multicast algorithms are discussed in [3]-[4].

10.2 PATH SEARCHING ALGORITHMS

10.2.1 Single Path and Standard Path Switch Fabrics

There is only one connecting path between any pair of input-output terminals in *single path switch fabrics*. When the switch fabric is nonblocking, the control algorithm needs only to change the states of the appropriate switches to set up (release) a connecting path. For relatively small switch fabrics, the control algorithm may use a state table with all possible states. When the new call arrives, the algorithm reads out how to set up the switches directly from this table. Depending on the switch fabric architecture, the state table may contain information on how to set up each switch separately, or how a group of these switches may be controlled together by one signal. In the latter case, the number of control signals is less than the number of switches. The state of switches and control signals may also be deduced directly from input and output terminals addresses.

An example of the single path switch fabric is the tree architecture discussed in chapter 9. In this architecture, switches placed in one stage of an active splitter or an active combiner can be controlled by one signal. The switch may be in one of two states: *cross* or *bar*. In the cross state input 1 is connected to output 2 while input 2 is connected to output 1. In the bar state the connection pattern is input 1 to output 1 and input 2 to output 2. Let cross and bar states of the 2×2 switch be activated by control signals 0 and 1, respectively. Let us assume that a new call from input terminal x to output terminal y is to be set up. This call will be denoted by $<x, y>$. Let also $x_{n-1}...x_0$ and $y_{n-1}...y_0$ be binary representations of the addresses of x and y, respectively (input and output terminals are numbered from 0 to $N-1$). The stages of switches in active splitter x are controlled by $y_{n-1}...y_0$, while the stages in active combiner y are controlled by $x_0...x_{n-1}$. This is shown in Figure 10-3 for call $<6, 5>$ in an 8×8 switch fabric (only the specific splitter and combiner involved in this connection are shown).

When some connecting paths use the same switch(s) in a single path fabric then a conflict occurs between connections. Examples are switch fabrics composed of 2×2 switches arranged in $\log_2 N$ stages (banyan, baseline). Therefore, the control algorithm has to check whether the connecting path for a new call is free. Example of an 8×8 baseline switch fabric is shown in Figure 10-4. There is exactly one connecting path between any pair of input-output terminals. This switch fabric is blocking. When call $<0, 0>$ is set up (bold line) and the new call $<1, 2>$ arrives (dashed line), the latter is blocked since the interstage link between switch 0 of stage 1 and switch 0 of stage 2 is already occupied by call $<0, 0>$.

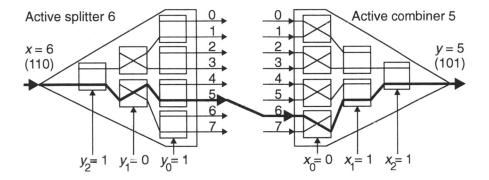

Figure 10-3. Path setup in the tree architecture

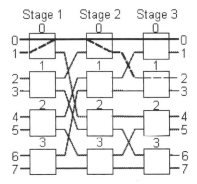

Figure 10-4. The 8×8 baseline switch fabric

In the *standard path switch fabric*, there may be several connecting paths between an arbitrary pair of input-output terminals. However, only one path, called the standard path, is always used to connect this input-output pair in order to preserve some fabric characteristics (for instance, nonblocking). A good example is the crossbar fabric composed of directional couplers, which is shown in Figure 10-5. The states of switches are stored in the state matrix, with rows and columns corresponding to input terminals and output terminals, respectively. Initially, all directional couplers are in the cross state. To set up the connecting path for the call $<x, y>$, the state of the directional coupler *x-y* should be changed to bar state by writing a "1" into row *x* and column *y* of the state matrix. Figure 10-5 shows a 4 × 4 crossbar example with call <1, 1>. In general, there are several possibilities to setup a call in this architecture. For instance, <3, 3> can be setup by changing the state of switch 3-3 or by changing states of switches 3-4, 4-4, and 4-3 (dashed lines in the figure). However, the latter setup results in the blocking of <4, 2> and <2, 4>.

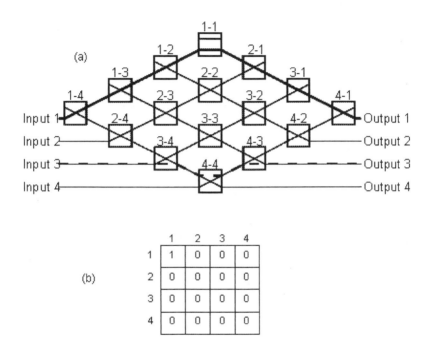

Figure 10-5. (a) A 4 x 4 crossbar with call <1, 1> marked by bold line (b) State matrix

10.2.2 Multi-Path Networks

In multi-path switch fabrics there are several connecting paths to connect any input terminal with any output terminal. Clos networks and stacked switch fabrics [5], for which several control algorithms are proposed, are examples of multi-path switch fabrics [3]. We describe here algorithms used in Clos networks. With slight modifications, these algorithms can also be used in other multi-path fabrics, such as multi-$\log_2 N$ fabrics [6], which are obtained by stacking $\log_2 N$ fabrics in parallel.

In the Clos network depicted in Figure 10-6, usually denoted by $C(m, n, r)$, a center stage switch is to be found for the new call. Let input terminals and output terminals of $C(m, n, r)$ be numbered from 0 to $N-1$,

$N = n \times r$, let the first stage switches (with n inputs each) and the third stage switches (with n outputs each) be numbered from 1 to r, and let center stage switches be numbered from 1 to m. Let $<x, y>$ denote the new call from input terminal x of the first stage switch I to output terminal y of the third stage switch O. The connecting path from input terminal x to output terminal y spans input switch I, center-stage switch M, output switch O and the inter-stage interconnect links among them.

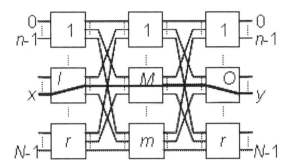

Figure 10-6. A three-stage Clos network, $C(m, n, r)$, with the call $<x, y>$

When the new call is to be set up, the control algorithm has to find the center stage switch M with free links to switches I and O. The first algorithm is a *Sequential algorithm*. It checks center stage switches sequentially starting from the center stage switch 1, and chooses the first available switch. The *Quasi-Random algorithm* also checks center stage switches sequentially but starting from switch $M_p + 1$, where M_p denotes the center stage switch used by the previous call. In the third algorithm, called the *Beneš algorithm*, the new call is set up through the first available switch with the biggest number of calls already routed through it [7]. These algorithms are the most popular and have been used in electromechanical and electronic switching systems. When the switch fabric is strict-sense nonblocking[1], it does not matter which algorithm is used. However, when the switch fabric is blocking, the blocking probability is affected by the control algorithm. The lowest blocking probability is achieved with the Beneš algorithm. It always tries to set up the new call through the most loaded center stage switch. Therefore, if this fails, the probability that it will find a path in one of less loaded switches is larger. In the Sequential algorithm switches with lower indexes are more loaded than those of higher indexes and the overall blocking probability is higher than in the Beneš algorithm. When switches are built using usage-sensitive technology (i.e. the more often a switch is

---

[1]   A connecting path can always be found without rearrangements. See discussion of blocking characteristics in section 9-5.

used, the higher the probability that it will break down), the sequential algorithm results in earlier failures of switches with lower indexes. The Quasi-Random algorithm spread connections equally through all center stage switches. Therefore all switches are roughly equally used, but the blocking probability is therefore higher than in the previous two algorithms. The time complexity of these algorithms is $O(m)$, i.e. in the worst case, any algorithm has to check all center stage switches before finding the connecting path. The operation of the Beneš algorithm is explained in the following example.

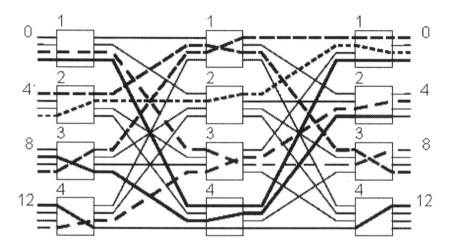

Figure 10-7. The state of C(4, 4, 4) in Example 10-1

Example 10-1:

Let us consider $C(4, 4, 4)$ with the state presented in Figure 10-7 where existing connections are shown in bold lines (solid and dashed). Let us assume that a new call <8, 4> arrives to the switch fabric, and that connections are set up using the Beneš algorithm. The algorithm will first check center stage switch 4 since it is already used for three calls. Since the link from the first stage switch 3 is occupied, the algorithm examines switches 1 and 3, each of which carries two connections. However, both are inaccessible for the new call <8, 4>. Finally, center stage switch 2 with one connection only is considered. Since both inter-stage links to this switch are accessible for the new call, the connecting path is set up through this switch. The number of calls set up through center stage switches may be calculated during the execution of the algorithm or may be stored in a table arranged according to switch loads (i.e. with the most loaded on top). In the second case, after a call has been set up or disconnected, the table is updated by a sort procedure, which moves switches up or down as needed. Now, let us consider the same state of this switch if the new call is <14, 1>. In this situation, all center stage switches are inaccessible for the new call. As a result the call is blocked.

10.3 REARRANGEMENT ALGORITHMS

10.3.1 Single Connections

Call <14, 1> in the switch fabric of Figure 10-7 can be set up by changing some of the existing connecting paths. We can for instance re-route call <15, 5> from switch 3 to switch 2. This will free center switch 3 for setting up <14, 1>. Rearrangement algorithms are used to determine which connecting paths should be re-routed and how. Slepian [8] and Paull [9] proposed several rearrangement algorithms. All of them use *Paull's matrix* $M_{r \times r}$ for representing the state of the switch fabric. Each row (column) in this matrix corresponds to one first stage (third stage) switch. An entry $M[I; O] = A$ means that the connection from the first stage switch I to the third stage switch O is set up through the center stage switch A. We will denote this connection by (I, O) to differentiate it from call $<x, y>$. For example, call <14, 1> is denoted by $(4, 1)$ since input terminal 14 is in the first stage switch 4 and output terminal 1 is in the third stage switch 1. It should be noted that there can be more calls (but not more than n) between switches I and O, so an entry $M[I; O]$ may contain more than one center stage switch. The Paull's matrix for the switch fabric of Figure 10-7 is shown in Figure 10-8. In this matrix calls <2, 8>, i.e. connection $(1, 3)$, and <3, 3>, i.e. connection $(1, 1)$, are represented by entries $M[1; 3] = 3$ and $M[1; 1] = 4$, since they are set up through center stage switches 3 and 4, respectively. The remaining calls are also represented in the matrix in the same manner. The Paull's matrix has the following properties:

1) In each row, there can be no more than n different elements since each first stage switch has n inputs and only one link to every center stage switch.

2) In each column, there can be no more than n different elements since each third stage switch has n outputs and only one link from every center stage switch.

Let us assume that the new connection is (I, O) and this connection is blocked. The input data for the algorithm are input switch I, output switch O, and Paull's matrix M representing the current state of the switch fabric. First we have to determine which switches will be used for rearrangements. An example of Paull's matrix is given in Figure 10-9(a) where it is assumed that $I = 1$ and $O = 1$, (I, O) is blocked and connections set up through center stage switches A and B are shown. These center-stage switches will be used for rearrangement. The algorithm (hereafter called *Paull algorithm*) and its operation is as follows [9]:

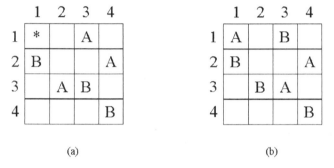

Figure 10-8. The state matrix of C(4, 4, 4); connection (4, 1) is   blocked

	1	2	3	4
1	*		A	
2	B			A
3		A	B	
4				B

(a)

	1	2	3	4
1	A		B	
2	B			A
3		B	A	
4				B

(b)

Figure 10-9. The Paull's matrix before (a) and after (b) rearrangements

1. Find the center switch not present in column O; i.e. A. The center stage switch A has the free link to the third stage switch O.
2. Find the center switch not present in row I; i.e. B. The center stage switch B has the free link from the first stage switch I.
   We have chosen two switches, A and B, which will be used to rearrange the state of the switch fabric. Now we take randomly one of these switches to set up the new connection, say switch A.
3. Set up (I, O) through A, i.e. $M[I; O] := A$; $i := I$; $o := O$. In other words, we put $M[1; 1] := A$ in Figure 10-9(b)[2].
   We now have A present two times in I. Note that variables i and o represent the currently considered row and column, respectively. At the beginning these two variables have values I and O.
4. In row i find such column o` that $M[i; o`] = A$; if there is such o`, then replace $M[i; o`] = A$ with $M[i; o`] = B$, set $o := o$` (i.e. move $(i, o`)$ from A

---

[2] The operator ":=" means assignment of value statement.

to B, and set $o\grave{}$ as the currently considered column), and go to step 5, otherwise the call is unblocked and the algorithm ends.

In the matrix of Figure 10-9(a), A is in column 3 of row $i = 1$, so we replace $M[1; 3] = A$ with $M[1; 3] = B$ and set $o: = o\grave{}\ (= 3)$. After this replacement we have to check if B is present in another row of column 3, so we go to step 5.

5. In column o find such row $i\grave{}$ that $M[i\grave{}; o] = B$; if there is such $i\grave{}$, then replace $M[i\grave{}; o] = B$ with $M[i\grave{}; o] = A$, set $i := i\grave{}$ (i.e. move $(i\grave{}; o)$ from B to A, and set $i\grave{}$ as the currently considered row), and go to step 4, otherwise the call is unblocked and the algorithm ends.

In the matrix of Figure 10-9(a), switch B is in row $i\grave{} = 3$ of column $o = 3$. So the new value of i is now 3, we replace $M[3; 3] = B$ with $M[3; 3] = A$ and we have to check if there is another column in row 3 containing A. Therefore we go back to step 4. In row $i = 3$, A is already in column $o\grave{} = 2$. We replace $M[3; 2] = A$ with $M[3; 2] = B$, update $o\ (o = o\grave{} = 2)$, and go to step 5 to check whether B is now present in another row of column $o\grave{} = 2$. Since there is no such row, the algorithm ends. The matrix after rearrangements is shown in Figure 10-9(b).

The algorithm always ends with success by step 4 or 5 where there is no switch to consider further. In the matrix of Figure 10-9(a) the algorithm ends on position $i = 3$ and $o = 2$. There is no another entry B in column 2, and when we look on the matrix we can see that there is no such row where B could be placed along this column. It cannot be in row $i = 1$, since in step 2 we chose B because it was not present in this row. It cannot also be in other rows (i.e. 2 and 4) since it is already present in other columns of these rows, and according to matrix properties, each symbol can be present in each column and in each row only once. More formal proof can be found in [9].

Example 10-2:
We now turn our attention back to the switch fabric in Figure 10-7, the Paull's matrix of which is given in Figure 10-8. Connection (4, 1) is blocked since in row 4 we have already numbers 3 and 4, while numbers 1, 2 and 4 are in column 1. Number 3 is not in column 1. Thus in applying the steps listed above, we put $A = 3$. In row 4 we have two numbers available, 1 and 2; we choose any of them, let say $B = 1$. We now set up the new call through switch 3 ($M[4; 1] := 3$ – step 3) and look for column $o\grave{}$ containing 3 in row 4. We have $M[4; 2] = 3$, so $o\grave{} = 2$ and we change $M[4; 2] = 3$ with $M[4; 2] = 1$ (step 4). In the next step we check if there is another row with 1 in column 2. Since there is no such row, the algorithm is ended and the new connection is established with one rearrangement. The matrix after this rearrangement is shown in Figure 10-10.

Figure 10-10. The state matrix of C(4, 4, 4) after rearrangement for connection (4, 1)

In the Paull algorithm we choose switches for rearrangements randomly. Paull proposed different modifications, which resulted in a lower number of rearrangements needed to unblock new calls. For instance, we can set up the new call through switch B (not A) or we can check all possible pairs of switches (A and B) and finally use the pair which results in the lowest number of rearrangements [9]. It was proved that using the Paull algorithm not more than $2r - 3$ rearrangements would be needed. When modified algorithms are used, the number of rearrangements needed is reduced to $r - 1$ [7]. These algorithms can also be applied for stacked switch fabrics [10]-[11].

The Paull algorithm can also be used in switch fabrics composed of more than three stages, which are obtained by recursively replacing center stage switches with other three-stage fabrics. The example of a five-stage 8×8 switch fabric, obtained from C(2, 2, 4) by replacing two center stage 4×4 switches with C(2, 2, 2) switches is shown in Figure 10-11(a). Calls <0, 1>, <3, 0>, <4, 5>, and <7, 6> are shown in bold lines. Paull's matrices for this switch fabric are given in Figure 10-11(b). Entries in matrix M represent the three-stage switch fabric (1 or 2) used for respective connections. Matrices $M1$ and $M2$ correspond to states of three-stage switch fabrics 1 and 2, respectively. In this switch fabric call <2, 4> (connection (2, 3)) is blocked, since we cannot put either 1 or 2 in position (2, 3) of matrix M (blocking is denoted by the asterisk in Figure 10-11(b)). We use the Paull algorithm to rearrange the state of the switch fabric. Connection (2, 3) is set up through the center three-stage switch fabric 2 and connection (3, 3) is moved to the switch fabric 1 (matrix M is shown after this rearrangement in Figure 10-11(c)). Now we have to set up these two connections ((2, 3) and (3, 3)) in the center switch fabrics. Connection (3, 3) can be set up through switch a of the center switch fabric 1 ($M1$ in Figure 10-11(c)), while connection (2, 3) is blocked in the center switch fabric 2 (asterisk in $M2$ of Figure 10-11(c)). We can now use the Paull algorithm for $M2$. All the matrices after rearrangements are shown in Figure 10-11(d) and connections are shown in Figure 10-11(e), where the new call is in dot line, and rearranged calls are in dashed lines. By analogy, the Paull algorithm can be used in switch fabrics composed of seven stages, nine stages, etc.

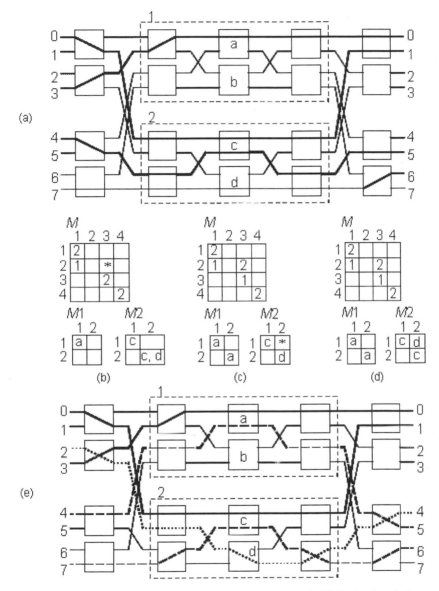

Figure 10-11. Rearrangements in a five-stage 8×8 switch fabric: (a) State before rearrangements, (b) Paull's matrices before rearrangements, (c) After rearrangements in M, (d) After rearrangements in $M2$, and (e) State after rearrangements and unblocking

10.3.2 Simultaneous Connections

In simultaneous connections we have a set of compatible calls, i.e. one input terminal requests a connection with exactly one output terminal, and one output terminal is requested by exactly one input terminal. Such a set of calls is also called *permutation* or *maximal assignment* if there is a request in each input terminal, and it is usually written as:

$$\Pi = \begin{pmatrix} 0 & 1 & 2 & \cdots & N-1 \\ \pi(0) & \pi(1) & \pi(2) & \cdots & \pi(N-1) \end{pmatrix}. \tag{10.1}$$

This means that input terminal 0 is to be connected to output terminal $\pi(0)$, input terminal 1 is to be connected to output terminal $\pi(1)$, etc. Similarly, the inverse permutation Π^{-1} denotes that output terminal 0 is to be connected to input terminal $\pi^{-1}(0)$, output terminal 1 is to be connected to input terminal $\pi^{-1}(1)$, and so on. Now, we have to choose the center stage switch for each call. Several approaches and algorithms were proposed to do this simultaneously. In the following, we discuss some of these approaches.

10.3.2.1 Matrix Decompositions

In algorithms based on matrix decompositions a maximal assignment is presented in the matrix $H_{r \times r}$. As in Paull's matrix, rows and columns correspond to the first and the third stage switches, respectively. In $H_{r \times r}$, an entry $H[I; O] = a$ means that the number of connections between switches I and O is equal to a, i.e. a maximal assignment contains a connections (I, O). When $a = 0$, there is no such connection. Since each first stage switch has n inputs, we have n calls from this switch in a maximal assignment. Therefore, when we add all elements in any row, we obtain n. The same is true for any column, since each third stage switch has n outputs. These properties of matrix H can be denoted by equations:

$$\sum_{i=1}^{r} H[i; O] = n \quad \text{and} \quad \sum_{j=1}^{r} H[I; j] = n. \tag{10.2}$$

Consider now a matrix for which the sum of any row or column is unity. It is often called an elementary permutation matrix, and is denoted by E. Neiman [12] has shown that the control of the rearrangeable switch fabric can be interpreted as a procedure of finding a set of E matrices which can be subtracted, one at a time, from some given H, and calls corresponding to each of these E matrices can be set up through one center stage switch. We say also that matrix H is decomposed into n matrices E. Let us consider $C(3, 3, 3)$ and the permutation

$$\Pi = \begin{pmatrix} 0 & 1 & 2 & 3 & 4 & 5 & 6 & 7 & 8 \\ 1 & 6 & 0 & 5 & 8 & 7 & 4 & 2 & 3 \end{pmatrix}. \qquad (10.3)$$

We have $H[1; 1] = 2$ (calls <0, 1> and <2, 0>) and $H[1; 3] = 1$ (call <1, 6>). There is no call between switches 1 and 2, so $H[1; 2] = 0$. Matrices H, E_1, E_2, and E_3 are shown in Figure 10-12(a), while connecting paths in $C(3, 3, 3)$ are shown in Figure 10-12(b) (in bold, dashed, and dotted lines for center stage switches 1, 2, and 3, respectively).

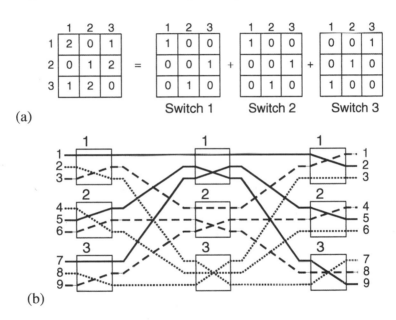

Figure 10-12. Simultaneous connections in $C(3, 3, 3)$: (a) Matrix H and its decomposition, (b) State of the switch fabric

Several algorithms have been proposed for decomposing matrix H into matrices E. The main drawbacks are time complexity and number of iterations. Neiman proposed an algorithm with the time complexity $O(r^2m^2)$ [12]. Some modifications in this algorithm, which resulted in fewer iterations, were proposed in [13]-[15]. Another algorithm with time complexity $O(nr^2)$ was proposed in [16]. Parallel algorithms as well as algorithms which realize only some of all possible permutations were also considered [17]. More efficient algorithms use graph coloring.

10.3.2.2 Matching and Graph Coloring Algorithms

10.3.2.2.1 Terminology of Graph Theory

Before we discuss graph coloring algorithms we introduce some of the basic terminology used in graph theory. Intuitively, a *graph* consists of *nodes* (called also *vertices*) and lines joining pairs of nodes called *edges*. A set of nodes is usually denoted by V, a set of edges is denoted by E, and a graph is denoted by $G(V, E)$ or simply by G. Graphs are generally used to model and solve numerous problems in different and independent fields of knowledge. It can be used to examine the interdependence between a number of elements in a given set. Graph algorithms have applications in communications and transportation networks and are used to solve many problems in these fields, including:

– Finding the shortest path in a network. For example, a path with the minimum delay;
– Finding the minimum spanning tree. For example, the least expensive way to join all nodes in a network;
– Finding the maximum throughput, or flow, in a network. For example, a path with the maximum speed at which information can be sent from a source to a destination.

Graph theory is also successfully applied to solve the maximal assignment problem in switching systems.

In graph theory we can distinguish *directed graphs* (*digraphs*) and *undirected graphs* (or just *graphs*). In digraphs the set E is the set of *ordered pairs* of nodes from V whereas in undirected graphs (graphs) the set E consists of *unordered pairs* of distinct nodes. For example, in Figure 10-13(a) the directed graph $G(V,E)$ is shown, where $V=\{1, 2, 3, 4\}$ and $E=\{(1,2), (1,3), (1,4), (2,1), (4,2), (4,3)\}$. The undirected graph is shown in Figure 10-13(b) where $V=\{1, 2, 3, 4\}$ and $E=\{(1,2), (1,3), (2,3), (3,4)\}$.

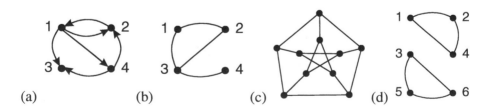

Figure 10-13. Graphs: (a) directed, (b) undirected, (c) 3-regular, and (d) not connected

The number of nodes of G is called the *order* of G. Two nodes joined by an edge are said to be *adjacent*, and this edge is said to be *incident* to these nodes. Two edges of G incident to the same node are called *adjacent edges*. When two nodes are joined by two or more edges, then these edges are called *multiple edges*. A graph containing multiple edges is called a *multigraph*. The *degree of a node*, denoted by $d(v)$, in a graph G is the number of edges of G incident to this node whereas in digraphs we distinguish between *in-degree of a node, $d^+(v)$* (i.e. the number of edges coming in a node) and *out-degree of a node, $d^-(v)$* (i.e. the number of edges coming out of a node). The degree of a node in a directed graph is the sum of in-degree and out-degree of the node. For example, for the graph presented in Figure 10-13(b) we have: $d(4) = 1$, $d(1) = d(2) = 2$ and $d(3) = 3$. In the directed graph (Figure 10-13(a)) the in-degree and out-degree of nodes are as follows: $d^+(1) = d^+(4) = 1$, $d^+(2) = d^+(3) = 2$, $d^-(1) = 3$, $d^-(2) = 1$, $d^-(3) = 0$, and $d^-(4) = 2$. The maximum degree of all nodes in G is called the *degree of the graph G*. When all nodes in G have degree n, a graph is called *n-regular*. For example, 3-regular graph is shown in Figure 10-13(c).

A *path* in a graph (digraph) $G(V, E)$ from node s to node t is a sequence of nodes $<v_0, v_1, v_2, ..., v_k>$ such as $s = v_0$, $t = v_k$, $(v_{i-1}, v_i) \in E$. For instance edges (2,3) and (3,4) form the path $<2, 3, 4>$ in the graph of Figure 10-13(b). The number of edges k of the path is called the *length of the path*. A graph is *connected* if every node can be reached from any other node, i.e. if there is a path between any two nodes. Graphs of Figure 10-13(a), (b), (c) are connected, while the graph of Figure 10-13(d) is not connected. A path containing at least two edges forms a *cycle* in a graph if $v_0 = v_k$. The path $<1, 2, 3, 1>$ containing edges (1,2), (2,3), and (3,1) forms a cycle in the graph of Figure 10-13(b). When a cycle traverses every edge of the graph exactly once, the cycle is called an *Euler cycle*. The problem of finding an Euler cycle in a graph is commonly seen in the form of puzzles where you are to draw a given figure without lifting your pencil from the paper, starting and ending at given points. A graph G has an Euler cycle if and only if it is connected and all its nodes are of even degree. This property of a graph will be used later on in this chapter.

An *edge coloring* of G is an assignment of colors to the edges of G so that adjacent edges are colored with different colors. For instance the graph of Figure 10-13(b) can be colored using three colors in the following way: (1,2) – blue, (1,3) – red, (2,3) – green, and (3,4) – blue. When all edges in G can be colored using n colors, the graph G is said to be *n-colorable*. The graph of Figure 10-13(b) is 3-colorable. A *spanning subgraph* of G is a graph H, which contains all nodes and a set of edges of G.

10.3.2.2.2 Bipartite Graph Coloring

A maximal assignment can be represented by a multigraph called a bipartite multigraph $G(V, E)$. A *bipartite graph* is a graph the nodes of which can be partitioned into two sets in such a way that each edge joins a node of the first set to a node of the second set. A *bipartite multigraph* is a bipartite graph with multiple edges. In the bipartite multigraph $G(V, E)$ the first set of nodes corresponds to the first stage switches, while the second set of nodes represents the third stage switches. Each call is represented by an edge joining respective nodes. Such graph representing maximal assignment has $2r$ nodes (r nodes in each set), N edges, degree n, and is n-colorable, or we can find n perfect matches. A *perfect matching* in G is a set of edges of G no two of which are adjacent and which include every node in G.

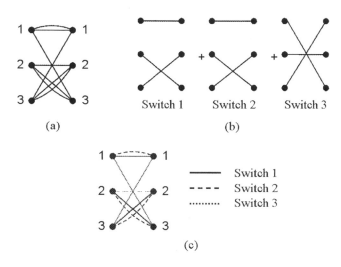

Figure 10-14. Simultaneous connections in C(3, 3, 3): (a) a bipartite multigraph, (b) perfect matches, and (c) graph coloring

A bipartite multigraph G for the calls in the C(3,3,3) of Figure 10-12 is shown in Figure 10-14(a). The graph G has $2r = 6$ nodes and $N = 9$ edges. Its degree is equal to 3 and it is 3-regular since each node in one set is joined to nodes in another set by exactly three edges. Perfect matchings in G and G colored with three colors are shown in Figure 10-14(b) and 10-14(c), respectively. It is clear that calls corresponding to edges in each perfect matching or colored with the same color can be set up through one center stage switch.

An efficient algorithm for finding perfect matching in bipartite graphs was proposed in [18]. This algorithm uses Euler cycles, therefore we first show how to find such cycles in the graph G. We will use two heaps. Nodes visited by the algorithm are stored in Heap_{temp}. Nodes stored in Heap_{CE} will constitute the Euler cycle [19]. The algorithm works as follows:

1) Put an arbitrary node of G on Heap_{temp}.
2) Repeat steps 2.1 to 2.3 until Heap_{temp} is not empty.
    2.1) $v :=$ the first node from Heap_{temp}.
    2.2) If v has any adjacent node then:

- $u :=$ a node adjacent to v;
- put u on Heap$_{temp}$;
- remove edge (u, v) from G;
- $v := u$.

2.3) If v has no adjacent nodes then move v from Heap$_{temp}$ to Heap$_{CE}$.

3) Draw Euler cycle traversing through successive nodes from Heap$_{CE}$ and assign successive numbers to respective edges.

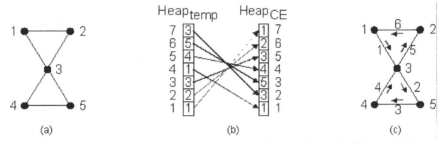

Figure 10-15. (a) The graph G, (b) Heaps contents during the run of the algorithm, and (c) the Euler cycle

Example 10-3:

Let us consider the graph G shown in Figure 10-15(a). All nodes are of even degree, so the Euler cycle exists in this graph. We start from node 1 by putting it on Heap$_{temp}$ in position 1 (see Figure 10-15(b)). The Heap$_{temp}$ has only one node at the beginning, so $v = 1$. Node 1 has two adjacent nodes (2 and 3). We choose one of them, say 2, put it on Heap$_{temp}$, and remove edge (1,2) from G. We have now $v = 2$. Node 2 has one adjacent node which is node 3 (1 is not adjacent since (1,2) is already removed). We put 3 on Heap$_{temp}$ and remove (2,3). Node 3 has three adjacent nodes, namely 1, 4, and 5. We choose 1, put it on Heap$_{temp}$, and set $v = 1$. Node 1 has no adjacent nodes so we move it from Heap$_{temp}$ to Heap$_{CE}$ on position 1. We start again with the first node in Heap$_{temp}$. At the top there is node 3, so we set $v = 3$ (step 2.1). In the next steps of the algorithm nodes 4, 5, and 3 will be put on Heap$_{temp}$. Node 3 has no adjacent nodes so it is moved to Heap$_{CE}$ (step 2.3). Similarly other nodes in Heap$_{temp}$ have no adjacent nodes, so all will be moved to Heap$_{CE}$ one by one, until Heap$_{temp}$ is empty. The Euler cycle can be drown starting from the node in position 1 of Heap$_{CE}$. This cycle is shown in Figure 10-15(c) (arrows show the direction of traversing nodes and respective numbers are assigned to edges).

Now we can move back to finding a perfect matching in bipartite multigraphs. The algorithm uses the property saying that any $2k$-regular multigraph H can be split into two k-regular spanning subgraphs H_1 and H_2. A perfect matching is a 1-regular spanning subgraph. To obtain a perfect matching of graph H by splitting it t times, it should be 2^t-regular. Therefore, the bipartite multigraph representing maximal assignment must be firstly converted to a 2^t-regular bipartite multigraph. This conversion is done by replacing each edge in G with $\alpha = \lfloor 2^t/n \rfloor$[3] multiple edges, and by adding $\beta = 2^t - n\alpha$ copies of edges in an arbitrary perfect matching P, where t is the nearest integer satisfying $2^t \geq N \, (= nr)$, i.e. $t = \lceil \log_2(rn) \rceil$. The edges of P and their copies are called bad edges. The algorithm works as follows:

1) Calculate parameters: $t = \lceil \log_2(2rn/2) \rceil$, $\alpha = \lfloor 2^t/n \rfloor$, and $\beta = 2^t - n\alpha$.
2) Create an arbitrary perfect matching P in G. This P does not necessarily consist of edges of G.
3) Create the graph H^1 by replacing each edge in G with α multiple edges and by adding β copies of each edge in P (called bad edges). Graph H^1 is 2^t-regular.
4) Starting from $s = 1$ to t do: split graph H^s into H_1^{s+1} and H_2^{s+1} edge disjoint 2^{t-s}-regular spanning subgraphs in the following way:
   - put $\lfloor \alpha/2^s \rfloor$ copies of each edge to H_1^{s+1}, and $\lfloor \alpha/2^s \rfloor$ copies to H_2^{s+1}; remove these edges from H^s;
   - put $\lfloor \beta/2^s \rfloor$ copies of each bad edge to H_1^{s+1}, and $\lfloor \beta/2^s \rfloor$ copies to H_2^{s+1}; remove these edges from H^s;
   - find an Euler cycle in each connected component of the remaining subgraph of H^s, move odd numbered edges to H_1^{s+1} and even numbered edges to H_2^{s+1}.
5) Choose a subgraph with the lower number of bad edges as the graph H^{s+1}.
6) A graph H^{t+1} is a perfect matching.

The total running time of this algorithm is $O(Nt) = O(N\log N)$. The following is an example of how it may be used.

Example 10-4:
Let us consider the example of Figures 10-12 and 10-14. In step 1, the parameters of the algorithm are calculated. Parameter t is equal to 4 (ceil of $\log_2 9$), and it means that the algorithm will perform step 4 four times. In steps 2 and 3, two additional graphs are created. Graph P is presented in Figure 10-16(a), while H^1, created from G (Figure 10-14(a)) and P, is shown in Figure 10-16(b). It consists of $\beta = 1$, copies of edges of P (bad edges are shown in dashed lines) and $\alpha = 5$ edges replacing each edge of graph G. So graph H^1 is still a regular graph but of degree $2^t = 16$. The first execution of step 4 results in two subgraphs presented in Figure 10-16(c). They were obtained by placing $\lfloor \alpha/2 \rfloor = 2$ copies of each edge in H_1^2 and H_2^2. Since $\beta = 1$, there is no bad edges in H_1^2 at this stage. In the remaining graph of H

---

[3] For any real z, $\lfloor z \rfloor$ denotes the largest integer less than or equal to z, while $\lceil z \rceil$ denotes the lowest integer greater than or equal to z.

we have $\beta = 1$ copy of bad edges and $\alpha - 2\lfloor \alpha/2 \rfloor = 1$ copy of each edge of G. For this remaining graph we look for an Euler cycle and add odd numbered edges to H_1^2 and even numbered edges to H_2^2. The subgraph H_2^2 contains only one bad edge, so it is chosen as H^2. Its degree is $2^{t-1} = 8$. Now, step 4 is performed for the subgraph H^2. It is divided into two subgraphs of degree 4 (graphs H_1^3 and H_2^3 in Figure 10-16(d)). Graph H_2^3 has no bad edges, so this is the graph taken into consideration in the next iteration of step 4. When s is equal to 3, graph H^3 is divided into two subgraphs presented in Figure 10-16(e). Either of subgraphs can be chosen as graph H^4 since neither of them has bad edges. The chosen graph can be divided into two subgraphs where each of them is a perfect match as per Figure 10-14. Thus $H^{t+1} = H^5$ is returned by the algorithm as its final result.

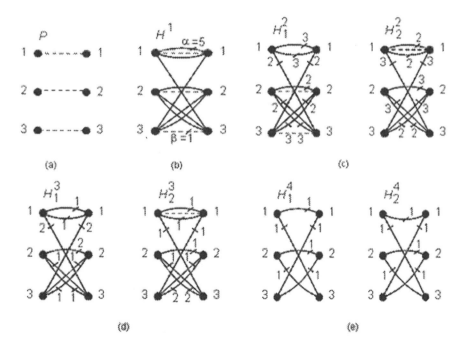

Figure 10-16. Algorithm perfect matching in use

10.3.2.3 The Looping Algorithm

This algorithm has been briefly discussed in chapter 9 in the context of Beneš networks. In this section we discuss it in more detail and examine its deployment in dilated Beneš networks. The looping algorithm was originally proposed to route simultaneous connections in switch fabrics composed of 2×2 switches [20]. It was later extended to switch fabrics with $2^t \times 2^t$ switches [21]. When the switch fabric is composed of 2×2 switches, two inputs (outputs) of the same first (last) stage switch are called *dual*. The dual of input x (output y) is denoted by $\sim x$ ($\sim y$). As discussed in section 9.8, and recalled in Figure 10-17, the algorithm is used for Beneš networks as follows:

1) Find a non-connected input terminal x and set $x_0 = x$. If all input terminals are connected, the algorithm is ended.
2) Connect x_0 to $\pi(x_0)$ through the upper switch fabric.
3) Find $\sim\pi(x_0)$.
4) Find an input terminal x, $\sim\pi(x_0)$ to be connected to.
5) Connect x to $\sim\pi(x_0)$ through the lower switch fabric.
6) Find $\sim x$; if connected, go to step 1; otherwise, set $x_0 = \sim x$ and go to step 2.

One run of steps 1 to 6 is shown in Figure 10-17. Arrows from left to right (from right to left) mark the connection set up in step 2 (5) through the upper (the lower) switch fabric. After all input terminals have are connected, the algorithm repeats itself.

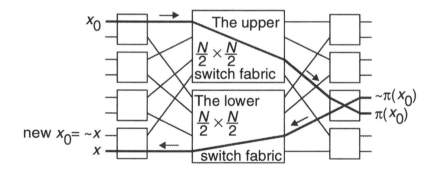

Figure 10-17. Operation of the looping algorithm in a Beneš switch fabric

We now examine how the lopping algorithm works in the dilated Beneš switch fabric [21]. Let $x_{m-1}x_{m-2}\ldots x_0$ be a binary representation of an input terminal x, whereas $\sim x$ is represented by $x_{m-1}x_{m-2}\ldots\bar{x}_0$, where $\bar{x}_0$ is a complementation of x_0. For instance 0 and 1 or 2 and 3 are duals. We use two tables: $MI[N;2]$ and $MO[N;2]$. An entry $MI[x;1] = y$ denotes that input terminal x is to be connected to output terminal y and $MO[y;1] = x$ denotes that output terminal y is to be connected to input terminal x. $MI[x;2] = MO[y;2] = 0$ means that the call $<x; y>$ is to be set up through the upper switch fabric, and $MI[x;2] = MO[y;2] = 1$ means that the lower switch fabric is used. Other entries, for instance $MI[x;2] = MO[y;2] = 2$, denote that this connection is not set up yet. The algorithm works as follows:

1) Set matrices MI and MO: for $i = 0$ to $N - 1$ (i.e. for all input and output terminals)

$MI[i;1] := \pi(i)$ (input terminal i is to be connected to $\pi(i)$);
$MI[i;2] := 2$ (input terminal i is not connected yet);

$MO[i;1] := \pi^{-1}(i)$ (output terminal i is to be connected to $\pi^{-1}(i)$);
$MO[i;2] := 2$ (output terminal i is not connected yet).
2) Find a non-connected input terminal x and set $x_0 := x$.
   If all input terminals are connected the algorithm is ended.
3) Connect x_0 to $\pi(x_0)$ through the upper switch fabric:
   $MI[x_0;2] := 0$;
   $MO[\pi(x_0);2] := 0$;
4) Find $\sim\pi(x_0)$.
5) Find an input terminal x, $\sim\pi(x_0)$ is to be connected to: $x := \pi^{-1}(\sim\pi(x_0))$
6) Connect x to $\sim\pi(x_0)$ through the lower switch fabric:
   $MO[\sim\pi(x_0);2] := 1$;
   $MI[x;2] := 1$;
7) Find $\sim x$; if connected go to step 2; otherwise, set $x_0 = \sim x$ and go to step 3.

Example 10-5:
An example of how the looping algorithm works in an 8×8 dilated Beneš
switch fabric is shown in Figure 10-18 assuming that the following
permutation is to be set up:

$$\Pi = \begin{pmatrix} 0 & 1 & 2 & 3 & 4 & 5 & 6 & 7 \\ 4 & 1 & 5 & 7 & 0 & 2 & 6 & 3 \end{pmatrix}. \tag{10.4}$$

Matrices MI and MO are initialized in step 1. We start from input
terminal $x_0 = 0$ (step 2). This input will be connected to output terminal 4 by
the upper switch fabric. After step 3 we have $MI[0;2] = MO[4;2] = 0$. For
output terminal 4 we have $\sim 4 = 5$ (step 4) and $x = \pi^{-1}(5) = 2$ (step 5), so
call <2, 5> is to be set up through the lower switch fabric, and after step 6
we have $MO[5;2] = MI[2;2] = 1$. At the end of step 6 we have $x = 2$. The
dual of 2 is 3. Since input terminal 3 is not connected yet, we put $x_0 = 3$ (step
7) and go back to step 3. The remaining part of the procedure is shown in
Figure 10-18. The upper and the lower switch fabrics at this level are
numbered 0 and 1. Connections to these inner switch fabrics are shown in
Figure 10-18(a) in bold lines. Each outer stage switch has only one input
(output) active. Permutation Π is now divided into two permutations Π_0 and
Π_1, where x in permutation Π is replaced by $\lfloor x/2 \rfloor$ in permutations Π_0 and Π_1.
We have:

$$\Pi_0 = \begin{pmatrix} 0 & 3 & 4 & 7 \\ 4 & 7 & 0 & 3 \end{pmatrix} \Rightarrow \begin{pmatrix} 0 & 1 & 2 & 3 \\ 2 & 3 & 0 & 1 \end{pmatrix}; \quad \Pi_1 = \begin{pmatrix} 1 & 2 & 5 & 6 \\ 1 & 5 & 2 & 6 \end{pmatrix} \Rightarrow \begin{pmatrix} 0 & 1 & 2 & 3 \\ 0 & 2 & 1 & 3 \end{pmatrix}$$

$$\tag{10.5}$$

Now, the looping algorithm can be used once again for these permutations. For permutation Π_0 the upper and the lower switch fabrics are numbered 00 and 01, while switch fabrics numbered 10 and 11 are used by Π_1. Matrices *MI* and *MO* for this run of the algorithm are shown in Figure 10-18(c) and connections are shown in bold lines.

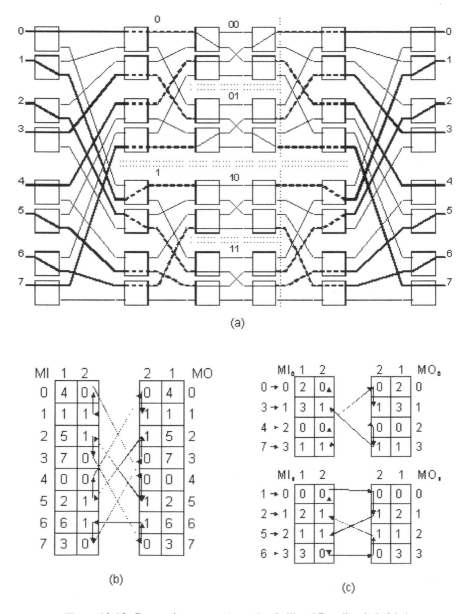

Figure 10-18. Connections set up in an 8 × 8 dilated Beneš switch fabric

10.4 OTHER CONSIDERATIONS FOR CONTROL ALGORITHMS OF OPTICAL FABRICS

We discussed control algorithms for space switching fabrics, whether based on optical or electronic technologies. Wavelength Division Multiplexing (WDM) adds the dimension of wavelength to that of space in optical switches and networks. Therefore, optical-switch control algorithms must manage switching operations at the wavelength granularity. Optical switching systems can also embrace wavelength conversion and some other optical processes. Therefore, control algorithms may have to control wavelength converters, tunable devices, optical amplifiers, and other optical components.

One of the important issues in designing control algorithms is their time complexity and whether, or not, they have to set up calls in real time. Different switching methods demand different requirements in this regard. Many of the control algorithms discussed in this chapter may be developed, or adapted, for circuit-switched OXCs. Optical packet switching will demand special control algorithms with time limits, like electronic packet switching, and certain optical technology considerations. Algorithms based on graph coloring are too complex and slow to fulfill these requirements on a packet by packet basis. Different approaches have been proposed for finding matches (not necessary perfect) in electronic packet switches using hardware solutions. Similar approaches and implementations may have to be pursued in the optical domain for optical packet switching.

REFERENCES

[1] G. Maier and A. Pattavina, "Design of photonic rearrangeable networks with zero first-order switching-element-crosstalk," IEEE Trans. Commun., Vol. 49, No. 7, pp. 1268-79, July 2001.

[2] A. Jajszczyk and G. Jekel, "A new concept—repackable networks," IEEE Trans. Commun., Vol. 41, No. 8, pp. 1232-37, Aug. 1993.

[3] F. K. Hwang, "The Mathematical Theory of Nonblocking Switching Networks," World Scientific Publishing, 1998.

[4] W. Kabacinski and G. Danilewicz, "Wide-sense and strict-sense nonblocking operation of Multicast Multi-$\log_2 N$ Switching Networks," IEEE Trans. Commun., Vol. 50, No. 6, pp. 1025-36, June 2002.

[5] C. Clos, "A study of non-blocking switching networks," The Bell System Tech. J., Vol. 32, No. 2, pp. 406-24, 1953.

[6] C.-T. Lea, "Multi-$\log_2 N$ networks and their applications in high-speed electronic and photonic switching systems," IEEE Trans. Commun., Vol. 38, No. 10, pp. 1740-9, Oct. 1990.

[7] V. E. Beneš, "Mathematical Theory of Connecting Networks and Telephone Traffic," Academic Press, 1965.

[8] D. Slepian. "Two theorems on a particular crossbar switching network," Unpublished memorandum, 1952.

[9] M. C. Paull, "Reswitching of connection networks," The Bell System Tech. J., Vol. 41, No. 3, pp. 833-55, May 1962.

[10] D.-J. Shyy and C.-T. Lea, "Rearrangeable nonblocking $Log_2^d(N, m, p)$ Networks," IEEE Trans. Commun., Vol. 42, No. 5, pp. 2084-6, May 1994.

[11] W. Kabacinski and M. Zal, "A new control algorithm for rearrangeable mulit-log_2N+m switching networks," Int. Conf. on Telecommunication, ICT 2001, Bucharest, Romania, Vol. 1, pp. 501-6, June 2001.

[12] V. I. Neiman, "Structure et commande optimales des réseaux de connexion sans blocage," Annales des Télécomunication, pp. 639-43, July-Aug. 1969 (in French).

[13] N. T. Tsao-Wu, "On Neiman's algorithm for the control of rearrangable switching networks," IEEE Trans. Commun., Vol. COM-22, No. 6, pp. 737-42, June 1974.

[14] A. Jajszczyk, "A simple algorithm for the control of rearrangeable switching networks," IEEE Trans. Commun., Vol. COM-33, No. 2, pp. 169-71, Feb. 1985.

[15] C. Cardot, "Comments on 'A simple algorithm for the control of rearrangeable switching networks'," IEEE Trans. Commun., Vol. COM-34, No. 4, p. 395, April 1986.

[16] H. Y. Lee, F. K. Hwang, and J. D. Carpinelli, "A new decomposition algorithm for rearrangeable Clos interconnection networks," IEEE Trans. Commun., Vol. 44, No. 11, pp. 1572-8, Nov. 1996.

[17] F. K. Hwang, "Control algorithms for rearrangeable Clos networks," IEEE Trans. Commun., Vol. COM-31, No. 8, pp. 952-4, Aug. 1983.

[18] N. Alon, "A Simple algorithm for edge-coloring bipartite multigraphs," Information Proc. Lett., Vol. 85, No. 6, pp. 301-2, March 2003.

[19] R. Sedgewick, "Algorithms in C Part 5. Graph Algorithms," Addison Wesley, 3rd edition, 2002.

[20] D. C. Opferman and N. T. Tsao-Wu, "On a class of rearrangeable switching networks," The Bell System Tech. J., Vol. 50, No. 5, pp. 1579-618, May-June 1971.

[21] S. Andresen, "The looping algorithm extended to base 2^t rearrangeable switching networks," IEEE Trans. Commun., Vol. COM-25, No. 10, pp. 1057-63, Oct. 1977.

[22] K. Padmanabhan and A. N. Netravali, "Dilated networks for photonic switching," IEEE Trans. Commun., Vol. COM-35, No. 12, pp. 1357-65, Dec. 1987.

Chapter 11

OPTICAL SWITCHING IN COMMUNICATIONS NETWORKS

Tarek S. El-Bawab

The success of optical fiber as a transmission medium throughout the period from the mid 1970s to late 1980s, and after, lead to large interest in the potential of fiber-optic communication systems. Researchers in this field have tapped the tremendous bandwidth of fiber by utilizing the wavelength and time domains besides the space domain. As a result, wavelength division multiplexing (WDM), optical time division multiplexing (OTDM), and optical code division multiplexing (OCDM) have all been pursued and systems based upon each were proposed. Some research has also investigated systems which exploit more than two domains. However, the combination of space and wavelength domains has clearly attracted most of the attention, due to the ease of accessing the optical wavelength domain, and to the huge success and widespread deployment of WDM.

WDM enabled true utilization of the huge fiber bandwidth and created the case for optical switching. These developments paved the way for the birth of the engineering and scientific discipline of optical networking. For decades, communications networks had been based only on electrical and electronic technologies. Optical networks presented a proposition that potentially facilitated great advances and benefits compared to networks based on these technologies.

Optical fiber is characterized by very large bandwidth and very low attenuation. WDM optical networking exploits these qualities and utilizes new network elements to realize communications networks with several distinctive attributes, such as huge information-carrying capacity, transparency to signal type and format, reconfigurability, scalability, flexibility of design and operation, and cost effectiveness.

WDM has been deployed for point-to-point transmission links to interconnect SDH/SONET network elements and rings. In order to improve their networks in terms of scalability, manageability, and cost-effectiveness, carriers upgraded their SDH/SONET rings to WDM rings. Meanwhile, WDM mesh solutions that are more bandwidth efficient emerged too. Mesh architectures can offer several diverse restoration paths in the event of a failure without dedicated protection capacity. Therefore WDM mesh networks have been receiving considerable research interest.

WDM created an abundance of bandwidth and motivated research in advanced optical transmission, switching and networking. In terms of switching methods, optical switching, like its electronic counterpart, may be performed at the granularities of circuits or packets leading to optical circuit switching (OCS) and Optical Packet Switching (OPS), respectively. In recent years, burst switching, where switching takes place at the intermediate granularity of bursts, was also proposed and optical burst switching (OBS) gained popularity among the research community.

In this chapter, we discuss WDM optical communications networks and the role and applications of optical switching in these networks. Our use of the term *communications networks* embraces telecommunications, computers, and cable-TV networks; or networks that carry voice, data, video, and mixtures thereof. The main focus in this chapter and in chapter 12 is on OCS as the strongest candidate for practical deployment of optical switching in the near future. Chapter 13 is dedicated to OPS and OBS.

This chapter puts optical switching into the context of optical networks. To do this, we begin by reviewing the progress in optical networking research from infancy until today. We explore the optical network elements which form the basis of the optical layer. We focus our attention on network elements which utilize optical switching technologies. Emerging networking techniques, which are being developed for these elements and networks, are discussed. Finally, we discuss the current status of optical switching deployment.

11.1 EARLY OPTICAL NETWORKING RESEARCH

In the early days of optical networking research, networks employing optical switching were not generally investigated. Strong interest in optical switching technologies and devices was not fully matched by a perception that optical switching was intrinsic to optical networks. This was probably due to two main reasons. First, the nature of optics may have implied that different techniques, such as those based on passive optics and tunable devices, may replace the traditional networking methods of electronics leading to communications architectures that are free of switching in the traditional sense. Optics was not well suited, for example, to mimic large-scale telephone switching centers. Second, optical switching technologies at that time were not mature enough to build reliable and commercially viable products. Some research in areas such as digital optics was very optimistic, but did not deliver. It is possible, ironically, that uncalculated optimism was counterproductive and did not help the cause of turning optical switching into deployment reality.

As interest in optical networks research emerged, new terminologies were coined to describe techniques that were devised to realize these networks. The most common of these were the *Broadcast and Select* (B&S) and *Wavelength Routing* (WR) techniques. The terminology of *single-hop* versus *multihop* networks was also common.

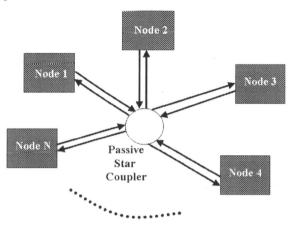

Figure 11-1. Passive-star-coupler based Broadcast-and-Select (B&S) optical network

Broadcast-and-Select (B&S) networks represent the first attempt by the research community to realize WDM optical networks [1]-[13]. This technique was also adopted for interconnect applications by computing researchers [14]. The B&S principle is depicted in Figure 11-1 using a star topology, based on a passive star coupler. Other topologies, such as bus and ring, and other optical devices have also been used. In Figure 11-1 the network interconnects N nodes and an $N \times N$ coupler. Each node is connected to the coupler by two fibers, one for transmitting to the coupler and one for receiving from it. At any given time, all nodes may transmit provided each of them uses a different wavelength than the others. All transmissions are broadcast to all nodes by virtue of the power-splitting and distribution feature of the coupler. On the receiving side, each node receives all WDM (multiwavelength) transmissions and selects only the wavelength that is intended for it. Different forms of this networking technique can be realized depending on design parameters such as the use of fixed and/or tunable transmitters/receivers and the type of control and management schemes deployed.

One of the first known experimental demonstrations of B&S optical networks was LAMBDANET [7]-[10]. This testbed was based on the topology of Figure 11-1. Each node was equipped with a fixed-wavelength transmitter, a demultiplexer, and up to N parallel receivers. Each node transmitted over a unique wavelength and received from all other nodes over their wavelengths simultaneously and asynchronously. Incoming WDM signals to any node were demultiplexed and each wavelength (*lightpath*) was directed to a receiver. The network had a control scheme based on a dedicated wavelength channel, which was time-slotted with fixed slot size. Therefore, each node was equipped with an additional transceiver to exchange control information with the other nodes. A special node might be designated as a centralized master for synchronization and monitoring purposes. While data and control signals were transmitted among nodes optically, signal reception, processing, and grooming were performed electronically.

With a combination of point-to-point, multicasting, and broadcasting capabilities, the non-blocking LAMBDANET was proposed as backbone for distribution of services among clusters of nodes, such as telecommunications switching centers, video distribution centers, and large business sites. Hierarchical networks and T1-based virtual private networks were envisioned using the LAMBDANET concept. The network was demonstrated experimentally with a 16×16 star and 16-18 wavelengths. It was suggested that incorporation of optical switches could improve signal routing, and optical cross-connection was identified as possible area of investigation. However, no work was reported later in this direction.

One of the problems of B&S networks is that the number of nodes is linearly related to the number of wavelengths. Also, the power budget of these networks can suffer large losses, which increase with increasing the number of nodes and with enlarging the physical size of the network. In the network of Figure 11-1, optical power launched by any node is split equally among all nodes. Moreover, it encounters coupler's excess/insertion loss (a logarithmic function in N) and fiber transmission loss. Hence, B&S networks are not scalable and are not particularly suitable for core/long-haul networks. Optical amplifiers (OAs) can be used to balance out loss. However, the use of amplifiers introduces noise onto the lightpath, which means that only a limited number of amplifiers may be cascaded. In B&S networks, each node gets the information destined to all others. Therefore, there is a security concern too. The design of B&S networks must take these issues into consideration.

Several research groups carried out work in the area of B&S local and metropolitan networks. Many efforts focused on medium-access-control (MAC) protocols to co-ordinate the access of nodes to the shared (broadcast) transmission medium [15]. Many projects and experimental initiatives/testbeds are reported as well [16]-[21]. For example, Rainbow [16] was deployed for several years.

Clearly, plain B&S networks do not embrace optical switching, at least not in the traditional telecommunications sense. Although signals are transferred from some space port to another, this is done over a broadcast medium where transmissions of all are received by all, without directivity control as found in switching systems.

An extension of the B&S approach for optical local area networks was proposed in [22]-[24]. This technique was packet based and used successive opto-electronic conversions to relay data packets from one node to another. Each node transmits at specific fixed wavelengths and receives at specific fixed ones. Wavelength assignment is made in such a way that although nodes may not all be able to communicate directly with each other (in a *single-hop* manner), all of them can ultimately connect by successive relaying of packets through intermediate nodes in a multihop fashion. Therefore, this approach was known as the *multihop* approach. In this arrangement each node acts as a repeater. It receives packets over one of its receiving wavelengths, converts it to the electronic domain. If the packets are destined to this node, they are delivered locally. If not, they are used to modulate a transmitting laser and sent out to another node. The process repeats itself until packets reach their destinations. When the connectivity provided by the wavelength assignment scheme is based on generalization of the well-known perfect shuffle configuration, the network is referred to as the ShuffleNet [24]. The physical topology, on the other hand, can be bus, tree, star, or other topologies. In this approach, the network enables concurrent optical transmissions among nodes at the expense of some bandwidth while getting around the technology limitations of the time, such as slow wavelength tuning speeds and immaturity of Lithium Niobate ($LiNbO_3$) Electro-Optic (EO) switches. However, with excessive optoelectronic conversions on a packet by packet basis, this approach was unpractical and expensive. Also, it did not provide the means to truly utilize the vast optical bandwidth.

The B&S technique, which can be used to interconnect network nodes, was also adopted by some research groups as a technique to actually build optical nodes and hubs with some switching capability (B&S in a box). One of the first examples of this abstraction is the parallel λ-switch [25]. This switch relied on extensive use of power splitters/combiners with tunable filters and wavelength converters (WCs) to build an optical hub node. The complexity and cost of this kind of nodes can be high. They are also not scalable beyond some limits.

Although the attention of the research community had focused on B&S networks for some time, the technique has enjoyed so far only little deployment. In part, this is because WDM has not penetrated the access/local area. Deployment of WDM in past years has mainly focused on core and regional metro networks for which B&S is not particularly suitable.

Another optical networking technique that is broadly discussed in the literature is *Wavelength Routing* (WR). A wavelength routed network is one where the wavelength itself serves as an identifier defining how routing/switching decisions for lightpaths are made. The wavelength-routed lightpath can be fixed (hard-wired) or reconfigurable (switched). Hence, unlike B&S, WR embraces optical switching. While early WR research mainly focused on fixed paths, later work turned its attention to optical-switching based WR. The concept of WR networks is depicted in Figure 11-2 assuming a network of N nodes. Both the fixed and switched principles are shown. Each node in the network is shown with its transmitter portion on the left and receiver portion on the right. In the case of fixed WR, N wavelengths are used in the network ($N-1$ are required for each node to communicate with all others). If optical switching is utilized, fewer wavelengths may be used at the expense of allowing an acceptable, customary, blocking probability.

WR routing enables wavelength reuse in the network; that is multiple lightpaths can use the same wavelength at the same time as long as they do not overlap with this wavelength on any link. Wavelength reuse makes it possible to have more lightpaths with fewer wavelengths. The combination of wavelength routing and wavelength conversion can enhance the connectivity and performance of WR networks[1].

One of the first demonstrations of the fixed WR concept was reported in [26]. In this demonstration a multiplexer/demultiplexer (mux/demux) based optical hub is used at the center of a star topology to provide optical cross-connection among electronic switching centers with optical terminations. The hub was optical-switching free, but it was suggested that an enhancement of its functionality could be provided by Lithium Niobate optical switches.

Other WR research activities were reported using passive optics with limited optical switching functionality and involving lightpath add/drop capability [27]-[28]. The complexity of telecommunications switching

---

[1] Some studies in the literature classify optical networks into wavelength translating (WT) and wavelength routing (WR) depending on whether wavelength conversion is used, or not, respectively. We treat wavelength conversion, which is an important feature, as an option in WDM networks. It can be used also in architectures based on the B&S concept.

centers was increasing with the rising bit rates. WR was viewed as an opportunity to relax this complexity by routing high-bit-rate data streams in the optical domain.

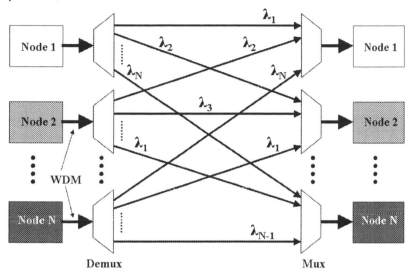

(a) Fixed (Static) WR

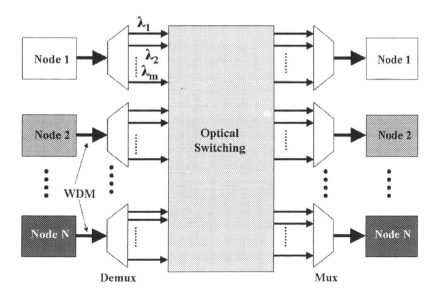

(b) Switched (Dynamic) WR

Figure 11-2. Concept of wavelength routing (WR), fixed and switched versions.

WR, which we have just described at the network level, can also be the basis for building optical switching nodes (WR in a box). Indeed, optical cross-connects (OXCs), which are discussed later in this chapter, use the principles of WR and networks based upon them are wavelength routed.

Until the end of 1980s optical switching research was largely focused on component and device issues. Little attention was paid to putting optical switches into the perspective of the progressing field of optical networking research [29]-[31]. This situation soon changed by the beginning of the following decade [32]-[38].

11.2 PATHWAY TOWARDS MATURITY

Optical networking and optical switching research enjoyed considerable progress and popularity during the 1990s. Industry and academia put their efforts together into several collaborations to investigate optical networking issues. For example, a pre-competitive consortium of major US industrial and academic partners was formed to study the technology, architecture, and applications of OCS networks with electronic control [39]. This proposal embraced a mix of the B&S and WR approaches. A proposal to extend the multihop approach to wide area networks by utilizing Acousto-Optic (AO) switching and opto-electronic (OEO) wavelength conversion was also reported [40]. This network consists of an OCS core surrounded by user access nodes. The core is based on OXCs and optical amplifiers (OAs) under common network control for dynamic configurability. Access nodes provided user interface, packet processing and switching, and wavelength conversion. An ATM-based data-communication network was proposed to host optical-network management systems [41]. Several other research projects were proposed and/or carried out.

Meanwhile, a layered approach emerged in optical networking research. This approach put optical networking into the same perspective as modern telecommunications and data networking. The SDH/SONET transport network comprises three main sub-layers, namely a circuit layer, a path layer, and a transmission media layer, as depicted in Figure 11-3. Optics is used only for point-to-point transmission and thereby confined to the transmission media layer. A new layer based on optical technologies was proposed at the path level by several research activities and was positioned as the cornerstone in future transport architectures. This optical-switching based layer became to be known as the optical path layer, the optical transport layer, or, simply, the optical layer.

Figure 11-4(a) depicts a special switch that is typically composed of passive couplers and 1×2 planar thermo-optic (TO) devices. This switch was named the delivery and coupling (D&C) switch [42]-[55]. An optical cross-connect (OXC) was proposed based on this fabric and is shown in Figure 11-4(b)[2]. The system also comprises an electrical digital cross-connect system (DCS) to interface with the add/drop ports and to provide switching at smaller granularities (not shown in figure). This design facilitates node compactness while increasing switch flexibility and capacity.

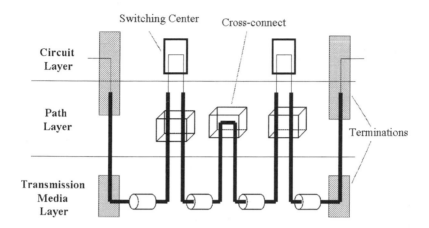

Figure 11-3. Functional layers of a transport network

The D&C-based OXC was designed for 16 space ports, with 8 wavelengths each, and 2.5 Gb/s transmission rate per wavelength leading to a switching capacity of 320 Gb/s. Nevertheless, 16-32 wavelengths ports and other system configurations were considered as well. In partial demonstration of this OXC, system switching time of 14 msec was reported. As part of this work, an SDH-compatible interface was proposed to accommodate client data into optical transport payloads, and vice versa. Performance monitoring, centralized management and an operation and maintenance (OAM) system were also reported. The management and control system was based on Telecommunication Management Network (TMN) concepts with a network processor and element processors attached to a dedicated data communication network via standardized interfaces. Inter-element management used proprietary interfaces.

The D&C based OXC demonstrated several favorable optical-switching attributes for the first time. Meanwhile, it was criticized as being optimized for many links populated with few wavelength channels (low fiber utilization). It can be argued that this OXC is only modular and scalable up to certain limit. There is also concern about its loss characteristics and cascadeability.

---

[2] In the beginning, OXCs were also named optical path cross-connects.

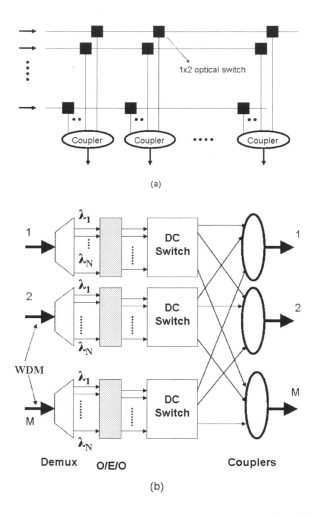

Figure 11-4. Switching architecture based on the delivery and coupling (D&C) switch: (a)
The D&C switch, (b) Optical cross-connection based on this switch

For the last two decades, research sponsored by the European community/union made several contributions to optical networking. Theoretical work in the COST[3] 239 project assessed the potential range and capacity of a European-scale multiwavelength optical network [56]-[57]. The RACE[4] program (1988-1995) had a number of optical-networking projects including OSCAR and MWTN [58]-[62].

The RACE I Optical Switching systems, Components, and Applications Research (OSCAR) project proposed a cross-connect node architecture

---

[3] COoperation in the field of Scientific and Technical research.
[4] Research and development in Advanced Communications technologies in Europe

where a Lithium Niobate based OXC was combined with an electronic DCS. The former assumed the responsibility of transporting large bandwidth pipes, while the latter handled smaller traffic granularities and performed traffic grooming, sorting, and signal regeneration. Experiments demonstrated the application of optical switching in provisioning and protection [59].

The RACE II "Multi-Wavelength Transport Network" (MWTN) project utilized wavelength-selective components, optical space switches, and optical amplifiers (OAs) to build an OXC [60], [62]. The node also comprised a DCS. More than one optical switching technology was tested in this project including Lithium Niobate Acousto-Optic (AO) switches and Semiconductor Optical Amplifier (SOA) gate arrays. The project embraced a centralized network management approach, where the management system architecture consisted of three main hierarchical processor levels: the network operation-system processor level on top, the network-element mediation processors level in the middle, and the device/component processors level at the bottom. Processors communicated with each other over dedicated administrative data-communication networks. Meanwhile, pilot tones were used to supervise lightpaths. At one transmitting end, a distinct electrical pilot tone in the kHz range was added to each lightpath. This tone was extracted at intermediate OXC nodes by tapping off a small proportion of the optical power into a monitoring module. The tone was electrically filtered out and used to obtain power level and channel quality information that were used by the management system for fault management. Using this approach, MWTN demonstrated the viability of managed optical networks and survivable optical rings.

The mid 1990s saw the shift from the RACE program to the ACTS[5] program (1995-1999) in Europe, where special focus was placed on field trials. The practicality of European-scale WDM networking was addressed by PHOTON (Pan-European Photonic Transport Overlay Network) and OPEN (Optical Pan-European Network) projects using analysis, laboratory work, and field trials [63]. Field trials were used to implement experimental optical networks over existing fiber infrastructures, crossing borders between countries. This provided challenging transmission platforms which are inhomogeneous in terms of fiber type, repeater spacing, and wavelength grids. Other projects, such as METON (METropolitan Optical Network), MEPHISTO (Management of Photonic Systems and Networks), MOON (Management Of Optical Networks) and PELICAN made other contributions within the framework of ACTS.

PHOTON [64]-[65] featured an OXC in a WR WDM network trial demonstration spanning 520 km between Germany and Austria. Up to 10 Gbit/s per wavelength channel was demonstrated in a system design for 8

[5] Advanced Communications Technologies and Services

wavelengths per port. The core part of the OXC was an opto-mechanical 16×16 non-blocking switch module and the system had 2-3 multi-wavelength bidirectional input/output ports and 2 single-channel bidirectional add/drop ports. Other modules in the system included a variable optical attenuator module for output power management[6] and a monitoring module that comprises a 16×1 fiber switch, meters, and spectrum analyzer. The system also incorporated a wavelength reference module, control module, and a basic OAM functionality. An optical loop of twice the transmission target (1050 km), passing through the OXC twice, was successfully demonstrated at 10 Gb/s transmission rate.

OPEN dealt with a pan-European optical network with wavelength conversion capabilities [66]. The OPEN OXC was based on a B&S architecture and consisted of three stages: a B&S stage, a wavelength-selection stage, and, finally, a wavelength conversion and multiplexing stage. The B&S switching fabric itself was an interconnection of splitters, clamped-gain SOA (CG-SOA) gates, and combiners in a perfect shuffle configuration. SOA-based Mach-Zehnder interferometers were used as wavelength converters (WCs). A monitoring board and a basic centralized management system were also devised. Partial implementation of a 4×4 version of the OPEN OXC was demonstrated. Transmission tests in a Norway-Denmark field trial demonstrated cascadeability of three OXCs over more than 1000 km of fiber with satisfactory performance. Simulation and laboratory studies suggested that up to ten OXCs may be cascaded and that chromatic dispersion and WC-induced jitter would be the primary limiting factor beyond this. It was claimed that the signal regeneration capability of WCs had a favorable effect on OXC cascadeability. Without wavelength conversion, transmission performance would have been primarily limited by noise and fiber non-linearities. Although SOAs compensates for part of the power losses, Erbium-Doped Fiber Amplifiers (EDFAs) were necessary in the OXC and in optical transmission lines. This was due to the B&S power budget. The Norway-Denmark trial ran for three months under harsh environmental conditions without hardware failure.

The project METON investigated metro WDM rings. As part of this work, the issue of optical networks scalability was examined and some guidelines were developed for using signal regeneration [67]. The ACTS' MEPHISTO and MOON projects focused on network management [68].

---

[6] It is necessary to equalize WDM signal powers. Different wavelengths may have different power levels because of diverse path lengths (in-line and within switch fabrics) and because OA gains are wavelength dependent. Equalization ensures that stronger signals do not dominate the others in subsequent links. It also controls crosstalk (which becomes worse with large signal-power differences).

Functional modeling and management-information modeling of an OXC were developed.

Today, EU funded research is focused on broadband access. There is appreciation that widespread deployment of affordable broadband services, wireless communications, next-generation cable services, and IP based applications depends heavily on optical networking and switching. Therefore, research in optical networks and technologies, represents strategic component of the ongoing IST[7] Program (1998-2006) [69]. For instance, the project LION (Layers Interworking in Optical Networks) has investigated the use of Generalized Multi-Protocol Label Switching (GMPLS)[8] to integrate the optical and IP layers [70].

In the COST program, meanwhile, action 266 (Advanced Infrastructure for Photonic Networks) has also investigated optical network control issues [71].

One of the most well-known optical network research programs is MONET (Multiwavelength Optical NETworking) [72]-[77]. MONET (1994-1999) was established by US industrial partners with collaboration of government agencies and laboratories and supported by the Defense Advanced Research Projects Agency (DARPA). The program was launched before commercial WDM deployment to research multiwavelength optical networks of national scale. Two eight-wavelength field demonstrations were carried out within the framework of MONET, one in New Jersey and the other in Washington D.C. The program defined several optical network elements including several types of OXCs and incorporated management and control functionalities in these elements.

The MONET New Jersey field trial interconnected three testbeds at three different locations: a Local-Exchange-Carrier (LEC) testbed, an OXC testbed, and a long-distance transmission testbed [73]. The LEC testbed was a survivable WDM ring with two dual 2×2 liquid-crystal (LC) based cross-connects and two Optical Add/Drop Multiplexers (OADMs). The latter used 2×2 opto-mechanical switches to either add/drop individual wavelengths channels or to pass them through. Optical switches were also used to demonstrate a 15-msec optional automatic protection switching and used for performance monitoring related measurements. In the OXC testbed, the switch fabric was structured in such a way that a space matrix was designated to each wavelength. Thus, it comprised eight 4×4 Lithium-Niobate wavelength-specific modules, where any of the four inputs on each module can be connected to any of its four outputs. A control unit was devised to configure the switch and monitor its performance. Finally, the long-distance testbed emulated a national-scale WDM transmission system.

---

[7] Information Society Technologies
[8] GMPLS is discussed in section 11.5

The MONET Washington DC network demonstration consisted of two interconnected rings (east and west) with wavelength conversion [77]. Six OXC and OADM network elements (nodes) were placed on the rings along with OAs.

Network elements on the east ring were all based on a principal research-prototype network-element architecture, which can be configured as an OXC, OADM, or OA. The functional difference between an OXC and an OADM in this demonstration was that the former supports four multi-wavelength interfaces whereas the latter supports only two. The percentage of add/drop the network element would perform, as opposed to pass-through, was another way to differentiate an OXC from an OADM. The main switching fabric in the OXC was a 48×48 matrix. This was made of eight 6×6 wavelength-specific non-blocking tree architectures. Additional 4×8 optical switching matrices were used to interconnect client signals to the main switch fabric. All optical switches were based on Lithium-Niobate EO units which can switch in nanosecond time scales [75]. System switching times in the microseconds range were reported [77]. Lithium-Niobate switch demonstrations in this field trial were large in terms of device quantity and system complexity.

The west ring was based on commercially available equipment including cross-connects and terminal multiplexers that were adapted to provide OADM functionality. The cross-connects in this ring were based entirely on electronic switching fabrics (OEO systems) and did not involve any optical switching. One of the goals of the Washington DC trial was to examine the interoperability of two WDM rings, the equipment of which were provided by two different vendors.

Optical network transparency was demonstrated in the New Jersey trial by concurrent transmission of analog FM-Video over one wavelength and digital SONET signals (OC-48) over the remaining seven wavelengths for more than 200 km. Other SONET transmission rates (OC-3 through OC-192) were transmitted concurrently in other experiments. Several other transmission experiments were carried out within the program. For example, the effect of propagation of optical power transients through optical switching fabrics was shown to be problematic.

Like many of the research initiatives of its time, the MONET management system architecture was hierarchical and comprised a network level, a network-element (mediation) level, and a component level. However, the architecture appeared more distributed than its predecessors and management was CORBA[9]-based. Management and control systems were hosted by ATM networks with an ATM switch assigned to each network element. The ATM network carried control signals among elements.

[9] Common Object Request Broker Architecture

Performance monitoring was carried out per wavelength and at the wavelength multiplex level. Measurements were performed inside network elements and at their edges in order to isolate faults. Interaction of different management systems, and administrative domains, to set up and tear down two-way 8-wavelength interconnections across the testbeds was demonstrated.

By the time the MONET program concluded, WDM systems with 32 wavelengths, and more, were commercially available. WDM had made impressive inroads into carrier networks. Also new optical switching technologies, such as optical MEMS, were developed. Therefore, the results of MONET are sometimes debated. However, the program did contribute to progress in optical networking research like many of the other projects discussed above.

Recently, a research program named OptIPuter was also reported [78]-[79]. The program is so named for its use of Optical networking, IP, and computer storage, processing and visualization technologies. OptIPuter can be thought of as an array of PC processors connected to an array of graphics cards and disks via an optical network. It is a virtual parallel computing system in which individual processors are widely distributed clusters. The backplane is delivered over reconfigurable 1-10 Gb/s lightpaths, and the mass storage system are large distributed data repositories which are fed by scientific instruments as peripheral devices. The interconnection operates in near real-time. Reconfiguration is facilitated by optical MEMS switching systems. Middleware is reported to control this switching equipment over an optical network that is sponsored by the State of Illinois (USA).

While large-scale research projects by consortia and long-term industry programs made significant contributions to optical networking research and represent milestones in this regard, several other activities by individual groups and organizations also made valuable contributions and lead to several developments in the 1990s (for example, see [80]-[103]).

By the turn of the century, optical networking and switching had made considerable progress, and had actually evolved from being an interesting area of research into a field of large investments, aggressive product developments, and standardization activities. The worldwide telecommunications market was booming and optical networking was the center of a great deal of attention. As WDM became established in carrier transmission systems [104]-[105], optical switching based OXCs emerged in product portfolios, and the attention of the industry, academia, and standardization communities turned to optical layer issues. Although the telecommunications market has undergone a sharp downturn since mid 2001, the fundamental vision of the optical layer remains sound and logical. Therefore, research in this area remains active (examples: [106]-[129]).

11.3 THE VISION OF THE OPTICAL LAYER

In sections 11.1 and 11.2 we discussed the evolution of optical networking research. Initially, the focus was on developing topologies, techniques and systems to enable concurrent optical transmission over multiple static channels per fiber. As the field progressed, the focus shifted to developing wavelength-routed network architectures where these channels can be allocated dynamically among several users with various service demands. By the mid 1990s, this led to the vision of the optical layer which can only be fully enabled by optical switching.

On the other hand, we have discussed in chapter 1 the evolution of switching in communications networks and the evolution of the architectures of these networks in recent decades. We saw how the growth of the Internet impacted both. Ultimately, the enormous demand for bandwidth, which is largely credited to the Internet, led to extensive deployment of WDM. The success of WDM coupled with progress in optical space switching technologies increased the interest of the telecommunications community in optical networks and in the optical layer during the second half of the 1990s [116].

The optical layer is expected to facilitate a transport infrastructure that is characterized by abundance of bandwidth, transparency, reconfigurability, manageability, scalability, reliability, and cost-effectiveness. It is expected to be capable of provisioning and managing lightpaths to serve the communication needs of clients such as IP routers, ATM switches, SDH/SONET network elements, and others. Transparency means that the optical network and its elements do not care whether the traffic they carry is composed of SONET/SDH frames, ATM cells, IP packets or Ethernet frames. Figure 11-5 is an elaboration of Figure 1-1 (chapter 1) where today's WDM transmission layer is empowered by optical switching and innovative network control to evolve into an agile optical layer. Several arbitrary paths are sketched. The path between routers A and C is a traditional path where both SDH/SONET and ATM equipment are involved. The path between routers B and E is based on the packet over SDH/SONET scheme and the path between C and F is where IP packets are carried directly over the optical layer. These alternatives have been discussed in chapter 1. None-IP paths are of course possible too and the figure shows as an example the path between two SDH/SONET network elements (G and H). The figure also depicts an entry strategy for the optical layer where it serves existing layers in today's network. As the network evolves, simpler architecture and models will be introduced. The roles of the IP and optical layers are expected to become even more important.

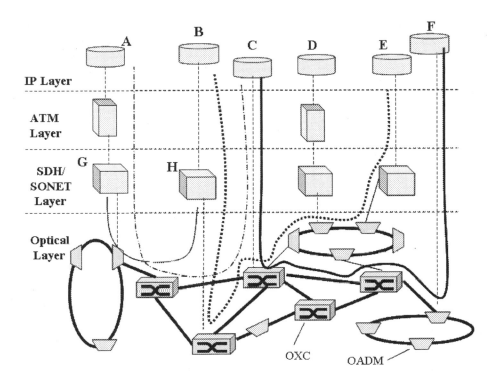

Figure 11-5. The vision of the optical layer as server layer for IP, ATM and SDH/SONET clients

A lightpath extends between two optical network elements (nodes) and can be routed through multiple intermediate elements. We envision several types of lightpaths. A lightpath can be an end-to-end fiber path including all wavelength channels carried by a fiber or a chain of fiber links. It can be a subset of these wavelength channels or a multiplex of channels sharing the same route. Finally, a lightpath can be an end-to-end single wavelength channel. We use the term wavelength channel in this classification to refer equally to fixed-wavelength channels and to channels where data may experience wavelength conversion(s) en route[10].

Lightpaths can be set up and torn down as required by the clients of the optical network. This results in a reconfigurable optical network.

---

[10] The terms wavelength path (WP) and virtual wavelength path (VWP) are used in the literature to classify single-wavelength lightpaths. A WP is a lightpath where data is transported over a fixed wavelength. A VWP is a lightpath where the wavelength can change from one link to another. A VWP-based network may require fewer wavelengths than a WP based network, especially when restoration capacity is taken into consideration. This is achieved at the expense of wavelength-conversion cost and complexity. Our use of the term lightpath embraces WPs, VWPs, multiple-wavelength paths, and fiber paths.

Reconfiguration is the ability of the network to provision lightpaths and topologies automatically and to alter them in response to changing traffic patterns, bandwidth demands, restoration requirements, and maintenance needs. Reconfigurability can also be exploited to design and evolve carriers' networks in a cost-effective way, to upgrade network links and elements without service disruption, and to provision optical virtual private networks (VPNs). Reconfigurability requires the network elements to be capable of rapid switching of lightpaths in a hitless manner, without traffic loss, or with minimum loss.

One of the important features of an optical network is the independence of its logical lightpath topologies and physical fiber topology, thanks to the passive nature of the fiber medium. This feature renders rich connectivity and provides a great deal of flexibility in design and operation. Several logical topologies can be established, and can co-exist, over a given physical topology. These logical topologies represent the optical network topologies as seen by various clients.

As such, the wavelength based optical network requires protocols and methods for Routing and Wavelength Assignment (RWA) in order to optimize its design, operation and minimize its cost. This is an active area of optical networking research at the time of writing, which is beyond our scope of discussion.

This vision of the optical layer implies a great deal of change to existing carriers' networks. Therefore, today's networks may evolve gradually, over time, toward this end. In the short term, network reconfiguration rates are expected to be slow. Once a lightpath is set up, it may remain for months or even years. In the future, this is expected to change leading to networks that are more dynamic in terms of handling larger arrival rates of service requests, having faster reconfiguration speeds, and shorter topology lifetimes.

11.4 OPTICAL NETWORK ELEMENTS

There are three main technological ingredients of optical networks. These are WDM transmission, optical switching, and optical-network control and management. Optical networks are composed of network elements (NEs) that utilize these ingredients and are interconnected by fiber links, cables and accessories. Owing to their role in configuring lightpaths and topologies, there are three main optical network elements:

1) Optical Line Terminal (OLT)

2) Optical Add/Drop Multiplexer (OADM)

3) Optical Cross-Connect (OXC)

OLTs, OADMs and OXCs usually incorporate optical amplifiers (OAs) as pre-amplifiers and/or post-amplifiers. An OA boosts the power of lightpaths, but is not in charge of switching and configuring them. An arbitrary example of an optical network is depicted in the bottom layer of Figure 11-5. In order to simplify the figure, only OADMs and OXCs are shown. However, clients can be connected to the optical network via OLTs, OADMs, or OXCs. OAs and OLTs are already used in today's WDM transmission systems. They do not utilize optical switching. Most OADM in use today do not utilize optical switching. However, interest in reconfigurable OADM (R-OADM), which can exploit optical switching, has been increasing over the past two years. OXCs utilize optical switching and have started to be deployed slowly during the past few years.

11.4.1 Optical Line Terminals

We only discuss OLT briefly. In this discussion, we mainly follow the approach of reference [130]. Figure 11-6 depicts a typical design for an OLT. OLTs are used currently in WDM point-to-point transmission systems. They are deployed at both ends of these systems, as shown in Figure 11-7. At one end, a number of wavelengths are multiplexed onto one fiber for outgoing transmission. At the other end, individual wavelengths are separated from each other and converted to the electrical domain where they are either delivered locally, or passed over to another OLT where they are regenerated, converted to optical domain, wavelength multiplexed, and transmitted over another link, and so on.

As depicted in Figure 11-6, a generic OLT is composed of transponders, a mux/demux pair, and OA(s). The role of a transponder is to adapt input signals from client layers to the requirements of the optical layer, and vice versa. The adaptation process usually involves wavelength conversion, to adapt proprietary wavelengths to ITU-standardized wavelengths, and network control/management functionalities. OLTs also extract the *optical supervisory channel* (OSC) which is used by the network for performance monitoring, OAM, and control. The current state of the art dictates that control, signal processing and wavelength conversion are performed by electronics. Therefore, transponders are OEO devices. The number of transponders in an OLT can be as large as the number of lightpaths it handles concurrently. However, transponders contribute significantly to the cost, size, and power consumption of OLTs. Therefore, it is often desirable to reduce their number.

After adaptation by transponders (Figure 11-6), signals are wavelength multiplexed onto one fiber. Several multiplexing technologies are commercially available today, the most important of which are gratings;

bulk, arrayed waveguide gratings (AWGs) and fiber Bragg gratings (FBG), and filtering technologies, such as dielectric thin-film filters. In principle, passive combiners and splitters may replace the mux/demux pairs in OLT. However, the use of mux/demux modules is advantageous since they introduce lower loss and because they can reject amplified spontaneous emission (ASE) noise.

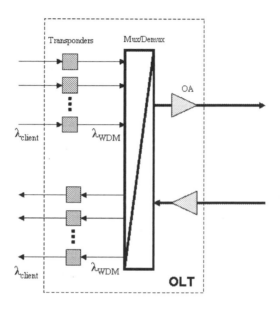

Figure 11-6. Optical line terminal (OLT) equipment

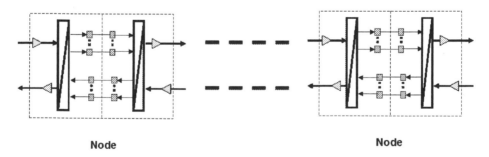

Figure 11-7. OLTs in a point to point WDM transmission link between two nodes

After multiplexing, signals may undergo amplification, if needed. EDFAs are typically used for this purpose. At the other end of the transmission link, the lightpath may also be amplified if needed and the reverse processes take place.

In an optical network, OLTs provide the interface between WDM transmission links and OXCs/OADMs. Figure 11-8 depicts an arbitrary example in this regard. In principle, OLTs can be included as part of OADMs and OXCs and this design approach has been considered by many equipment vendors. Today, they are usually treated as separate network elements and this helps carriers get around interoperability problems (due to lack of standardized WDM interfaces) and enable them to use equipment from different vendors.

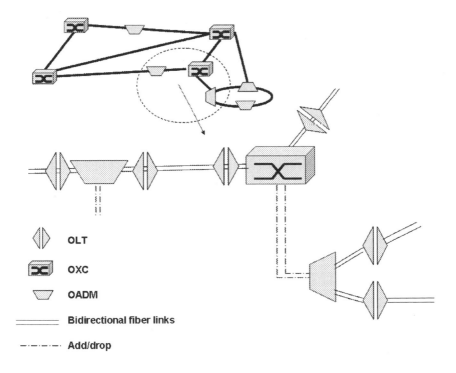

Figure 11-8. Zoom on a part of an arbitrary optical network topology

11.4.2 Optical Cross-Connects

OXCs are best suited for network locations where extensive bandwidth management is required, such as when several WDM transmission lines and many digital highways converge. OXCs are advanced optical network elements that are capable of switching lightpaths, passing them through, adding and dropping locally generated and terminated client-layer traffic, and configuring optical network topologies. They can also groom traffic and carry out sophisticated wavelength management tasks where complex network topologies and large number of wavelengths are involved. They are particularly useful for mesh topologies and to interconnect several WDM rings in core and metro networks. They also enable the optical network to reconfigure to meet client layer needs, to get around node and link failures, and to be upgraded and maintained [102], [116], [130].

11.4.2.1   Classification of Optical Cross-Connects

As indicated in chapter 1, some equipment vendors have developed OXC systems that are based on OE input ports, EO output ports, and electronic switching fabrics. This type of cross-connects is outside our scope, since we restrict our discussion of OXCs to those based on optical switching. OXCs may be classified according to their switching capability into four main types:

1) The Fiber Cross-Connect (FXC). This OXC switches all the wavelengths of an input fiber port to an output fiber port. Compared to the other three types, the FXC is the least complex and the least expensive. However, it is the least flexible in operation. In effect, it functions as an automated fiber patch panel. Since all wavelengths are switched together, and depending on transmission system design, the need for OLT equipment can be reduced or avoided. The FXC can be used for protection purposes when fiber cuts, as opposed to node failures, are of primary concern [102]. The principle of operation of the FXC is described in Figure 11-9.

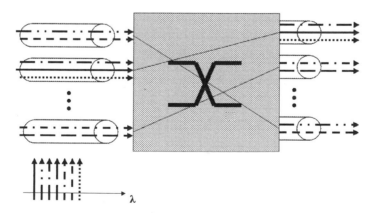

Figure 11-9. Principle of operation of the Fiber Cross-Connect (FXC). Different wavelengths are indicated by different input/output line styles.

2) The Wavelength-Band (wave-band) Cross-Connect (WBXC). In this OXC, lightpaths which share the same end points and have adjacent wavelengths on the transmission spectrum are grouped into bands that are switched together. In this case, special input/output port structures are needed to sort lightpaths into distinctive bands. The WBXC is more complex and more expensive, but more flexible, than the FXC. The

strength of this cross-connect is that it can switch wavelengths at a relatively low cost compared to a WSXC or a WIXC, which are explained next. The principle of operation of the WBXC is shown in Figure 11-10.

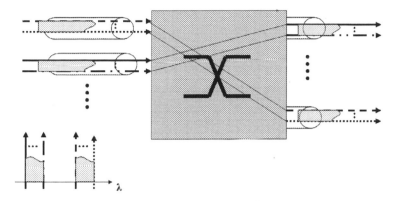

Figure 11-10. Principle of operation of the Wavelength-Band (or waveband) Cross-Connect (WBXC). Different wavebands are indicated by shaded areas that are bound by different line styles (wavelengths). For simplicity, only two bands are shown.

3) The Wavelength Selective Cross-Connect (WSXC). This OXC has the capability of switching individual wavelength channels, simultaneously, from any input port to any output port. This type uses single-wavelength ports. It offers much more switching flexibility than the above two types, permits wavelength-based services and is also more able and resource-efficient in provisioning and restoration. However, these advantages are gained at the expense of extra complexity and cost. The principle of operation of the WSXC is shown in Figure 11-11.

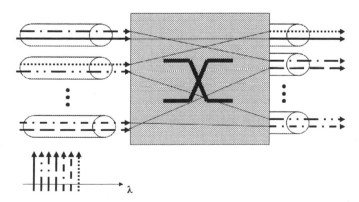

Figure 11-11. Principle of operation of the Wavelength Selective Cross-Connect (WSXC). Different wavelength are indicated by different input/output line styles.

4) The Wavelength Interchanging Cross-Connect (WIXC). This OXC has the same switching capability of a WSXC and adds to it the capability of wavelength conversion. This provides more flexibility, superior connectivity, and optimal blocking characteristics. A WIXC also uses single-wavelength ports. However, this cross-connect has higher complexity than the other three types and is obviously the most expensive to implement. Figure 11-12 depicts the principle of operation of a WIXC.

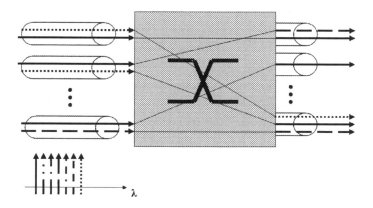

Figure 11-12. Principle of operation of the Wavelength Interchanging Cross-Connect (WIXC). Different wavelengths are indicated by different input/output line styles. Wavelength assignments and interchanges are arbitrary. The control unit determines whether, or not, a wavelength needs to be changed depending on the input/output ports involved and on existing wavelengths at any given time.

FXCs, WSXCs, and WIXCs have been around for some time and their generic requirements are defined by Telcordia, along with the requirement of hybrid OXCs composed of numerous mixtures of these three types [131]. The WBXC has emerged more recently as a research subject (see, for example, [96], [100], [112], and [120]). Since wavelengths are grouped together into bands which are handled all together, fewer optical components are required and the optical designs of the switch can be less stringent.

Multi-granularity OXCs, which combines switching at the fiber, waveband and wavelength granularities, have been discussed in the literature. They represent an interesting research topic and involve several issues. They may require extra control and complexity to retain the flexibility and reconfigurability of OXCs with one switching granularity. Designing a suitable switch architecture for a cross-connect of this kind while maintaining acceptable transmission performance can be a challenge.

11.4.2.2 Optical Cross-Connect Architectures

There are three main functional blocks of an OXC. These are input/output ports, a switch fabric, and a control unit. In the following discussion, we assume a WSXC as an example.

Some of the OXC ports are connected to ports of other optical network elements and others are connected to client ports. Today, SDH/SONET rates are the de facto units of transport. ATM-switches and IP-routers interfaces are set at the SDH/SONET rates. Therefore, OXCs have to accommodate these rates.

The switch fabric is the core of the OXC. In principle, it can be based on any of the technologies we have discussed in Part II. Depending on the application, however, and on the required specifications of the OXC, certain technologies would be more suitable than others. Switching devices are interconnected using various interconnection strategies to form the switch fabric. Many of these strategies have been discussed in chapter 9. The fabric can take the form of a single stage crossbar matrix or a multistage fabric based on interconnection of smaller switches. The interconnection can be based on Clos, Beneš, banyan, tree, or any other architecture. Depending on the design, and on the switching technology, it is possible also to build the switch as one wavelength-independent fabric or to have a wavelength-partitioned architecture where a relatively small space switch is provided for each wavelength. These two options are shown in Figure 11-13.

At the network level, an end-to-end control mechanism is necessary to set up and tear down lightpaths. The control unit of an OXC communicates with its peers in the network to establish lightpaths and to tear them down. The control unit also uses the lightpath set-up information to generate local commands to configure the switch fabric. Several standard and proprietary algorithms can be used to configure switch fabrics. Chapter 10 has discussed this subject and demonstrated many of the issues involved therein. End-to-end lightpath control is an advanced networking topic that is outside our scope. We only cover this topic briefly in section 11.5.

Typically, an OXC should be equipped with a monitoring module to measure optical powers, wavelengths and signal to noise ratio (SNR). A small proportion of a lightpath power can be tapped into such a module to check these parameters. This process is sometimes referred to as *bridging*. Passive splitters may be used to tap signal power and optical switching devices can be involved to perform this functionality for several channels in a dynamic manner. Chapter 12 discusses a typical configuration in this regard. Other techniques, such as looping the signals back for monitoring purposes can be used too.

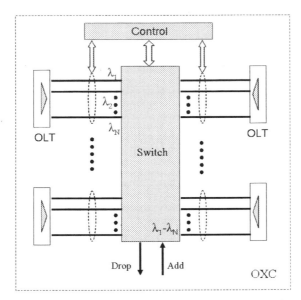

(a) OXC with multiwavelength switch fabric

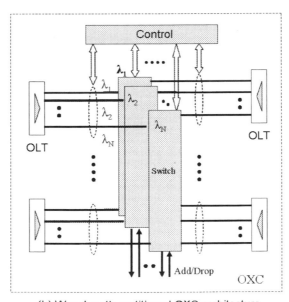

(b) Wavelength partitioned OXC architecture

Figure 11-13. (a) OXC with one wavelength-independent switch fabric. (b) Wavelength partitioned OXC, with wavelength-specific fabrics. For simplicity, only two OLTs are shown. Typically, an OXC would handle many more. Add/drop management, digital cross-connection and other details are not shown.

Some DCS capabilities are typically incorporated in OXC architectures. This is to enable the cross-connect to handle traffic at smaller granularities via its add/drop ports and to serve the needs of client layers more efficiently.

Boards, modules, and building blocks of telecommunications network elements are linked to each other using interconnects and backplanes, which are mainly based on electrical conductors. However, electrical interconnects have performance limitations and are not ideally suited for optical network elements. Advanced optical interconnects and backplanes are under research [132]-[135] and development and will improve the performance of OXCs.

11.4.2.3 Design Considerations for Optical Cross-Connects

There are a number of features to seek in an OXC in terms of blocking characteristics, complexity, transparency, reconfigurability, switching speed, modularity, upgradeability, scalability, cascadeability, reliability, OAM capability, support of multicast, and operation in multi-vendor environments. All these features are strongly related to the optical design of the OXC, and to many of its performance measures, such as, insertion loss, crosstalk, wavelength dependence, polarization dependence, extinction ratio, signal-to-noise ratio (SNR), and performance monitoring capabilities.

The choice of a switching technology and switch architecture embraces a number of tradeoffs among the desirable features of the OXC as a network element, and between them and the optical design and performance of the cross-connect. The size of the OXC and its cost are important in this regard. Decisions made concerning these design tradeoffs are critical in evaluating the cross-connect for various applications. In the following, we discuss some of the compromises involved in designing OXCs.

An OXC must be nonblocking and should have as low a complexity as possible. Although rearrangeably nonblocking switches utilize fewer switching devices (crosspoints) than other nonblocking architectures and are therefore less complex, rearrangement of high-bandwidth lightpaths can be problematic to carriers, even if it takes a very little time. Therefore, it is preferable that the OXC be wide-sense nonblocking or strictly nonblocking. Low complexity also means that actuators, where applicable, should be as few and simple as possible. Control circuitry needs to be as simple and reliable as possible. Wavelength conversion is very desirable because it improves the blocking performance and scalability of the OXC. However, it increases its complexity and cost. Several studies have demonstrated that WCs need not be incorporated in each and every node/port in the network. Also some schemes have been proposed for sharing wavelength converters among OXC ports.

Transparency equips the network with outstanding features and capabilities and is a favorable feature in optical networks. Due to optical transmission impairments, end to end transparency may not always be possible in long haul links. These impairments become larger with increasing transmission distances, number of wavelengths per fiber, and bit rate per wavelength. Thus, transparency can only be realized within some design limits. Transparency also requires that the transmission plant be designed for the worst case (e.g., highest bit rate), which means high startup cost. Performance monitoring is another issue in transparent optical networks. Although lightpaths carry digital data, they are treated as analog carriers since the optical medium is transparent to data bits and the optical network can not monitor them to detect errors and locate faults. Therefore, monitoring parameters in transparent optical transmission systems are optical power, wavelength, and optical SNR. However, an optical signal with adequate SNR could still be experiencing errors caused by chromatic dispersion or optical nonlinearities. Therefore, optical networks continue to rely on electrical performance monitoring via 3R regeneration.

3R regeneration is a process of Re-amplification, Re-shaping and Re-timing of signals. Transparent networks are 1R networks, with re-amplification only, and are therefore not scalable beyond some design limit. Scalability demands 2R regeneration, re-amplification and re-shaping, or 3R regeneration and the latter is usually required [67]. Today, 3R regenerators must rely on OEO conversion. Transponders of an OXC perform 3R regeneration among other functionalities. Therefore, they are needed in large scale networks. They enable performance monitoring and fault management, handle transmission impairments, facilitate wavelength conversion and multi-vendor inter-operability. On the other hand, transponders disrupt transparency and are themselves expensive. Therefore, it is not desirable to use them extensively.

Hence, the design of an OXC needs to strike a balance between transparency and opaqueness while minimizing cost and ensuring credible performance monitoring.

OXC are expected to be intelligent network elements that are capable of provisioning lightpaths with speed and agility, sparing carriers the slow and error prone task of setting up connections manually. While optical switching times in the milliseconds range, and shorter, are possible with devices based on several technologies, the overall system switching time of an OXC is determined by many of these devices, the switch architecture, and the configuration algorithm. Also, the set up time of an end-to-end lightpath is determined by the switching speed of several OXC systems. In the near term, it is reasonable to expect that the provisioning time of lightpaths will range from sub-seconds to minutes. In the longer term, this time could be in the milliseconds range. It must be noted however that the need for very large switching speeds is sometimes overemphasized. The fact of the matter is that not all applications require them, or even require frequent switching. Even for restoration, switching time in the millisecond range is sufficient.

The combination of transparency and fast reconfigurability arms the OXC with a great deal of flexibility.

It is desirable that the insertion loss of the switch fabric be as low as possible. Insertion loss depends on the switching technology, the switch architecture and its size. Switch fabrics with large loss require sensitive and expensive transceivers, and demand more OAs to compensate for loss. However, there are more into optimizing the loss performance of an OXC than having low insertion loss. The variation of loss among different paths across the fabric must also be as low as possible[11]. Return loss, where applicable, must be as high as possible. It is also desirable that the performance of the optical switch be wavelength and polarization independent. Polarization dependent loss (PDL) and polarization mode dispersion (PMD) should be as low as possible.

Switching technology, port count, insertion loss, and switching speed are interrelated. Today, guided-wave technology can achieve small to moderate port counts (less than about 128×128) with moderate to high insertion loss and rapid switching speeds. Free-space technologies are more likely to achieve larger port counts (256 and higher) with low loss and slower switching speeds. Thus, waveguide based technologies can be a choice for applications that require switching time in the range of sub-millisecond or less (down to nanoseconds). For applications that require large port count, free-space technologies are the more likely candidates.

The higher the extinction ratio of the switching device, the better it is. SNR is related to the extinction ratio and is also function of the size and the architecture of the fabric. The design of an OXC has to consider also the issue of whether the switch need to be latching. One of important characteristics of the switching device is its optical response. For instance, digital switches are preferred over interferometric ones.

An OXC must have modular design, be easily upgradeable and scalable in terms of the number of space ports, number of wavelengths per port, and switch fabric. Also, OXCs must exhibit satisfactory transmission performance when cascaded. This requires the use of optical switches, components and designs which can result in low loss, low crosstalk, negligible error probability, and negligible passband misalignments.

---

[11] This is sometimes expressed in terms of *differential loss* which is defined as the difference between the worst-case and best-case insertion loss across the fabric. It is desirable that differential loss be as low as possible. Differential loss depends on the switching technology, switch architecture, and the manufacturing processes. Variable optical attenuators (VOA) can be used to balance out differential loss and equalize signal powers.

Passband misalignments can arise from manufacturing imperfections, temperature variations and environmental changes.

In order to ensure all these features, high-precision fabrication of optical components is essential. High density electronic and optical packaging is important because it influences the size (footprint) and reliability of the OXC. Adequate packaging also enables the switch to be insensitive to external mechanical and environmental effects.

Like other telecommunications equipment, OXCs are required to attain high levels of reliability, availability, and survivability. Therefore, careful choices of technologies and components are necessary. Certain applications demand high reliability even if the switching devices remain inactive for years. Redundancy (1+1) of critical equipment building blocks, such as switch fabric, power supply and control, can enable the OXC to attain the required level of carrier-grade performance.

Last, but not least, OXCs should provide the minimum possible startup investment while supporting future growth at reasonable incremental cost. The case for OXCs is even stronger by demonstrating how their technical merits are matched by economic and service benefits for carriers and their customers. A number of case studies have been performed in this regards [111], [117]. The results of these studies suggest that by sharing protection capacity, OXC based mesh networks can lead to considerable savings in bandwidth (from 22% to more than 50% depending on network topology and traffic distribution). Also, provisioning time, and all its associated cost, can be cut by more than half. The results also suggest that network-wide reconfiguration brings a great deal of benefit in dealing with traffic churn and can retrieve capacity that is provisionally taken by traffic which is not routed optimally. This form of reconfigurability can lead to up to 8.5% more savings in network capacity. All in all, these studies suggest that OXCs represent meaningful capital investments and demonstrate that they introduce considerable savings in operation cost, potential revenue opportunities for carriers, and new services for their customers.

11.4.3 Reconfigurable OADM

Like OXCs, OADMs perform wavelength management tasks (Figure 11-11). However, they are simpler network elements than OXCs. While OXCs manage several WDM optical transmission lines and have a large number of bidirectional space ports, a typical OADM handles one line and has one bidirectional port. Thus, OADMs have smaller size and smaller functional responsibilities. They are used where some wavelength channels need to be added and/or dropped (terminated) at some node, while others are to pass through (transit). They are suitable for linear chain (point to point) and ring architectures in metro and core networks.

Multiplexing and filtering technologies, such as those used in OLTs, are the primary components in an OADM. OADMs may be differentiated depending on the number of wavelengths they support, the proportion of these wavelengths that can be added/dropped, and whether there are any restrictions in this regard.

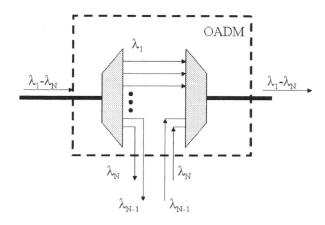

(a) Fixed (static) OADM

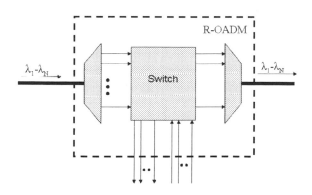

(b) Reconfigurable (dynamic) OADM

Figure 11-14. Optical add/drop multiplexers (OADMs): (a) Fixed OADM, (b) Optical-switching based R-OADM

There are two principle categories of OADMs, as shown in Figure 11-11, namely, the fixed (static) type and the reconfigurable (dynamic) type. Fixed OADMs do not utilize optical switching technologies. They have static add/drop and pass-through configurations.

Reconfigurable OADM (R-OADM), on the other hand, have the ability to arbitrarily select wavelengths, on the fly, to add/drop, or pass through transparently. This is done automatically without manual intervention and without disruption of any operational channels. As such, R-OADMs have more design and operation flexibility. This reconfigurability can be realized by using optical switches, as shown in Figure 11-11(b). The R-OADM in this figure resembles a small OXC with one bidirectional port and flexible add/drop capability.

R-OADMs share many of the requirements of OXCs, with some adaptation to their specific functionality. For instance, while R-OADMs must be able to pass lightpaths through the switch fabric transparently, the fabric should also provide full connectivity between input and drop ports and between add and output ports. R-OADMs should also be characterized by modularity, scalability, cascadeability, high switching speed, support of multicast (drop and continue), reliability, OAM capability, and should operate in multi-vendor environments. They should have low insertion loss, low crosstalk, and high SNR. Polarization independence and possession of performance monitoring capabilities are important too.

11.5 OPTICAL LAYER CONTROL PLANE

As indicated earlier, the control unit of an optical-switching system, such as an OXC or a R-OADM, generates the necessary commands to configure the switch fabric in such a way as to connect certain inputs to certain outputs at any given moment. This is done in accordance with the topology of the network. The control unit also works with its peers in the optical layer and in client layers to establish end-to-end lightpaths. These two tasks are obviously interrelated. In chapter 10, we discussed switch-fabric control. We now briefly highlight some technologies which have been developed to control optical networks and set-up/tear-down lightpaths.

Two general approaches have been adopted for optical network control by the International Telecommunication Union- Telecommunication Standardization Sector (ITU-T) and the Internet Engineering Task Force (IETF). The work of the ITU-T is within the framework of the *Automatically Switched Optical Network* (ASON) architecture [136] whereas the IETF work is based on *Generalized Multi-Protocol Label Switching* (GMPLS) [137]-[141]. The ITU-T community is dominated by the telecommunication industry. Therefore the ASON concepts follow the tradition of many telecommunication standards. The IETF, on the other hand, is primarily dominated by the Internet and data-communication community and GMPLS is an IP centric technology.

Both the ASON and GMPLS approaches adopt the notion of a distributed *control plane* for the optical layer. This plane has to be capable of tracking network topology and resources, and is responsible for fast provisioning of lightpaths. A centralized connection management system has been used for similar purposes in many transport networks worldwide. In principle, this approach may be extended for the optical layer. However, provisioning and recovery of connections using centralized network management systems (NMS) poses problems in multi-vendor multi-carrier environments. Carriers who want to deploy equipment from different optical networking vendors will have to create an umbrella management system and will find it difficult to establish a connection across separate administrative domains since these domains do not provide proprietary information (such as topology, link capacities, element capabilities) to each other.

The ASON architecture[12] has been developed as an Optical Transport Network (OTN) [143] with dynamic configuration capability. This capability depends on a control plane that undertakes the responsibility of setting up and tearing down connections (lightpaths) and maintaining them. The ASON architecture is based on three planes, namely, a *transport plane*, a *management plane* and the *control plane*. The control plane facilitates route determination and signaling, and supports connections resulting from a client request (which is referred to as *switched connection*) and/or a management request (*soft permanent connection*). Connection state information (e.g. signal quality, fault information) is detected by the transport plane and provided to the control plane. The control plane distributes link status information (e.g. adjacency, available capacity and failures) to support provisioning and restoration. Detailed fault management and/or performance monitoring information is transported within the transport plane (via overhead/OAM) and via the management plane (using data communication channel).

ASON is not a protocol or a suite of protocols. It is an architecture which mainly defines the components in the control plane and the interactions both among these components and between them and the transport and management planes. The architecture also identifies which of those interactions will involve vendor-interoperability issues, and therefore require standardized protocols. The ASON adopts the ITU-T approach in defining system and component architectures from scratch and specifying their requirements. Then, protocols that are capable of fulfilling these requirements can be embraced by the ASON.

---

[12] ITU-T has also standardized the Automatic Switched Transport Network (ASTN) [142] architecture. ASTN is technology independent and is therefore open to different technology implementations. ASON is focused on optical technology.

GMPLS, on the other hand, is a generalization of the concept of Multi-Protocol Label Switching (MPLS) [144]. The latter works with IP routing to provide high-speed packet forwarding with Quality of Service (QoS). MPLS uses a technique known as label swapping to forward data through the network. A label is added to each data packet when it enters the IP/MPLS network and the packet is routed based on this label and on the input interface through which it enters the router. The packet is forwarded to an output interface with a new label that identifies the next hop for the packet, and so on. Therefore, a router in an MPLS-based IP network is referred to as a Label-Switched Router (LSR) and the path of the packet across the network is termed a Label Switched Path (LSP).

The work of the IETF in distributed control planes for the optical layer started at the end of the 1990s with the development of the concept of MPλS based on extensions of MPLS Traffic Engineering (TE). MPλS evolved to Generalized MPLS (GMPLS) which is largely considered as an acceptable basis for optical-layer control. GMPLS is a suite of IP-centric protocols which extend MPLS-TE from supporting only Packet Switching Capable interfaces, for which MPLS was designed, to other classes of interfaces. Four classes with different interface/switching capabilities have been defined: Layer-2 switching capable, Time-Division Multiplex (TDM) capable, Lambda (λ) switching capable, and fiber switching capable. As such, GMPLS should enable a variety of electronic and optical switching systems to interoperate. GMPLS also extends the MPLS definition of an LSP from being a path that merely connects LSRs to one that may connect pairs of layer-2 switches, SDH/SONET network elements, or a lightpath that interconnects optical network elements [116].

Using the same framework of MPLS-TE, GMPLS allows the client and optical layers to inter-operate while potentially reducing network complexity. In the IP over optics model, an LSR is connected to its peer(s) via the optical layer (Figure 11-15). Three different interconnection models are outlined for organizing the control plane while accommodating all possible carrier deployment scenarios. These are the overlay model, the augmented model and the peer model.

In the overlay model, the client (IP) layer and the server (optical) layer are completely separate, with independent control planes and without exchange of routing or topology information. The separation is made at the User-Network Interface (UNI), as shown in Figure 11-15(b), and at the optical Network-Network Interface (NNI). Independent addressing schemes are used in both layers. The overlay model is of interest to large carriers, optical-transport providers and the service providers who lease optical-transport from them [116].

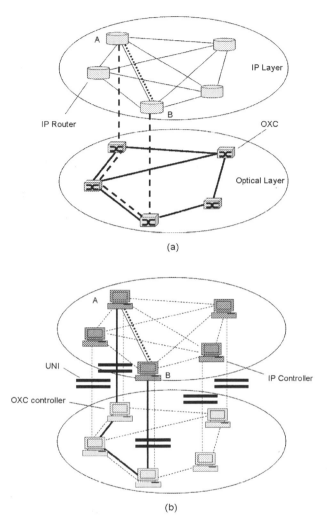

(a)

(b)

Figure 11-15. An arbitrary example of an IP over optics architecture: (a) Optical network physical topology, IP network logical topology(s), and a typical connection between two routers (A and B) across the optical network. The diagram shows the actual physical path data may take through the optical network (dashed) along with the logical path seen by routers (dotted). (b) The corresponding control network involving the control plans of both the IP and optical layers.

The augmented and peer models deal with the case when the IP and the optical layers are slightly, or tightly, integrated together, respectively. Hence, they can be of interest to service providers who are also optical-transport providers. These models follow common addressing schemes and the optical layer elements become IP addressable devices. In the augmented model, reachability information is exchanged between the control plane instances of the two layers. Because only reachability information is passed between layers, integration is incomplete and additional routing information processing is required at the border between the IP and optical networks (at the boarder router) [116].

The peer model is a unified service model with peer interconnection of the control plane instances in the client and server layers. Typically, the routing information exchange is not restricted to reachability but also includes topology and traffic-engineering attributes of the optical layer. This information is meaningful for the whole domain through which they are available, not only at the border between the IP and optical networks. This implies tight integration between the control plane instances in the two layers. The peer model enables full integration between the optical and IP layers and full exploitation of GMPLS benefits. However, it requires open exchange of proprietary network information among service providers [116].

Another standardization group that is also involved in control-plane issues is the Optical Internetworking Forum (OIF). The OIF activities blend some of the work of IETF with that of the ITU-T and the main contributions of this forum have been in the area of user-network interface.

The ASON and GMPLS approaches are not to be regarded as competitors although this is how it seems sometimes. The two standardization efforts should complement each other, especially since the desired outcome for both efforts has evolved toward having the ASON reference the various protocol specifications in the GMPLS suite. The ASON reference architecture, on the other hand, is intended to be broad in scope, as complete as possible, and future proof.

11.6 DEPLOYMENT OF OPTICAL SWITCHING: STATUS UPDATE

By the turn of the century, optical networking was in a period of high growth and OXCs were already undergoing several field trials in carrier networks. The downturn of the telecommunications industry, by mid 2001, has obviously put a lot of progress on hold, especially in carrier networks[13]. Nevertheless, deployment of optical switching equipment went on slowly, but steadily. Worldwide national research and education networks have taken the lead in deploying optical switching. Usually, these are non-profit government-sponsored computing/networking research consortia and initiatives. Internet2 is an example in this regard. It is a combined effort of more than 200 US universities working with industry and government. Internet2 has announced in July 2005 that it will use optical switching in five

---

[13] This subject is discussed in chapter 14.

national network nodes. This is part of its Hybrid Optical and Packet Infrastructure (HOPI) project.

Other examples include SuperSINET and JGN II in Japan. SuperSINET is a networking initiative that promotes advanced science, technology, and research and is operational since 2002. JGN II is an open testbed network environment for research and development. It is an expansion of Japan Gigabit Network (JGN, 1999-2004). Optical switching, based on MEMS and new polymer technologies, is inherent part of these initiatives. Other optical networking and switching initiatives include OptIPuter[14] (US), TransLight (US/Europe), UltraLight (US), and others [145].

There is a trend to use optical networking to support emerging *grid* applications. These are advanced distributed parallel-processing systems. High bandwidth applications in fields such as visualization, computing and bio-informatics are driving this move forward. Optical networks can facilitate reconfigurable high-bandwidth lightpaths that are needed for grid applications. The goal is also for a grid application to be able to perform network-aware functions such as storage management and scheduling. Optical switching is key enabler of these applications.

The National LambdaRail (NLR) initiative in the USA is expected to promote optical switching as well. NLR is a national fiber optic infrastructure designed to foster the advancement of next-generation networks and their application in science, engineering and medicine. It is a partnership of U.S. research universities and private sector technology companies. It supports many networks which are independent from each other. Also, the Global Lambda Integrated Facility (GLIF) is an international virtual organization whose members support data-intensive scientific research and middleware development through the international LambdaGrid. LambdaGrid is expected to bring together worldwide efforts to architect an international research infrastructure using advanced optical networking and switching technologies.

Another initiative to be mentioned in regards to optical networking is CANARIE, which is Canada's advanced Internet development organization. Part of CANARIE's mission is to facilitate the widespread adoption of next generation optical technologies. CA*net 4 is expected to demonstrate the concept of a customer-empowered network where dynamic allocation of network resources is placed in the hands of end users to enable them to develop innovative network-based applications.

Finally, several experimental demonstrations of GMPLS controlled optical-switching systems are reported in literature and/or in professional media. Some of these testbeds feature innovative applications, such as wide-area distributed storage services [139].

---

[14] We have discussed this project in sec. 11.2

Recently, the market has taken several positive steps in regards to metro R-OADMs. North American Multiple System Operators (MSOs) have taken the lead in this respect to secure support for cable TV and data services. Japanese carriers are starting to deploy R-OADMs too.

Finally, optical switching is being proposed today for several other applications, and is starting to be introduced in many of them. The list include shared access to transport bandwidth; automated fiber management in central offices, data centers, and large enterprises; Fiber to the Home (FTTH); remote testing and monitoring; and fiber lab automation [146].

Thus, there has been progress in optical switching deployment during the past few years and in the midst of the downturn of the telecommunications industry. Up to 320×320 OXC systems are said to be installed today [145]. In chapter 12, we discuss the applications of optical switching in protection and restoration. These applications, in particular, represent a driver for deployment of optical switching in carrier networks in the future. Optical switching research has focused on protection and restoration since long time and lightpath recovery times in the order of milliseconds were reported in the early stages of this research.

Throughout previous chapters, we have demonstrated that the vision of the optical layer in telecommunications remains sound, despite the current downturn. We see that optical switching has finally reached the stage of maturity where carriers will start to invest in it within few years. Provided the telecom business conditions improve, it is a matter of time for OXCs and R-OADMs to become important cornerstones in transport networks.

Our focus throughout this book is mostly on optical switching in communications networks and systems. It must be noted however that optical switching has other applications. These include measurement and instrumentation systems, sensor networks, photonic surveillance and security systems, military applications, and many others.

REFERENCES

[1] T. Hermes, B. Hoen, J. Saniter, and F. Schmidt, "LOCNET-a local area network using optical switching," IEEE J. Lightwave Technol., Vol. LT-3, No. 3, pp. 467-71, June 1985.
[2] T. Shibagaki, I. Hiroyuki, and O. Takeshi, "Video transmission characteristics in WDM star networks," IEEE J. Lightwave Technol., Vol. LT-3, No. 3, pp. 490-5, June 1985.
[3] H. Kobrinski, "Applications of coherent optical communications in the network environment," Proc. of SPIE, Vol. 568, pp. 42-9, San Diego, Aug. 1985
[4] D. B. Payne and J. R. Stern, "Wavelength switched, passively coupled, single mode optical networks," In: IOOC-ECOC '85, 5th International Conference on Integrated Optics and Optical Fibre Communication and 11th European Conference on Optical Communication, Technical Digest, pp. 585-8, Venice, Italy, October 1985.

[5] S. S. Cheng, H. Kobrinski, and L. T. Wu, "A distributed star network architecture for interoffice applications," In: IOOC-ECOC '85, 5th International Conference on Integrated Optics and Optical Fibre Communication and 11th European Conference on Optical Communication, Technical Digest, Vol. 1, pp. 699-702, Venice, Italy, October 1985.

[6] D. B. Payne and J. R. Stern, "Transparent single mode fiber optical networks," IEEE J. Lightwave Technol., Vol. LT-4, No. 7, pp. 864-9, 1986

[7] M. S. Goodman, H. Kobrinski, and K. W. Loh, "Applications of wavelength division multiplexing to communication network architectures," International Communications Conference (ICC'86), Vol. 2, 29.4.1, pp. 931-3, 1986.

[8] K. D. Cooley, M. S. Goodman, M. Kerner, H. Kobrinski, and D. M. Simmons, "Wideband virtual networks," Conf. Record of Globecom'86, Houston, pp. 979-82, 1986.

[9] H. Kobrinski, R. M. Bulley, M. Goodman, M. P. Vecchi, C. A. Brackett, L. Curtis, and J. L. Gimlett, "Demonstration of high capacity in the LAMBDANET architecture: A multiwavelength optical network," Electron. Lett., Vol. 23, No. 16, pp. 824-6, July 30, 1987.

[10] M. Goodman, C. A. Brackett, R. M. Bulley, C. N. Lo, H. Kobrinski, and M. P. Vecchi, "Design and demonstration of the LAMBDANET system: A multiwavelength optical network," Proc. Of the IEEE/IEICE Global Telecommunications Conference 1987 (Globecom'87), 37.4.1-37.4.4, Tokyo, 1987.

[11] H. Bunning, M. Burmeister, T. Hermes, and F. Raub, "LOCNET-an experimental communications system using optical routing," IEEE J. Select. Areas Commun., Vol. 6, No. 7, pp. 1241-7, Aug. 1988.

[12] T.-H. Wu, D. J. Kolar, and R. H. Cardwell, "Survivable network architectures for broad-band fiber optic networks: model and performance comparison," IEEE J. Lightwave Technol., Vol. 6, No. 11, pp. 1698-709, Nov. 1988.

[13] B. S. Glance, J. Stone, K. J. Pollock, and P. J. Fitzgerald, "Densely spaced FDM coherent star network with optical signals confined to equally spaced frequencies," IEEE J. Lightwave Technol., Vol. 6, No. 11, pp. 1770-81, Nov. 1988.

[14] E. Arthurs, J. M. Copper, M. S. Goodman, H. Kobrinski, M. Tur, and M. P. Vecchi, "Multiwavelength Optical Crossconnect for Parallel-Processing Computers," Electron. Lett., Vol. 24, No. 2, pp. 119-20, Jan. 21, 1988.

[15] G. N. M. Sudhakar, M. Kavehrad, and N. D. Georganas, "Access protocols for passive optical star networks," Computer Networks and ISDN Systems, Vol. 26, No. 6-8, pp. 913-30, March 1994.

[16] F. J. Janniello, R. Ramaswami, and D. G. Steinberg "A prototype circuit switched multi-wavelength optical MAN," IEEE J. Lightwave Technol., Vol. 11, No. 5/6, pp. 777-82, May-June 1993.

[17] E. Hall, J. Kravitz, R. Ramaswami, and M. Halvorson, "The Rainbow-II gigabit optical network," IEEE J. Select. Areas Commun., Vol. 14, No. 5, pp. 814-23, June 1996.

[18] L. G. Kazovsky and P. T. Poggiolini, "STARNET: A multi-gigabit-per-second optical LAN utilizing a passive WDM star," IEEE J. Lightwave Technol., Vol. 11, No. 5/6, pp. 1009-27, May-June 1993.

[19] T-K Chiang, S. K. Agrawal, D. T. Mayweather, D. Sadot, C. F. Barry, M. Hickey, and L. Kazovsky, "Implementation of STARNET: A WDM computer communication network," IEEE J. Select. Areas Commun., Vol. 14, No. 14, No. 5, pp. 824-39, June 1996.

[20] T. S. El-Bawab, C. Vaishnav, A. P. Jayasumana, H. Temkin, J. R. Sauer, and H. A. Willebrand, "Robust wavelength division multiplexed local area networks," Fiber and integrated Optics, Vol. 16, No. 3, pp. 237-60, July 1997.

[21] T. S. El-Bawab and A. P. Jayasumana, "Modeling and performance analysis of a symmetric fast-circuit switched Robust-WDM LAN with the AR/LTP protocol," IEEE J.

Lightwave Technol., Vol. 17, No. 6, pp. 973-88, June 1999. See also Vol.17, No. 6, p. 1514, September 1999.

[22]   A. S. Acampora, "A multichannel multihop local lightwave network," Globecom'87, pp. 1459-67, 1987

[23]   A. S. Acampora and M. J. Karol, "An overview of lightwave packet networks," IEEE Network Mag., Vol. 3, No. 1, pp. 29-41, January 1989.

[24]   M. G. Hluchyj and M. J. Karol, "ShuffleNet: An application of generalized perfect shuffles to multihop lightwave networks," IEEE J. Lightwave Technol., Vol. 9, No. 10, pp. 1386-97, October 1991.

[25]   M. Nishio and S. Suzuki, "Photonic wavelength-division switching network using parallel λ-Switch," in Proc. of the International Topical Meeting on Photonic Switching, PS'90, Vol. PD14-B-9, Kobe, Japan, April 12-14, 1990. (Photonic Switching II, Springer-Verlag, pp. 286-90, 1990)

[26]   H. Kobrinski, "Crossconnection of wavelength-division-multiplexed high-speed channels," Electron. Lett., Vol. 23, No. 18, pp. 974-6, 27th Aug. 1987.

[27]   G. R. Hill, "A wavelength routing approach to optical communications networks," Proc. 7th Annual Joint Conference of the IEEE Computer and Communications Societies (IEEE INFOCOM), pp. 354-62, March 1988

[28]   P. J. Chidgey and G. R. Hill, "Experimental demonstration of wavelength routed optical networks over 52 Km of monomode optical fiber," European Conference on Optical Communications (ECOC'89), Paper MoA1-5, pp. 9-12, Gothenberg, Sweden, September 10-14, 1989.

[29]   S. F. Su, L. Lou, and J. Lenart, "A review on classification of optical switching systems," IEEE Commun. Mag., Vol. 24, No. 5, pp. 50-5, May 1986.

[30]   J. E. Midwinter, "Photonic switching technology: component characteristics versus network requirements," .IEEE J. Lightwave Technol., Vol. 6, No. 10, pp. 1512-19, Oct. 1988.

[31]   J. A. McEachern, "Gigabit networking on public transmission network," IEEE Commun. Mag., Vol. 30, No. 4, pp. 70-8, 1992.

[32]   C. A. Brackett, "Dense wavelength division multiplexing networks: principles and applications," IEEE J. Select. Areas Commun., Vol. 8, No. 6, pp. 948-64, Aug. 1990.

[33]   G. R. Hill, I. Hawker, and P. J. Chidgey, "Applications of wavelength routing in a core telecommunications network," IEE Conference on Integrated Broadband Services and Networks, pp. 63-7, London 15-18 October 1990.

[34]   M. Fujiwara, "Studies on optical digital cross-connect system using photonic switching matrices and optical amplifiers," ECOC '91 Vol. 1, pp. 97-100, Valbonne, France, 1991.

[35]   H. J. Westlake, P. J. Chidgey, G. R. Hill, P. Granestrand, L Thylen, G. Grasso, and F. Meli, "Reconfigurable wavelength routed optical networks: a field demonstration," 17th European Conference on Optical Communication & 8th International Conference on Integrated Optics and Optical Fibre Communication (IOOC-ECOC '91), Vol. 1, pp. 753-6, Valbonne, France, 1991.

[36]   I. Hawker, "Evolution of digital optical transmission networks," BT Technol. J., Vol. 9, No. 4, Oct. 1991.

[37]   I Chlamtac, A. Ganz, and G. Karmi, "Lightpath communications: an approach to high bandwidth optical WAN's," IEEE Trans. Commun., Vol. 40, No. 7, pp. 1171-82, July 1992.

[38]   T. G. Lynch, P. J. Chidgey, E. G. Bryant, P. Brown, and M. Greatbanks, "Experimental field demonstration of a managed multi-noded reconfigurable wavelength routed optical network," Proc. 18th European Conference on Optical Communication (ECOC'92), Th A12.4, pp. 609-12, Berlin, 1992.

[39] S. B. Alexander, R. S. Bondurant, D. Byrne, V. W. S. Chan, S. G. Finn, R. Gallager, B. S. Glance, H. A. Haus, P. Humblet, R. Jain, I. P. Kaminow, M. Karol, R. S. Kennedy, A. Kirby, H. Q. Le, A. A. M. Saleh, B. A. Schofield, J. H. Shapiro, N. K. Shankaranarayanan, R. E. Thomas, R. C. Williamson, R. W. Wilson, "A precompetitive consortium on wide-band all optical networks," J. Lightwave Technol., Vol. 11, No. 5/6, pp. 714-35, May/June 1993.

[40] C. A. Brackett, A. S. Acampora, J. Sweitzer, G. Tangonan, M. T. Smith, W. Lennon, K.-C. Wang, and R. H. Hobbs, "A scalable multiwavelength multihop optical network: a proposal for research on all optical networks," IEEE J. Lightwave Technol., Vol. 11, No.5-6, pp. 736-53, May-June 1993.

[41] G-K Chang, G. Ellinas, J. K. Gamelin, M. Z. Iqbal, C. A. Brackett, "Multiwavelength reconfigurable WDM/ATM/SONET network testbed," J. Lightwave Technol., Vol. 14, No. 6, pp. 1320-40, June 1996.

[42] K. Sato, S. Okamoto, and H. Hadama, "Optical path layer technologies to enhance B-ISDN performance," Conference Record, IEEE International Conference on Communications, (ICC'93), Vol. 3, pp. 1300-7, 1993.

[43] K. Sato, S. Okamoto, and H. Hadaroa, "Network performance and integrity enhancement with optical path layer technologies," J. Select. Areas Commun., Vol. 12, No. 1, pp. 159-70, January 1994.

[44] A. Watanabe, S. Okamoto, and K. Sato, "Optical path cross-connect node architecture with high modularity for photonic transport networks," IEICE Transaction on Communications, Vol.E77-B, No. 10, pp. 1220-9, Oct. 1994

[45] S. Okamoto, A. Watanabe, and K. Sato, "A new optical path cross-connect system architecture utilizing delivery and coupling matrix switch," IEICE Transaction on Communications, Vol. E77-B, No. 10, pp. 1272-4, Oct. 1994.

[46] N. Nagatsu, Y. Hamazumi, and K. Sato, "Optical path accommodation designs applicable to large scale networks," IEICE Transactions on Communications, Japan, Vol. E78-B, No. 4, pp. 597-607, April 1995.

[47] N. Nagatsu and K. Sato, "Optical path accommodation design enabling cross-connect system scale evaluation," IEICE Transactions on Communications, Vol. E-78-B, No. 9, pp. 1339-43, Sept. 1995.

[48] K. Sato, S. Okamoto, and A. Watanabe "Photonic transport networks based on optical paths," Int'l J. Commun. Sys., Vol. 8, No. 6, pp. 377-89, Nov. Dec. 1995

[49] M. Koga, Y. Hamazumi, A. Watanabe, S. Okamoto, H. Obara, K.-I Sato, M. Okuno, and S. Suzuki, "Design and performance of an optical path cross-connect system based on wavelength path concept," IEEE J. Lightwave Technol., Vol. 14, No. 6, pp. 1106-19, June 1996.

[50] S. Okamoto, A. Watanabe, and K.-I Sato, "Optical path cross-connect node architectures for photonic transport network," IEEE J. Lightwave Technol., Vol. 14, No. 6, pp. 1410-22, June 1996.

[51] A. Watanabe, S. Okamoto, M. Koga, K. Sato, and M. Okuno, "Packaging of 8x16 delivery and coupling switch for a 320 Gb/s throughput optical path cross-connect system," the 22nd European Conference on Optical Communication (ECOC'96), Vol. 4, ThD1.3, pp. 111-14, 15-19 Sept. 1996.

[52] Nagatsu, S. Okamoto, and K. Sato, "Optical path cross-connect system scale evaluation using path accommodation design for restricted wavelength multiplexing," IEEE J. Select. Areas in Commun., Vol. 14, No. 5, pp. 893-902, June 1996.

[53] K. Sato, "Photonic transport network OAM technologies," IEEE Commun. Mag., Vol. 43, No. 12, pp. 86-94, December 1996.

[54] S. Okamoto, M. Koga, H. Suzuki, and K. Kawai, "Robust photonic transport network implementation with optical cross-connect systems," IEEE Commun. Mag., Vol. 38, No. 3, pp. 94-103, March 2000.

[55] S. Aisawa, A. Watanabe, T. Goh, Y. Takigawa, M. Koga, and H. Takahashi, "Advances in optical path crossconnect systems using planar-lightwave circuit-switching technologies," IEEE Commun. Mag., Vol 41, No. 9, pp. 54-7, September 2003.

[56] M. J. O'Mahony, D. Simeonidou, A. Yu, and J. Zhou, "The design of a European optical network," IEEE J. Lightwave Technol., Vol. 13, No. 5, pp. 817-28, May 1995.

[57] T. S. El-Bawab, M. J. O'Mahony, and A. P. Jayasumana, "A European multiwavelength optical network," The 2nd IEEE Symposium on Computers and Communications (ISCC'97), pp. 568-73, Alexandria, Egypt, July 1-3, 1997.

[58] N. F. Whitehead and N. Parsons, "Network applications of photonic switching," European Conference on Optical Communications (ECOC'89), Paper WeA15-1, pp. 110-3, Gothenberg, Sweden, September 10-14, 1989.

[59] S. Johansson, M. Lindblom, P. Granestrand, B. Lagerstrom, and L. Thylen, "Optical cross-connect system in broad-band networks: system concept and demonstrators description," IEEE J. Lightwave Technol., Vol. 11, No. 5-6, pp. 688-94, May-June 1993.

[60] G. R. Hill, P. J. Chidgey, F. Kaufhold, T. Lynch, O. Sahlen, M. Gustavsson, M. Janson, B. Lagerstrom, G. Grasso, F. Meli, S. Johansson, J. Ingers, L. Fernandez, S. Rotolo, A. Antonielli, S. Tebaldini, E. Vezzoni, R. Caddedu, N. Caponio, F. Testa, A. Scavennec, M J. O'Mahony, J. Zhou, A. Yu, W. Sohler, U. Rust, and H. Herrmann, "A transport network layer based on optical network elements," IEEE J. Lightwave Technol., Vol. 11, No. 5-6, pp. 667-79, May-June 1993.

[61] L. Gillner and M. Gustavsson, "Scalability of optical multiwavelength switching networks: power budget analysis," IEEE J. Select. Areas in Commun., Vol. 14, No. 5, pp. 952-61, June 1996.

[62] S. Johansson, "Transport Network Involving a Reconfigurable WDM Network Layer-A European Demonstrator," IEEE J. Lightwave Technol., Vol. 14, No. 6, pp. 1341-8, June 1996.

[63] M. Berger, M. Chbat, A. Jourdan, M. Sotom, P. demeester, B. Van Caenegen, P Godsvang, B. Hein, M. Huber, R. Marz, A. Leclert, T. Olsen, TG. Tobolka, and T. Van den Broeck, "Pan –European Optical Networking Using Wavelength Division Multiplexing," IEEE Commun. Mag., Vol. 35, No. 4, pp. 82-8, April 1997.

[64] M. Lehdorfer, W. Plotz, M. Schreiblehner, K. Zhuber-Okrog, and O. Jahreis, "Optical cross-connect with supervision functions implemented in PHOTON," In: Technology and Infrastructure (NOC '98), pp. 27-34, Manchester, UK, 23-25 June 1998 (IOS Press: Amsterdam, Netherlands, 1998)

[65] M. Lehdorfer, M. Rasztovits-Wiech, D. Werner, B. Hein, and W. Mullner, "PHOTON-results from the WDM field trial network over 520 km with up to 10 Gbit/s per channel and optical cross-connecting," 24th European Conference on Optical Communication (ECOC'98), Vol. 1, pp. 569-70, Madrid, Spain, 20-24 September 1998.

[66] M. W. Chbat. E. Grard, L. Berthelon, A. Jourdan, P. A. Perrier, A. Leclert, B. Landousies, A. Ramdane, N. Parnis, E. V. Jones, E. Limal, H. N. Poulsen, R. J. S. Pedersen, N. Flaaronning, D. Vercauteren, M. Puleo, E. Ciaramella, G. Marone, R. Hess, H. Melchior, W. V. Parys, P. M. Demeester, P. J. Godsvang, T. Olsen, and D. R. Hjelme, "Toward wide-scale all-Optical transparent networking: The ACTS optical pan-European network (OPEN) project," IEEE J. Select. Areas Commun., Vol. 16, No. 7, pp. 1226-44, September 1998.

[67] P. Ohlen, E. Berglind, and L. Thylen, "Scalability issues in optical networks," IEICE Trans. on Commun., Vol. E82-B, No. 2, pp. 231-8, Feb. 1999.

[68] A. Rafel, J. Prat, J. Jude, J. Comellas, and G. Junyent, "A new functional model for management of optical transport network nodes," IEEE J. Lightwave Technol., Vol. 19, No. 6, pp. 810-20, June 2001.

[69] A. Houghton, "Supporting the rollout of broadband in Europe: optical network research in the IST program," IEEE Commun. Mag., pp. 58-64, September 2003.

[70] C. Cavazzoni, V. Barosco, A. D'Alessandro, A. Manzalini, S. Milani, G. Ricucci, R. Morro, R. Geerdsen, U. Hartmer, G. Lehr, U. Pauluhn, S. Wevering, D. Pendarakis, N. Wauters, R. Gigantino, J. P. Vasseur, K. Shimano, G. Monari, and A. Salvioni, "The IP/MPLS over ASON/GMPLS test bed of the IST project LION," Vol. 21, No. 11, pp. 2791–803, Nov. 2003.

[71] S. De Maesschalck D. Colle, B. Puype, Q. Yan, M. Pickavet, P. Demeester, T. Cinkler, S. Tomic, C. Mauz, M. Ljolje, M. Lackovic, R. Inkret, B. Mikac, M. Köhn, C. Gauger, D. Schupke, S. Sanchez-Lopez, X. Masip-Bruin, J. Sole-Pareta, and M. Mattiello, "Circuit/wavelength switching and routing. report of the achievement of the COST-action 266," 7th International Conference on Telecommunications (ConTEL 2003), pp. 769-73, June 11-13, 2003, Zagreb, Croatia

[72] R. E. Wagner, R. C. Alferness, A. A. M. Saleh, and M. S. Goodman, "MONET: multiwavelength optical networking," IEEE J. Lightwave Technol., Vol. 14, No. 6, pp. 1349-55, June 1996.

[73] L. Garrett, R. Derosier, A. Gnauck, A. McCormick, R. Tkach, R. Vodhanel, J. Chiao, J. Gamelin, C. Gibbins, H. Shirokmann, M. Rauch, J. Young, R. Wagner, A. Luss, M. Maeda, J. Pastor, M. Post, C. Shen, S. Wei, B. Wilson, Y. Tsai, G. Chang, S. Patel, C. Allyn, A. Chraplyvy, J. Judkins, A. Srivastava, J. Sulhoff, Y. Sun, A. Vengsarkar, C. Wolf, J. Zyskind, A. Chester, B. Comissiong, G. Davis, G. Duverney, N. Jackman, A. Jozan, V. Nichols, B. Lee, R. Vora, A. Yorinks, G. Newsome, P. Bhattacharjya, D. Doherty, J. Ellson, C. Hunt, A. Rodriguez-Moral, N. Srinivasan, W. Kraeft, and J. Ippolito, "The MONET New Jersey network demonstration," IEEE J. Select. Areas Commun., Vol. 16, No. 7, pp. 1199-219, Sept. 1998.

[74] H. Dai, G. K. Chang, B. Meagher, J. Young, W. Xin, S. J. Yoo, G. Ellinas, and W. T. Andersen, "Field deployment and evaluation of 1510 nm data communication network for MONET Washington DC network trial," Optical Fiber Communication Conference and the International Conference on Integrated Optics and Optical Fiber Communications (OFC/IOOC'99), Vol. 2, pp. 187-9, 21-26 Feb. 1999

[75] A. M. Gottlieb and S. R. Johnson, "MONET WDM network elements," Proc. of Military Communications Conference (MILCOM 1999), Vol. 2, pp. 963-7, 31 Oct.- 3 Nov. 1999.

[76] S. R. Johnson and V. L. Nichols, "Advanced optical networking- Lucent's MONET network elements," Bell Labs Technical Journal, Vol. 4, No. 1, pp. 145-62, Jan.-March 1999.

[77] W. Anderson, J. Jackel, G.-K. Chang, D. Hongxing W. Xin; M. Goodman, C. Allyn, M. Alvarez, O. Clarke, A. Gottlieb, F. Kleytman, J. Morreale, V. Nichols, A. Tzathas, R. Vora, L. Mercer, H. Dardy, E. Renaud, L. Williard, J. Perreault, R. McFarland, and T. Gibbons, "The MONET project- A final report," IEEE J. Lightwave Technol., Vol. 18, No. 12, pp. 1988-2009, December 2000.

[78] T. DeFanti, M. Brown, J. Leigh, O. Yu, E. he, J. Mambretti, D. Lillethun, and J. Weinberger, "Optical switching middleware for the OptIPuter," IEICE Transactions on Communications, Vol. E86-B, No. 8, pp 2263-72, August 2003.

[79] L. L. Smarr, A. A. Chien, T. DeFanti, J. Leigh, and P. M. Papadopoulos, "The OptIPuter [parallel optical network based computing]," Communications of the ACM, Vol. 46, No. 11, pp. 58-66, Nov. 2003.

[80]   K. J. Hood, P. W. Walland, C. L. Nuttall, L. J. St. Ville, T. P. Young, A. Oliphant, R. P. Marsden, J. T. Zubrzycki, G. Cannell, C. Bunney, J. P. Laude, and M. J. Anson, "Optical distribution systems for television studio applications," IEEE J. Lightwave Technol., Vol. 11, No. 5/6, pp. 680-7, May-June 1993

[81]   J. Sharony , K. Cheung, and T. E. Stern, "The wavelength dilation concept in lightwave networks-implementation and system considerations," IEEE J. Lightwave Technol., Vol. 11, No. 5, pp. 900-7, May-June 1993.

[82]   P. J. Smith, D. W. Faulkner, and G. R. Hill, "Evolution scenarios for optical telecommunication networks using multiwavelength transmission," Proceedings of IEEE, Vol. 81, No. 11, pp. 1580-7, November 1993.

[83]   G. Hill and A. McGuire, "Management and functionality of optical networks," Proc. European Conference on Optical Communication (ECOC'94), Vol./Issue 20, pp. 255-62, Florence, Italy, Sept. 25-29, 1994.

[84]   P. E. Green, "Optical networking update," IEEE J. Select. Areas Commun., Vol. 14, No. 5, pp. 764-79, June 1996.

[85]   R. Sabella E. Iannone, and E. Pagano, "Optical transport networks employing all-optical wavelength conversion: limits and features," IEEE Journal of Selected Areas in Communications, Vol. 14, No. 5, pp. 968-78, June 1996.

[86]   Y. Tada, Y. Kobayashi, Y. Yamabayashi, S. Matsuoka, and K. Hagimoto, "OA&M framework for multiwavelength photonic transport networks," IEEE J. Select. Areas in Commun., Vol. 14, No. 5, pp. 914-22, June 1996.

[87]   I. P. Kaminow, C. R. Doerr, C. Dragone, T. Koch, U. Koren, A. M. Saleh, A. J. Kirby, C. M. Ozveren, B. Schofield, R. E. Thomas, R. A. Barry, D. M. Castagnozzi, V. W. S. Chan, B. R. Hemenway, Jr, D. Marquis, S. A. Parikh, M. L. Stevens, E. A. Swanson, S. G. Finn, and R. G. Gallager, "A wideband all-optical WDM network," IEEE J. Select. Areas in Commun., Vol. 14, No. 5, pp. 780-99, June 1996.

[88]   N. Wauters and P. Demeester, "Design of optical path layer in multiwavelength cross-connected networks," IEEE J. Select. Areas Commun., Vol. 14, No. 5, pp. 881-91, June 1996.

[89]   C. A. Brackett, "Is there an emerging consensus on WDM networking," IEEE J. Lightwave Technol., Vol. 14, No. 6, pp. 936-41, June 1996

[90]   A. Jourdan, F. Masetti, M. Garnot, G. Soulage, and M. Sotom, "Design and implementation of a fully reconfigurable all-optical crossconnect for high capacity multiwavelength transport networks," IEEE J. Lightwave Technol., Vol. 14, No. 6, pp. 1198-206, June 1996.

[91]   G. Hill, "The emerging shape of optical networks: A European perspective," 22nd European Conference on Optical Communications (ECOC'96), MoA.2.1, 1. 3-6, Oslo, Sept. 1996.

[92]   P. A. Perrier, M. Sotom, A. Jourdan, and L. Berthelon, "Optical crossconnect systems and technologies for WDM transport network," IEE Colloquium, February 6, 1997.

[93]   G. Hill, A. McGuire, and E. Lowe, "Issues in the deployment of optical networks," the 1997 Digest of the IEEE/LEOS Summer Topical Meetings: Vertical-Cavity Lasers/Technologies for a Global Information Infrastructure/WDM Components Technology/Advanced Semiconductor Lasers and Applications/Gallium Nitride Materials, Processing, and Devices, pp. 30-1, New York, NY, USA: IEEE, 1997.

[94]   N. V. Srinivasan, "Add-drop multiplexers and cross-connects for multiwavelength optical networking," Optical Fiber Communication Conference (OFC'98) Technical Digest, Vol. 2, pp. 57-8, Paper TuJ1, San Jose, CA, USA, 22-27 Feb. 1998.

[95] N. A. Jackman, S. H. Patel, B. P. Mikkelsen, and S. K. Korotky, "Optical Cross Connects for Optical Networking," Bell Labs Technical Journal, Vol. 4, No. 1, pp. 262-81, Jan.-March 1999.

[96] K. Harada, K. Shimizu, T. Kudou, and T. Ozeki, "Hierarchical optical path cross-connect systems for large scale WDM networks," Optical Fiber Communication Conference and the International Conference on Integrated Optics and Optical Fiber Communications (OFC/IOOC'99), Vol. 2, pp. 356-8, 1999.

[97] E. L. Goldstein, L. Y. Lin, and R. W. Tkach, "Multiwavelength Opaque Optical-Crossconnect Networks," IEICE Transactions on Electronics, Vol. E82-C, No. 8, pp. 1361-70, Aug. 1999.

[98] X. Lin, Y. Wang, S. Poinat, and Y. J. Chen, "Smart optical cross-connect switch array with build-in monitoring functions," Conference on Lasers and Electro-Optics (CLEO 2000), Technical Digest, Vol. 39, pp. 255-6, Salem, MA, USA, 7-12 May, 2000.

[99] D. Cavendish, "Evolution of Optical Transport Technologies: From SONET/SDH to WDM," IEEE Commun. Mag., Vol 38, No. 6, pp. 164-72, June 2000

[100] E. Ciaramella, "Introducing wavelength granularity to reduce the complexity of optical cross connects," IEEE Photon. Technol. Lett., Vol. 12, No. 6, pp. 699-701, June 2000.

[101] J. Gruber, P. Roorda, and F. La Londe, "The photonic switch/cross-connect (PSX)- its role in evolving optical networks," National Fiber Optic Engineers conference (NFOEC), pp. 678-89, Denver Colorado, August 27-31, 2000.

[102] P. Perrier and S. Thompson, "Optical cross-connects: the newest element of the optical backbone network," Alcatel Telecommunication Review, 195-201, 3Q, 2000

[103] J. Lacey, "Optical crossconnects: their impact on optical networking," 13th Annual IEEE Lasers and Electro-Optics Society Meeting (LEOS 2000), Vol. 1, pp. 224-5, 13-16 Nov. 2000

[104] H. J. F. Ryan, "WDM: North American deployment trends," IEEE Commun. Mag., Vol. 36, No. 2, pp. 40-44, Feb. 1998

[105] E. Lowe, "Current European WDM deployment trends," Vol. 36, No. 2, pp. 46-50, Feb. 1998

[106] L. Zong; Y. Li, H. Zhang, X. Zheng, and Y. Guo, "Low crosstalk structure for integrated OXC/OADM in WDM optical transport networks," Optics Communications, Vol. 195, No. 1-4, pp. 179-86, 1 Aug. 2001.

[107] S. Kuroyanagi and T. Nishi, "A sub-grouped wavelength conversion switch architecture for scalable and large-capacity optical cross-connect," Proceedings 27th European Conference on Optical Communication (ECOC'01), Vol. 3, pp. 474-5, Amsterdam, 30 September- 4 October 2001.

[108] I. Lelic, "Large multi-stage OXC," Proceedings 27th European Conference on Optical Communication (ECOC'01), Vol. 4, pp. 540-1, Amsterdam, 30 September- 4 October 2001.

[109] S. Bruno, D. Domin, S. Ruggeri, M. Audoin, R. Balcon, and M. Benomar, "128*128 fully broadcasting, managed and rack-mounted optical cross-connect," Proceedings 27th European Conference on Optical Communication (ECOC'01), Vol. 4, pp. 542-3, Amsterdam, 30 September- 4 October 2001.

[110] C. F. Lam, M. Boroditsky, M. D. Feuer, N. J. Frigo, B. Desai, and J. Kang, "Demonstration of programmable optical multicasting in a regional/metro area network," IEEE Photon. Technol. Lett., Vol. 13, No. 11, pp. 1236-8, Nov. 2001.

[111] A. Antonopoulos, "Benefit and applicability analysis of OXC based solutions in transport networks," Proc. 3rd International Workshop on Design of Reliable Communication Networks (DRCN 2001), pp. 1-8, Budapest, Hungary, 2001.

[112] M. Lee, J. Yu, Y. Kim, C-H. Kang, and J. Part, "Design of hierarchical crossconnect WDM networks employing a two-stage multiplexing scheme of waveband and wavelength," IEEE J. Select. Areas Commun., Vol. 20, No. 1, pp. 166-71, January 2002.

[113] M. Birk, J. Hobson, P. Lin, S. Clark, M. Curry, R. Ramaswami, and S. Yadegar, "Evaluation of a large-scale photonic cross connect for carrier network applications," Optical Fiber Communications Conference (OFC 2002), Postconference Technical Digest, Vol. 1, pp. 227-8, Anaheim, CA, USA, 17-22 March 2002

[114] X. Qin and Y. Yang, "Nonblocking WDM switching networks with full and limited wavelength conversion," IEEE Trans. Commun., Vol. 50, No. 12, pp. 2032-41, Dec. 2002.

[115] C. Nuzman, J. Leuthold, R. Ryf, S. Chandrasekhar, C. R. Giles, and D. T. Neilson, "Design and implementation of wavelength-flexible network nodes," IEEE J. Lightwave Technol., Vol. 21, No. 3, pp. 648-63, March 2003.

[116] T. S. El-Bawab, A. Agrawal, F. Poppe, L. B. Sofman, D. Papadimitriou, and B. Rousseau, "The evolution to optical-switching based core networks," Optical Networks Magazine, Vol. 4, No. 2, pp. 7-19, March/April 2003.

[117] N. Geary, A. Antonopoulos, and J. O'Reilly, "Analysis of the potential benefits of OXC-based intelligent optical networks," Optical Networks Magazine, Vol. 4, No. 2, pp. 20-31, March-April 2003.

[118] O. Gerstel and H. Raza, "On the synergy between electrical and photonic switching," IEEE Commun. Mag., Vol. 41, No. 4, pp. 98-104, April 2003

[119] S. Sengupta, V. Kumar, and D. Saha, "Switched optical backbone for cost-effective scalable core IP networks," IEEE Commun. Mag., Vol. 41, No. 6, pp. 60-70, June 2003.

[120] X. Cao, V. Anand, Y. Xiong, and C. Qiao, "A study of waveband switching with multilayer multigranular optical cross-connects," IEEE J. Select. Areas Commun., Vol. 21, No. 7, pp. 1081-95, Sept. 2003.

[121] H. A. Jäger, A. D. Monitzer, R. Rieken, and B. Kracker, "Quasi-dynamic intelligent networks: A proposal to solve essential scalability and cost issues in photonic core networks," Optical Network Magazine, Vol. 4, pp. 219, September/October 2003.

[122] M. Vasilyev, I. Tomkos, M. Mehendale, J.-K. Rhee, A. Kobyakov, M. Ajgaonkar, S. Tsuda, and M. Sharma, "Transparent ultra-long-haul DWDM networks with "broadcast-and-select" OADM/OXC architecture," IEEE J. Lightwave Technol., Vol. 21, No. 11, pp. 2661-72, Nov. 2003.

[123] J. Strand and A. Chiu, "Realizing the advantages of optical reconfigurability and restoration with integrated optical cross-connects," IEEE J. Lightwave Technol., Vol. 21, No. 11, pp. 2871-82, Nov. 2003.

[124] P. Peloso, D. Penninckx, M. Prunaire, and L. Noirie, "Optical transparency of a heterogeneous pan-European network," IEEE J. Lightwave Technol., Vol. 22, No. 1, pp. 242-8, Jan. 2004.

[125] N. Hanik, A. Ehrhardt, A. Gladisch, C. Peucheret, P. Jeppesen, L. Molle, R. Freund, and C. Caspar, "Extension of all-optical network-transparent domains based on normalized transmission sections," IEEE J. Lightwave Technol., Vol. 22, No. 6, pp. 1439-53, June 2004.

[126] O. Gerstel and H. Raza, "Predeployment of resources in agile photonic networks," J. Lightwave Technol., Vol. 22, No. 10, pp. 2236-44, Oct. 2004.

[127] J. C. Attard, J. E. Mitchell, and C. J. Rasmussen, "Performance analysis of interferometric noise due to unequally powered interferers in optical networks," IEEE J. Lightwave Technol., Vol. 23, No. 4, pp. 1692-703, April 2005.

[128] S. D. Dods and T. B. Anderson, "Calculation of bit-error rates and power penalties due to incoherent crosstalk in optical networks using Taylor series expansions," IEEE J. Lightwave Technol., Vol. 23 , No. 4, pp. 1828-37, April 2005.

[129] E. Bouillet, J.-F. Labourdette, R. Ramamurthy, and S. Chaudhuri, "Lightpath re-optimization in mesh optical networks," IEEE/ACM Transactions on Networking, Vol. 13, No. 2, pp. 437-47, April 2005.

[130] R. Ramaswami and K. N. Sivarajan, "Optical Networks: A Practical Perspective," 2nd Edition, Morgan Kaufmann, Academic Press, 2002

[131] Telcordia GR-3009, Optical cross-connect generic requirements

[132] A. Varma, "Optical switched-star network for computer-system interconnection," in Proc. of the International Topical Meeting on Photonic Switching, PS'90, Vol. PD14-B-9, Kobe, Japan, April 12-14, 1990. (Photonic Switching II, Springer-Verlag, pp. 386-90, 1990)

[133] A. Husain and J. N. Lee (editors), Special Issue on Optical Interconnections for Information Processing, J. Lightwave Technol., Vo. 13, No. 6, pp. 985-1120, June 1995.

[134] R. A. Nordin, W. R. Holland, and M. A. Shahid, "Advanced optical interconnection technology in switching equipment," IEEE J. Lightwave Technol., Vol. 13, No. 6, pp. 987-94.

[135] M. Jonsson, "Optical interconnection technology in switches, routers, and optical cross connects," Optical Networks Magazine, Vol. 4, No. 4, pp. 20-34, July-Aug. 2003.

[136] ITU-T G.8080, Architecture for the automatically switched optical network (ASON)

[137] www.ietf.org. See also numerous white papers in several industrial web sites.

[138] G. Bernstein, B. Rajagopalan, and D. Saha, "Optical Network Control: Architecture, Protocols, and Standards," Addison Wesley Professional, 2003

[139] T. S. El-Bawab (Editor), "Generalized multi-protocol label switching (GMPLS): In quest of data and transport integration," Proc. of the Workshop on, Part of the 2004 IEEE Global Telecommunications Conference (GLOBECOM '04), Dallas, Texas, 3 December 2004.

[140] M. J. Morrow, M. Tatipamula, and A. Farrel (Editors), "GMPLS: The promise of the next-generation optical control plane," Special feature, IEEE Commun. Mag., Vol. 43, No. 7, pp. 26-74, July 2005

[141] A. Farrel and I. Bryskin, "GMPLS, Architecture and Applications," Morgan Kaufmann, 2006

[142] ITU-TG.807, Requirements for automatic switched transport networks (ASTN)

[143] ITU-T G.872, Architecture for optical transport networks (OTN)

[144] Bruce S. Davie and Yakov Rekhter, "MPLS: Technology and Applications," Morgan Kaufmann, 2005

[145] O. Jerphagnon, D. Altstaetter, G. Carvalho, and C. McGugan, "Photonic switches in the worldwide National Research and Education Networks," Lightwave, November 2004.

[146] T. S. El-Bawab (Editor), "Optical Cross-Connects: From Research to Practice," Proc. of the Workshop on, Part of the 2005 IEEE Global Telecommunications Conference (GLOBECOM '05), Saint Louis, Missouri, 2 December 2005.

Chapter 12

PROTECTION AND RESTORATION ARCHITECTURES

Alan McGuire

This chapter reviews the application of optical switching to protection and restoration in optical networks. In doing so, technologies, system requirements and network architectures are examined. The techniques developed for protection and restoration have striking similarities to those already being exploited in existing SDH/SONET networks. In many ways this is not surprising, they are all connection-oriented, circuit switched technologies. The most significant difference between optical networking and SDH/SONET networking is the need in the former to account for the accumulation of analog impairments along a concatenation of links and switches. Critical to the success of optical networking is the capability to handle changes to the lightpath that occur as a result of protection and restoration events. Whilst these issues have been successfully tackled in simple topologies such as protected optical line systems and rings they still represent major challenges for large scale optical networks.

12.1 SDH AND SONET – TODAY's OPTICAL TRANSPORT NETWORKS

Networks deployed using the Synchronous Digital Hierarchy (SDH) and Synchronous Optical Networking (SONET) can be considered as the first generation of optical networks based upon optical transmission and electronic digital switching. Together SDH and SONET form most of the core transport infrastructure covering the globe. SDH is the International version whilst SONET may be considered a North American variant. The major difference between the two is the multiplexing structure. Networks based on SDH/SONET add/drop multiplexer rings and cross-connects have been widely deployed. An excellent description of both SDH and SONET can be found in [1]. Both are based on a layered network structure, as shown in Figure 12-1, consisting of path layers providing end-to-end networking, including cross-connection, and section layers that support the path layers.

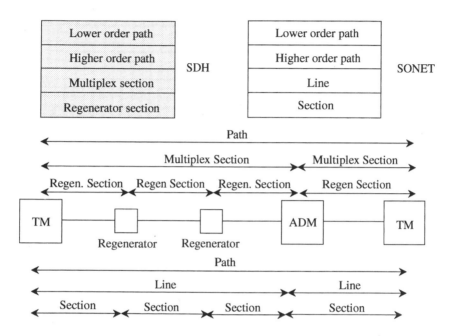

TM - Terminating mux, ADM - Add/drop multiplexer

Figure 12-1. SDH and SONET layers

Two forms of section layer exist, a multiplex section in SDH (line in SONET) between line terminals, add/drop multiplexers and cross-connects and a regenerator section (section in SONET), which is between every SDH/SONET network element and is terminated at digital regenerators. Each of these layers also transports overhead bytes in the frame structure to allow for operations, administration and maintenance functions. Protection switching is provided at either path or multiplex section (line) level.

SONET and SDH have both continued to evolve and have easily coped with the introduction of new optical technologies, starting with optical amplifiers, moving onto WDM line systems and are capable of interworking with optical networks with optical add/drop multiplexers (OADMs) and optical cross-connects (OXCs). Next generation SDH and SONET networking is introducing control plane technology, as discussed in chapter 11 and described later in this chapter, and support for the transport of a wide variety of data protocols. Both will continue to be deployed and used with the next generation of optical networks.

12.2 OPTICAL NETWORK ARCHITECTURE

The second generation of optical networking is intended to remove the digital switch (or rather place it more toward the edge) and route, or switch as much as possible in the optical domain. Such networks, though currently limited in size, (up to around 32 wavelengths 8 nodes and a distance of about 500-km) have begun to be deployed, mainly in the form of Dense Wavelength Division Multiplexing (DWDM) rings with OADMs. The phase after this is viewed by many as being the introduction of mesh networks using OXCs interconnected by DWDM line systems. Initially the inputs and outputs of the DWDM line systems will contain optical/electrical/optical (O/E/O) conversion. A longer-term view is that O/E/O conversions will eventually be eliminated paving the way to an end-to-end transparent, and possibly all-optical, network. The challenge lies in extending the reach, flexibility and connectivity of such networks as discussed later.

Of considerable interest in any form of optical network is the ability to provide survivability against failure by means of protection and restoration. Protection and restoration schemes can operate on single optical wavelengths or on aggregates of multiplexed wavelengths. To describe these new networks as well as their protection and restoration schemes it is useful to place them in the context of an optical network architecture. Although there are a number of ways of describing an optical network, the one described here allows us to reuse much of the existing terminology of protection and restoration schemes employed in SDH and SONET networks.

The optical transport network can be described as a series of layer networks as described in [2]. Three optical layer networks are defined as shown in Figure 12-2:

-   An optical channel (OCh) layer network providing end-to-end networking, including cross-connection, of optical channels for transparently conveying digital client information of varying formats. This layer network currently, due to limitations of technology, provides little in the way of operations, administration and maintenance functionality. To compensate for this the architecture defines preferred digital client layers – the optical transport unit (OTU), akin to an SDH section layer, and the optical data unit (ODU), akin to an SDH path layer, which provides frame structures for encapsulation of other protocols. Although other formats can be carried directly on optical channels they do not necessarily provide all of the functionality necessary to manage the optical network. The reason for this is that many existing protocols assume that transmission in the optical layer is point-to-point without any switching.

- An optical multiplex section (OMS) layer network that provides functionality for the transport of a multi-wavelength optical signal. An OMS of order n can support up to n optical channels.
- An optical transmission section (OTS) layer network that provides functionality for transmission of optical signals on optical fiber.

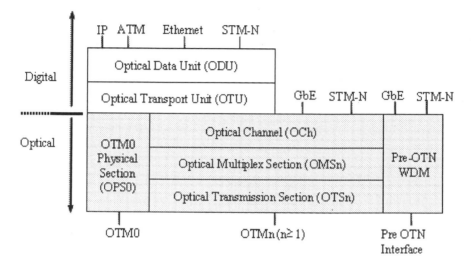

ATM - Asynchronous Transfer Mode, GbE - Gigabit Ethernet,
OTM- Optical Transport Module

Figure 12-2. Optical Transport Network (OTN) architecture

This network model is very similar to that of SDH with its path layers, multiplex section and regenerator section, and the SONET model of path, line and section. SDH, SONET and optical networking provide protection and restoration schemes at the path level (optical channel or path layers, optical multiplex section or SDH multiplex section, or SONET line). The protection and restoration schemes for the optical layers are remarkably similar to the schemes currently in use for SDH/SONET. As a result many existing techniques can be reapplied with only minor modifications.

12.3 PROTECTION

Protection is a mechanism for enhancing availability of a connection through the use of additional, assigned capacity. Traffic is automatically re-routed toward the additional assigned capacity in the event of a failure of the working capacity. Protection schemes can be applied on an end-to-end basis or in only part of the network. In both cases the aim is to provide a mechanism of restoring service so that service level agreements between the network operator (service provider) and the customer can be met.

In what follows we refer to working and protection in such a way that working capacity (fibers, sections or paths) carry traffic under normal conditions whilst the protection capacity provides an alternative path under failure conditions. To avoid a single-point of failure (or pinch-point) the working and protection capacity are routed diversely. The level of diversity can vary from fully diverse (no common ducts (conduits), cable, equipment or buildings at intermediate points) to some form of partial diversity (e.g. no common equipment or ducts, but equipment may be in the same building).

Protection may be dedicated, where each working entity (e.g. a path) has associated with it capacity for protection purposes which is only available to that particular working entity, or shared, where the protection capacity is shared amongst several working paths. In the latter case it is assumed that not all of the working paths will fail at the same time, so the required protection capacity can be less than the working capacity. The two most common forms of protection schemes are linear and ring protection. Both can be further categorized by the actual scheme used.

12.3.1 Linear Protection Schemes

Linear protection schemes are the simplest to design and implement. In the event of a fiber failure optical protection switching at the OMS level provides bulk protection with a single switching action. There are two main schemes based on 1+1 (one-plus-one) and 1:1 (one-for-one).

12.3.1.1 1+1 Optical Multiplex Section Protection

This form of protection has been deployed in both terrestrial and submarine WDM line systems. An example system is shown in Figure 12-3. The transponders provide optical-electrical-optical conversion to convert incoming optical signals from client systems (e.g. SDH equipment, ATM switches or IP routers) into optical signals that are compatible with a WDM line system. Each transponder has a well-defined individual wavelength on the line side. The wavelengths are commonly defined on a 50 or 100 GHz grid. Conversely transponders at the receive end of the line system convert the optical signal into a form that is suitable for the attached client network elements. At the transmit side of the line the outputs from the transponders are wavelength multiplexed via the multiplexer and then bridged onto two disjoint paths. The bridge is normally a simple passive 3dB optical splitter. The line terminals often include optical amplification, generally using erbium doped fiber amplifiers, on both outgoing fibers. Further amplification may be provided between the terminals at intermediate repeater stations. At the receiving terminal a 2x1 (or 2x2 using only one output) optical switch is used to select the best signal. Further optical amplification may also be provided on both inputs to this switch. The output is then wavelength demultiplexed via the demultiplexer (demux) and each wavelength directed toward the appropriate transponder. When a fiber break on the working path (or an amplifier failure) occurs the receiving terminal detects the failure and switches to the other path. This scheme provides for fast protection switching and avoids the need for any form of signaling protocol to the far end. There are two basic means of detecting failures: measurement of the incoming optical power, obtained by tapping of a small proportion of the incoming signal, or by comparing the inputs of all the active transponders for a signal fail condition. In the latter case, protection switching is initiated if all flag a failure.

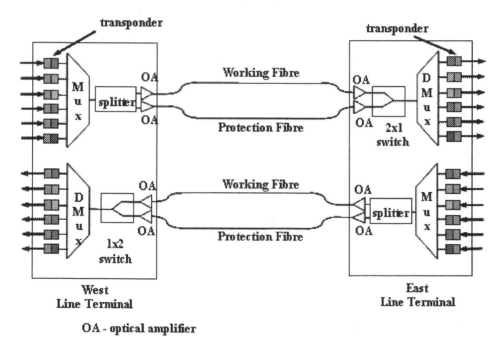

Figure 12-3. 1+1 Optical multiplex section protection

One mode of operation is unidirectional switching where each direction acts independently of the other. The most common form of optical switching technology used in this application is the 2x1 opto-mechanical switch, discussed in chapter 8. This technology provides a high level of reliability coupled with excellent switching characteristics as illustrated in Table 12-1. The switches may be of the moving fiber or moving deflector type.

Parameter	Value
Insertion Loss	0.5 –1.0 dB
Back reflection	≤ -45 dB
Polarisation dependent loss	≤ 0.1 dB
Cross-talk	-50 to –80 dB
Switching time	5-10 ms
Operating voltage	4.5 to 5.5 V
Durability (switch cycles)	$10^7 - 10^8$

Table 12-1 Typical characteristics of opto-mechanical switches

This overall performance sets a very high benchmark for competing technologies. Electro-optic switching technologies such as those based on lithium niobate, acousto-optic switches, thermo-optic switches and liquid crystal switches can all be pushed toward this benchmark only by using advanced device structures and complex manufacturing processes.

Switching speeds at the component level need only be of the order of a few milliseconds for protection. The reason for this is that the time required to allow for propagation delays, handshaking between nodes (where necessary) and complete switching is 50 ms. Note that from the perspective of technologies such as SDH/SONET the WDM line system is not visible, so protection mechanisms in SDH/SONET act independently of those in the optical layers. The detection time of failures in optical networking and SDH/SONET are of the same timescale so each initiates protection mechanisms autonomously. Speeding up the optical switching time at the component level below a few milliseconds provides no significant benefit.

It should be noted that the use of a switch in the 1+1 configuration represents a common point of failure in an active component for both the working and protection paths. Should power fail on an optical switching card the switch should maintain its current transmission state – in other words the switch should be latching. Some implementations avoid the use of a single active 2x1 switch, replacing it with a passive combiner (which has a higher reliability) with switching achieved using line side optical amplifiers feeding into the combiner. One of the amplifiers operates normally on the working path, whilst the other is "powered down" to operate in glow mode, thereby effectively isolating the protection path. Rapid switching in the order of a few milliseconds can be achieved by ramping up the gain, rather than starting with no power at all. Provided that optical amplification is actually required in the line terminals (and it isn't always, depending on distance) this is a more cost-effective solution than using switches. Bidirectional switching (where both directions go from working to protection fibers, even for single fiber failures) can also be invoked in this configuration. When a failure is detected on the incoming side it is flagged to the outgoing side in the same terminal. Then the terminal can power down its optical amplifier on the outgoing working fiber, thus imitating a failure and forcing the far end to switch. This avoids the need for any signaling protocol between terminals and has the advantage that network operators can ensure that working capacity for both directions can always be transported via the same route – even for single fiber cuts.

Where the differential path length between the working and protection routes is significant the power level of the optical channels incident upon the transponders is path dependent and can result in performance problems. To avoid this, "attenuation balancing" or equalization can be provided by means of varying the attenuation on the paths. This was generally achieved in earlier systems by optical measurements during installation and commissioning of the line system. Alternatively if the losses of the fiber links were known in advance the attenuation required could be calculated by means of software programs provided by some suppliers. Modern systems tend to provide automatic control using electronically controlled variable optical attenuators. This problem is not unique to optical line systems and has to be addressed in optical networks in general.

12.3.1.2 1:1 Optical Multiplex Section Protection

This is similar to 1+1 protection, but now the bridge at the transmit end is replaced with a switch. Consequently traffic is only transmitted over one fiber at a time. As with the previous example opto-mechanical 2x1 switches can be used for this application. For this scheme to work an Automatic Protection Switch (APS) protocol is needed to co-ordinate (handshake) between the switches at the two ends. The additional time to carry out the signaling procedure means that the overall protection switching time, from detection of a fault to completion of switching is slightly longer than that of the 1+1 scheme. In 1:1 schemes additional low priority traffic can be carried on the protection fiber until a cut occurs on the working fiber and then it is displaced in favor of the working traffic. To achieve this, the 2x1 switches must now be replaced with 2x2 switches. The same technology as used in 2x1 switches can be used for the 2x2 switches. Such a protection scheme may become more prevalent with the growth of best effort data services. One could conceive of providing such a form of protection for additional traffic using optical amplifiers along with splitters and combiners. However, the use of 2x2 switches (both terminals, both directions) is a more attractive solution than that of providing additional optical amplifiers for protection.

12.3.1.3 Other Linear Protection Schemes

In addition to providing optical protection at the OMS level, linear protection can also be provided at the optical channel level. This requires a 2x1 optical switch for each optical channel. It is also possible to implement

1+N or 1:N protection schemes where multiple working sections are protected by a single protection path [3]. The same technologies as used for 2x1 and 2x2 switches can be applied in these cases, whether opto-mechanical switching or gate arrays with splitters/combiners.

12.3.2 Ring Protection Schemes

The main advantages of rings are an inherent ability to provide protection and multiplexing of traffic onto a structure that allows simple interconnection of multiple sites. From the perspective of optical networking, beyond SDH/SONET, rings represent the second stage of deployment, the first stage being the introduction of WDM point-to-point line systems. Systems have already been implemented and are being deployed and will continue to evolve increasing in reach, the number of wavelengths, and the number of nodes supported in the ring.

By analogy with SDH and SONET which are constructed from digital add/drop multiplexers, optical network rings are constructed from optical add/drop multiplexers, OADMs. OADM architectures can range from very simple solutions that only allow a fixed subset of traffic to be added and dropped at each node to advanced solution with support for fully flexible add/drop multiplexing of between one and all optical channels. Different forms of OADM architecture are discussed in [4]. Optical rings can provide for protection at both the OCh and OMS levels. In this section we consider what are likely to be the most common forms of optical ring. Some of the high-level requirements for optical ring structures can be considered as follows:
- Provide healing of single points of failure.
- Accommodate where possible multiple failures.
- Provide healing quickly – of the same time period as SDH/SONET schemes.
- Coexist in harmony with possible client layer schemes (e.g. SDH/SONET). An example of this harmony is the capability to enable/disable the optical protection scheme on an individual OCh basis.
- Minimize the rerouting distance on the protection path (to avoid physical layer impairments on the signals).
- Avoid performance hits to traffic unaffected by the failure.
- Minimize the amount of signaling complexity required.

As indicated in the following discussion, different ring structures do not always meet all of these requirements – there is a trade-off between simplicity, complexity of signaling protocols, speed of response and efficient use of resources. Different ring structures are also required for different applications in outer core/access aggregation and inner core or long haul transport.

12.3.2.1 Optical Channel Subnetwork Connection Protection (SNCP)

Figure 12-4 shows the application of subnetwork connection protection in a unidirectional ring. This type of ring is attractive in applications where traffic is hubbed into a single exchange (central office) and is simple to implement.

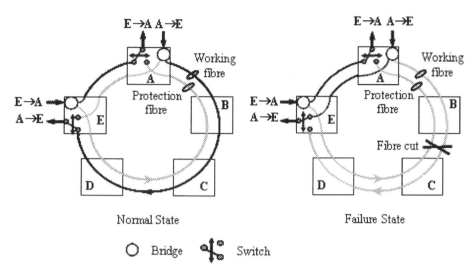

Figure 12-4. Optical channel subnetwork connection protection

Working traffic is carried around the ring in one direction, clockwise in the case of Figure 12-4. Hence traffic from A to E is routed in the same direction as traffic from E to A. At the entry point to the ring traffic is bridged (a simple passive splitter is sufficient) such that protection traffic is routed anticlockwise around the ring. Selection of the best path is made at the exit node via a switch. As it is a single-ended protection scheme there is no need for APS. It is similar to the linear 1+1 scheme described earlier. The ring can deal with failures of the transmitters or receivers or fiber failures. Such a ring is also referred to as a unidirectional line switched ring. The protection type is that of dedicated protection. The same switching technologies as described for linear protection schemes can also be used for this application. SNCP is not restricted to ring structures – it can be applied to any meshed subnetwork that offers two disjoint routings between source and destination. An alternative to using switches based on optical amplifiers is described in [5].

12.3.2.2 Bi-Directional Rings

A second form of ring is the bidirectional ring where working traffic is carried in both directions around the ring. Hence working traffic from A to E may be transported clockwise around the ring whilst traffic from E to A is transported anti-clockwise. In contrast to the unidirectional ring that is based upon dedicated protection, bidirectional rings use shared protection.

Bidirectional switching in rings requires the use of an APS protocol. In SONET and SDH these protocols are carried in the line (section) overhead of the frame. For optical networking an alternative mechanism can convey the optical APS protocol. An optical supervisory channel is proposed for this purpose among others. The optical supervisory channel is a wavelength (typical values are 1310 nm, 1480 nm, 1510 nm or 1620 nm) that allows for management of optical network elements. The APS protocols provide the means of access to the shared protection bandwidth. They are also complex and the form of failure encountered dictates their behavior. Details of existing APS protocols for SDH can be found in [6]. It is expected that optical layer APS protocols will have many similarities to their SDH/SONET counterparts.

In the following sections we consider bidirectional rings that switch at the OMS and OCh levels: the 4–fiber OMS Shared Protection Ring (OMS-SPRing/4), the 2-fiber OMS Shared Protection Ring (OMS-SPRing/2), and the Optical Channel Spring (OCh/SPRing).

12.3.2.2.1   OMS-SPRing/4

The OMS-SPRing is the optical networking equivalent of the SDH Multiplex Section shared protection ring (MS-SPRing) or the SONET bidirectional line switched ring.

Two forms of protection are provided in the form of span switching and ring switching. In the case of span switching if one or both working fibers fails between adjacent nodes (or associated transmitters or receivers) the traffic on the affected fiber(s) is simply rerouted to the protection fiber(s) between the same two nodes. Span protection is illustrated in Figure 12-5.

However, if both the working and protection capacity between two nodes fails due to cable failure, then ring, rather than span protection must be employed. This is shown in Figure 12-6. The traffic is now routed around the ring to the nodes adjacent to the failure on the protection fibers via the intermediate nodes.

Ring switching can also be used to deal with node failures. In this particular case nodes on either side of the failed node see the node failure as a link failure. Unfortunately if both adjacent nodes act independently and try to respond to what they believe is a single link failure using ring protection misconnections will occur. To avoid this happening adjacent nodes require the knowledge of the failure type, which is achieved by sharing information about link failures with other nodes. In the event of a node failure misconnections are avoided by examining tables in the nodes adjacent to the failure (each node in the ring has one) that indicate which connections begin and end on the failed node and then squelching (eliminating) these connections. For this to work properly the tables need to be updated whenever connections are setup or removed in the ring. On completion of squelching, messages are passed around the ring to allow protection switching to occur. The traffic originating and terminating at the failed node are naturally not protected, but transit traffic is.

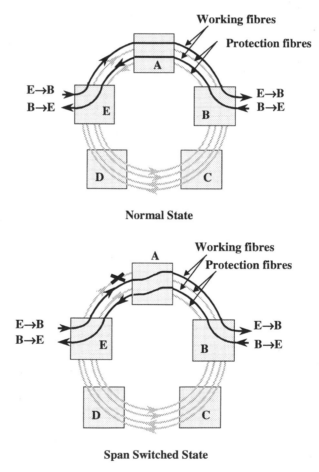

Figure 12-5. Span switching in OMS-SPRing/4. When the clockwise working fiber, between E and A, fails traffic is redirected to the protection fiber between E and A.

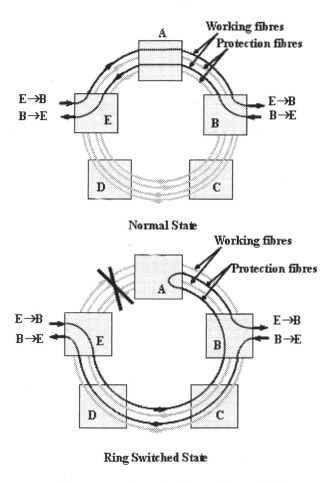

Figure 12-6. Ring switching in 4-fiber SPRING

12.3.2.2.2   OMS-SPRing/2

In this SPRing both fibers are used to carry working traffic in both directions around the ring. However, half of the capacity in each fiber, in this case half of the optical channels within an OMS, is reserved for protection capacity. In contrast to its 4-fiber cousin the 2-fiber SPRing can only provide ring switching. An example of ring switching in a 2-fiber ring is shown in Figure 12-7.

12.3.2.2.3   OCh-SPRing

OMS based protection is excellent for both fiber and node failures. In doing so it provides bulk protection, switching all optical channels in a single action. OCh-SPRing, on the other hand, is designed to provide protection against failure of individual optical channels, due for example to failure of a laser or a transponder at the edge of the ring. A four fiber OCh/SPRing can be considered in a similar way to the OMS-SPRing/4, where all N optical channels in an OMS are treated as a single entity for the purposes of protection, but now with each optical channel treated separately, and each working fiber carrying N working optical channels and each protection fiber carrying N protection optical channels. In effect the 4-fiber ring supports N independent OCh-SPRings.

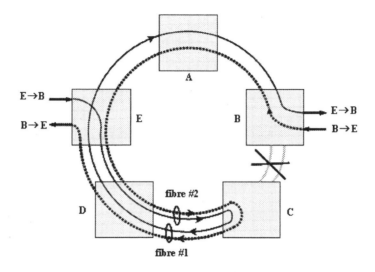

Figure 12-7. Ring switching in an OMS-SPRing/2

12.3.2.2.4   Transoceanic Rings

A disadvantage of the SPRing architecture when used in ring switching mode is that the traffic loops back upon itself. Consider Figure 12-6 where nodes E and A are on one side of an ocean, whilst B, C and D are on the other. Ring switching results in a lightpath that would cross the ocean three times, once between E and D and twice between A and B. This would result in a very long lightpath and a generally unacceptable increase in delay. To avoid this, a transoceanic switching protocol was defined in [6]. The transoceanic protocol ensures that in the event of switching occurring the path does not go over any span more than once. Consequently no optical channel need travel a distance longer than one circuit of the ring. Clearly this greatly simplifies transmission engineering and as such the transoceanic protocol may well find application in terrestrial rings as well. To avoid wavelength conversion the protection capacity on one fiber uses the same wavelengths as the working capacity on the other fiber.

12.4 OPTICAL MESH NETWORKS, CROSS-CONNECTS, AND RESTORATION

Restoration of a connection is the replacement of a failed connection by rerouting it using any available spare network capacity. In contrast to protection, some, or all, of the network resources used to support the connection may be changed during a restoration event. Restoration requires less overall capacity than protection as restoration capacity can be shared amongst a large number of potential connections rather than being dedicated. It offers network operators/service-providers the opportunity to improve network availability for unprotected traffic with the minimum of extra capacity. Restoration tends to be more beneficial in networks where there is a reasonable degree of meshing e.g. in optical cross-connect (OXC) networks, and can be applied at OXCs that switch at the optical channel or the OMS level. Eventually, this is an area where optical switching can find a lot of applications in future.

Chapter 11 has discussed several OXC architectures which may utilize numerous optical switching technologies. This discussion has also examined the characteristics of optical switching fabrics in terms of insertion loss, extinction ratio, signal to noise ratio, polarization sensitivity, wavelength dependence, switching speed (a few milliseconds is sufficient for restoration), and blocking. All these characteristics are determined by technology choice and fabric architecture. While taking all these factors into consideration, the design of an OXC needs to be optimized for high-performance cost-effective provisioning and restoration of lightpaths.

The performance tradeoffs for different switch fabric architectures have been discussed in chapters 9 and 10. Common architectures, such as crossbar, double cross-bar, permutation, 3-stage Clos (commonly used in electronic switching fabrics), Benes, Communicative splitter combiner, distributive splitter combiner and one sided, are all examined in literature too, for example in [7]-[9]. The analysis of [7] was carried out for lithium niobate based switches, but can be applied to any implementation based on 1x2 or 2x2 crosspoints. For large fabrics (32x32 and above) it is suggested that only the communicative and distributed splitter combiner architectures are suitable. Indeed, these strict sense non-blocking architectures with no differential loss[1] have received much attention. The active splitters or combiners can be based on 1xN or Nx1 switches. These can be constructed from 1x2, or 2x2 switches or gated optical amplifiers and passive splitters. Generally speaking however, switches based on bulk opto-mechanical, electro-optic, thermo-optic, liquid crystal, or SOAs, are limited to around

---

[1] The difference between the worst and best case insertion loss across the switch fabric

8x8 in size. Bubble-based waveguide switches can reach around 32x32. Whilst larger multi-stage switches can be created using these fabrics as building blocks factors such as total insertion loss may make it impractical.

MEMS technologies remove many of the drawbacks of waveguide based solutions. They also offer the potential of mass-produced low cost reliable switches along with excellent optical performance. 2-D MEMS are limited to around 32x32. As with waveguide based solutions larger fabrics could be built using multiple stages of smaller switches, but again optical losses make this unattractive. 3-D MEMS switches however have the potential to grow to sizes of between 256x256 and 1024x1024 with optical insertion losses of less than 10 dB [10]-[13]. The ability to support this number of optical wavelengths makes cross-connects of this size attractive for optical restoration schemes.

12.4.1 Restoration Schemes

A taxonomy for restoration schemes is shown in Figure 12-8. There are two basic architectures, a centralized control architecture or a distributed control architecture, each of which is described below. All of these can be used with optical switches.

In the centralized architecture a single control system, normally associated with a network management system, or operational support system (OSS), is linked to optical cross-connects via a data communications network (DCN). The DCN is essentially a router network that carries management and control information. For the purposes of resilience and to avoid single points of failure the DCN provides alternate paths to optical cross-connects and the central control system is often duplicated at two remote sites with the backup system operated in a hot standby mode. This architecture is illustrated in Figure 12-9(a).

The central control system contains a database of the transport network including the details of the network topology and the configuration of connections. When a network failure occurs alarm information is passed to the central control system. To ensure that all relevant information related to the event has been collected, sufficient time has to be allocated to allow for the alarms to be generated and propagated through the DCN. If the time period is too short, or the DCN congested, errors in the resultant restoration computation can occur. The alarms are analyzed to determine which resources have failed and the connections that are affected. Then, computation and selection of alternative routes follows based on the remaining free resources in the network. As part of this process individual connections may have a higher priority than others and the re-establishment of these connections is attempted first. Once the new route has been calculated configuration commands are passed to the cross-connects via the DCN. These configuration commands may be provided by network management protocols or via a signaling protocol. Examples of such signaling protocols are Private Network-Network Interface (PNNI), Resource Reservation Protocol with traffic engineering extensions (RSVP-TE) and Constraint Routed Label Distribution Protocol (CR-LDP). These signaling protocols may also be used to setup connections normally. A description of these protocols and their application to optical networking can be found in [14].

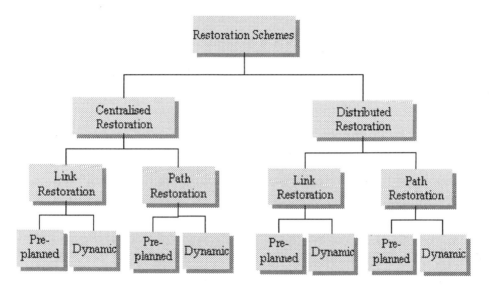

Figure 12-8. A classification of restoration schemes

The major limitation of the centralized architecture is the time taken for the control system to respond to a failure event. This is primarily dictated by the time taken to collect and analyze alarms, correlate this information with network connection data and then search for alternative routes. Restoration times can be of the order of minutes and performance is limited by the ability to collect and process data, which does not scale well with increasing size of network.

In the distributed architecture the controllers are located on each optical cross-connect and they communicate with each other via messaging channels embedded in the links between the cross-connects, as shown in Figure 12-9(b). Distributed architectures are based on control plane technology rather than network management. The control plane provides functionality such as signaling, route calculation, topology maintenance and resource discovery. Restoration is now achieved in seconds.

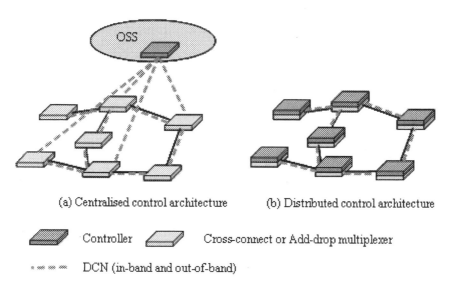

(a) Centralised control architecture   (b) Distributed control architecture

⬛ Controller   ⬛ Cross-connect or Add-drop multiplexer

˟ ▬ ▬ ▬ DCN (in-band and out-of-band)

Figure 12-9. Centralized and distributed architectures for restoration

Each optical cross-connect controller contains a routing table that contains routes from itself to other cross-connects. To ensure that the routing tables are up to date a routing protocol (such as Open Shortest Path First with traffic engineering extensions (OSPF-TE) or Intermediate System to Intermediate System (IS-IS)) exchanges information with other cross-connects and changes are then made to the routing tables as necessary. The information in the routing tables can be used to calculate routes for connections during set-up or to reroute connections as part of a restoration action.

As discussed in chapter 11, there are two basic models under development for the control plane of optical networks, the Automatic Switched Optical Network (ASON) as defined by the International Telecommunication Union– Telecommunications standardization sector (ITU-T) and Generalized Multi-protocol Label Switching (GMPLS) as defined by the Internet Engineering Task Force (IETF). Details of both can be found in [14]. ASON is supported by a number of different signaling and routing protocols whilst GMPLS is based on the control plane used in Multi-Protocol Label switching (MPLS)- an Internet Protocol (IP) based control plane.

12.4.2 Link and Path Restoration

Restoration can be provided at a path level, where the end-to-end optical channel (or lightpath) is restored or on a per link basis where all of the traffic routed on a failed link is routed around the link. Generally link restoration is faster and requires simpler rerouting than path restoration. However it is often less efficient at using network resources as it can lead to longer routes than necessary. Both restoration schemes can be implemented with either centralized or distributed architectures.

12.4.3 Preplanned and Dynamic Restoration

To improve restoration speed alternative routes between any two nodes can be precomputed for the majority of simple failure scenarios. For example, alternative routes can be calculated to cater for individual link and node failures. When a network failure occurs the alternate routes can be pruned to eliminate those that are no longer valid and a choice made from the remaining routes. This can be used in conjunction with a signaling mechanism known as crankback. If the attempt to signal the new route for a connection fails, the connection is cranked back and a new connection attempt is made based on the next viable precomputed route. Crankback is a useful mechanism in that the choice of a precomputed route may be based upon incomplete information regarding the fault, or, in the case of a distributed architecture discovering during the connection setup phase that the resource has already been taken. This improves the ability of preplanned restoration to respond to rapid changes in the network. Precomputation of routes can also be used to identify any failures that cannot be restored as a result of insufficient spare capacity.

In contrast, dynamic restoration schemes make their routing decision based on the state of the network at the time of failure. Whilst slower than preplanned restoration it does have the advantage of being able to respond to complex network failures, such as multiple failures, which are computationally difficult to calculate in advance. Crankback can also be used in these schemes.

Both preplanned and dynamic restoration schemes can be used with centralized or distributed architectures and both can be applied to both link and path restoration.

12.4.4 Aspects of Restoration Specific to Optical Networks

In any network the description of physical interfaces is of critical importance. In the SDH world the interfaces are broadly classified according to the distance (intra-office, short haul, long haul, etc.), fiber window (1310 nm or 1550 nm), bit rate (STM-N) and fiber type. Nominally each SDH section can be assumed to have the same transmission properties when viewed from the digital level. At the termination of each section the signal is regenerated and with the exception of jitter, degradations do not accumulate. Effectively each section can be treated independently of another. A consequence of this is that routing decisions tend to be based on simple metrics such as number of hops, quality, delay or implied cost in common with other digital technologies.

In contrast the cascading of optical transmission sections within an optical transport network presents a very different proposition. The result of analog transmission through a number of optical network elements such as optical amplifiers and optical cross-connects is that all of the properties of the signal at any point are dependent on what has gone before. Essentially it is the result of passing through a number of transfer functions. Although in principle it is possible to define the transfer function of optical equipment it does depend upon physical implementation (the transfer function of an equipment is dependent on the aggregation of the transfer functions of the components contained within it). As such it is manufacturer specific. It should be clear however that the transfer function can be simplified by choosing optical cross-connect architectures that minimize differential path effects. The cascading of such transfer functions is also complicated by the fact that the total transfer function will also depend on where the amplifiers and cross-connects are located in the network, in other words it is also network operator specific and path specific. This is very different to the digital world. Taken together these issues make it extremely difficult to define physical interfaces and to understand what such interfaces mean. Indeed it is difficult to envisage how optical specifications which would allow different suppliers equipment to be mixed and matched, can be achieved in the short term for anything other than the simplest systems. This is analogous to the SDH radio world, where it is argued that the rapid changes in component technology and the wide variety of modulation formats that can be implemented effectively rule out interoperable specification.

The consequence of this is that vendors and operators will interwork via a common digital client layer, such as an SDH path or an OTN ODU. Optical networks will effectively be provided as single vendor all-optical islands within which all the connections in the optical channel layer can be provided in the optical domain between terminating electronics. These optical islands are interconnected with other optical islands via electronic gateways which may be as simple as transponders that provide regeneration or as complex as digital cross-connects.

The absence of such specifications within an optical island and the need for bespoke engineering rules means that route calculation algorithms and extensions to routing protocols to disseminate information that permits the transmission properties of an optical channel to be calculated are likely to remain proprietary, at least in the foreseeable future. As a simple example, optical networks with wavelength conversion available at each OXC remove the constraint of calculating routes which have to have the same wavelength across the optical island.

Within a single optical island rerouting in response to a network failure can be achieved by creating a rerouting domain that is coincident with the boundary of the optical island. The cross-connects at the edges of the rerouting domains coordinate domain-based rerouting operations for all connections that traverse the rerouting domain. Where a connection is rerouted inside a rerouting domain, the domain-based rerouting operation takes place between the edges of the rerouting domain and is entirely contained within it.

The activation of a rerouting service is negotiated as part of the initial call establishment phase. For a single domain an intra-domain rerouting service is negotiated between the source and destination nodes within the rerouting domain. Once a connection has been established the rerouting services cannot be renegotiated. Within the rerouting domain routing information, including extensions that provide details of optical parameters is fully distributed. Any optical connection that fails between the end points is cleared back to the end points of the rerouting domain and rerouted via a new path between these end points. The advantage of this approach is that the restoration action is contained within individual optical islands. This avoids the need to reroute over several domains and speeds up the restoration process. The only restriction is that connections must continue to use the same source (ingress) and destination (egress) gateways nodes in the rerouting domain. Should a gateway node fail then restoration can be achieved by means of path restoration at the client level.

12.5 OPTICAL SWITCHING IN MONITORING SYSTEMS

Optical switches can also be used in optical network elements to provide access to dedicated or shared test resources as shown in Figure 12-10. This functionality can be provided for performance monitoring and as part of the mechanisms of survivability and recovery from failures. At each test point a passive tap (e.g. 90:10) removes a small portion of the optical signal. By attaching the outputs of these taps to a Nx1 switch it is possible to test each of the input ports on a round robin basis. This allows test facilities such as measurement of power per channel, channel equalization, optical-signal-to-noise ratio and wavelength stability to be shared amongst the inputs. It also allows parameters for individual signals to be tested at the same time. For the measurement of many parameters such a round-robin approach is adequate since time averaged measurements of the parameters are all that is required.

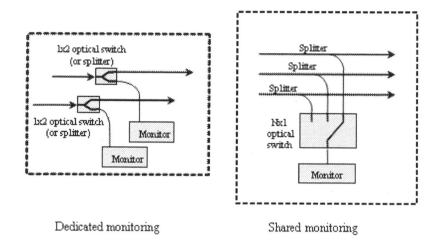

Dedicated monitoring   Shared monitoring

Figure 12-10. Application of optical switching to monitoring

This approach can be used for test facilities embedded in the network element or for the attachment of standalone diagnostic tools. A similar approach can also be adopted in reverse for the injection of test signals.

REFERENCES

[1] M. Sexton and A. Reid, "Broadband Networking: ATM, SDH and SONET," Artech House, 1997.

[2] ITU-T Recommendation G.872, Architecture of optical transport networks, 2001.

[3] Tsong-Ho Wu, "Fiber Network Service Survivability," Artech House, 1992.

[4] R. Ramaswami and K. N. Sivarajan, "Optical Networks: A Practical Perspective," Morgan Kaulman, 2002.

[5] R. Batchellor, "Optical layer protection: benefits and implementation," In Proceedings of National Fiber Optic Engineers Conference (NFOEC), Orlando, Florida, 1998.

[6] ITU-T Recommendation G.841, Types and characteristics of SDH network protection architectures, 1998.

[7] R. J. Reason, "Optical space switch architectures based on lithium niobate crosspoints," British Telecom Technology Journal, Vol. 7, No. 1, pp. 83- 91, 1989.

[8] J. Selvarajan and J. E. Midwinter, "Photonic switches and switch arrays on lithium niobate," Optical and Quantum Electronics, Vol. 21, No.1, pp. 1-15, 1989.

[9] R. A. Spanke, "Architectures for guided-wave optical space switching systems," IEEE Commun. Mag., Vol. 25, No.5, pp. 42-8, 1987.

[10] P. B. Chu, S. Lee, and S. Park, "MEMS: the path to large optical cross-connects," IEEE Commun. Mag., Vol. 40, No.3, pp. 80-7, 2002.

[11] P. D. Dobbelaere, K. Falta, S. Gloeckner, and S. Patra, "Digital MEMS for optical switching," IEEE Commun. Mag. Vol. 40, No. 3, pp. 88-95, 2002.

[12] R. Ryf, J. Kim, J. P. Hickey, A. Gnauck, D. Carr, F. Pardo, C. Boelle, R. Frahm, N. Basavanhally, C. Yoh, D. A. Ramsey, R. Boie, R. George, J. Kraus, C. Lichtenwalner, R. Papazian, J. Gates, H. R. Shea, A. Gasparyan, V. A. Muratov, J. E. Griffith, J. A. Prybyla, S. Goyal, C. D. White, M. T. Lin, R. Ruel, C. Nijander, S. Arney, D. T. Neilson, D. J. Bishop, P. Kolodner, S. Pau, C. Nuzman, A. Weis, B. Kumar, D. Lieuwen, V. Aksyuk, D. S. Greywall, T. C. Lee, H. T. Soh, W. M. Mansfield, S. Jin, W. Y. Lai, H. A. Huggins, D. L. Barr, R. A. Cirelli, G. R. Bogart, K. Teffeau, R. Vella, H. Mavoori, A. Ramiirez, N. A. Ciampa, F. P. Klemens, M. D. Morris, T. Boone, J. Q. Liu, J. M. Rosamilia, and C. R. Giles, "1296-port MEMs transparent optical crossconnect with 2.07 Petabit/s switch capacity," Post-deadline paper, Optical Fiber Communications Conference, Tech. Digest OFC, PD28, Anaheim, California, 2001.

[13] M. Birk, J. Hobson, and P. Lin "Evaluation of a large scale photonic cross-connect for carrier network applications," Optical Fiber Communications Conference, Tech Digest OFC, WH3, pp 227-228, Anaheim, California, 2002.

[14] G. Bernstein, B. Rajagopalan, and D. Saha, "Optical Network Control: Architecture, Protocols and Standards," Addison Wesley, 2003.

Chapter 13

OPTICAL PACKET SWITCHING AND OPTICAL BURST SWITCHING

Alan E. Willner
Reza Khosravani
Saurabh Kumar

Optical packet switching (OPS) and optical burst switching (OBS) enable switching of optical data at the granularities of packets and bursts, respectively. Switching/routing decisions in OPS and OBS are made based on packet/burst "header" information. Ideally, OPS is performed all-optically. However, due to numerous technological limitations, the research community has been investigating different forms of OPS where control, at least in part, is carried out by electronics. This includes header processing, synchronization and other functionalities. OBS, on the other hand, is based on electronic control. Thus, OPS and OBS switch fabrics are configured electronically using routing information derived from headers and locally managed routing tables.

OPS and OBS improve the utilization of optical network resources by statistical multiplexing of traffic streams, especially when traffic is dominated by data. They are more suited to bursty data traffic than Optical Circuit Switching (OCS), where switching is performed at the granularity of wavelength, at best.

In this chapter, we review some of the OPS and OBS technologies and the challenges associated with them. As discussed in chapter 1, packet networks handle traffic by storage, delay and contention resolution, and perform a great deal of signal processing operations. Therefore, we pay special attention to contention resolution techniques and optical signal processing in OPS, since these techniques and processes represent the major challenges of OPS. We describe the operation of OBS and discuss its techniques, potential merits and challenges.

13.1 OPTICAL PACKET SWITCHING

Figure 13-1 shows a block diagram of a typical OPS node. The node is composed of three main parts – input/output interface modules, an optical switch unit and a control unit [1]-[3].

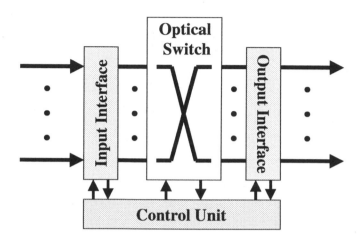

Figure 13-1. Schematic diagram of an optical packet switching node

The input/output interfaces are where data transmission lines are terminated and packets are prepared for switching. Many physical layer functionalities that are required for optical packet switching, such as synchronization and packet delineation, are realized at these interfaces. A number of other technologies are also housed at these interfaces and used for for contention resolution. This include optical buffers (Fiber Delay Lines, FDLs) and wavelength converters.

The switching unit is the main component of the node. It is especially important for OPS that the switching fabric be capable of fast switching of data, on a packet by packet basis. Only technologies with very high switching speed, such as Lithium Niobate Electro-Optic (EO) switching and Semiconductor Optical Amplifiers (SOAs), have been considered as candidate optical switching technologies for OPS.

The control unit is the intelligent part of a switching node, which retains information about network topology, forwarding table(s), scheduling, and buffering. The control unit instructs the switch fabric to configure as needed in the right times, and is in charge of resolving contentions among packets. Therefore, a significant amount of information processing is required in this unit. However, these functions are extremely difficult to accomplish in the optical domain. Indeed, current optical signal processing techniques are at a nascent stage compared to their electronic counterparts. Therefore, OPS control functionalities can be easily implemented in electronics.

It is important to note that the control unit in an OPS network detects and processes only the header of a packet (not the payload), and it is possible to transmit the header at a lower bit-rate to facilitate the processing. As a result, the presence of electronics at the control unit does not necessarily pose a limitation on the data transmission rate. Nevertheless, there have been efforts to implement some of the control unit functions (e.g. header recognition) in the optical domain in the hope of realizing pure OPS in the future. We will discuss some of these technologies later in this chapter.

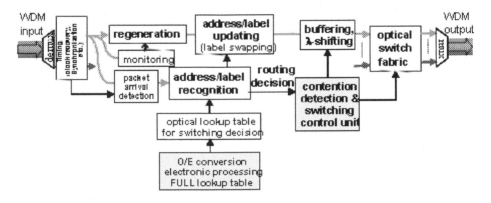

Figure 13-2. Functional process in a future OPS node [4]

Figure 13-2 shows some of the functional building blocks of an OPS node [4]. The functions involved in the process of routing a packet include:
1) Address/label recognition to determine the destination output port
2) Header updating or label swapping to prepare the packet for next node
3) Bit and/or packet synchronization to time the switching process
4) Routing-table caching as a reference for routing decisions
5) Contention resolution, mainly via buffering and/or wavelength conversion
6) Signal monitoring to assess its quality
7) Signal regeneration to recover from accumulated noise and distortions
8) Optical switching to direct packets to the appropriate output ports.

The electronics in Figure 13-2 is concerned with header processing, and the switching control unit may operate at the rate of transmission of header bits, which can be lower than the data transmission rate. Implementation of control and logical functions in electronics is easy, thanks to maturity of electronic processing technologies, devices and memories. Electronic random access memories (RAM) are used to make buffers in conventional routers. On the other hand, simple logic operations can be hard to implement in the optical domain. However, research in this area has been active for many years and a lot of progress has been established. In the next section, we discuss the current status of OPS contention-resolution technologies. Then we review progress in the areas of optical signal processing, synchronization and regeneration technologies.

13.2 CONTENTION RESOLUTION TECHNIQUES

Contention occurs when input packets from different sources are destined to the same output port at the same time. Conventional routers resolve the contention problem by buffering packets and managing their flow in such a way that no more than one packet is sent to any output port at a given time. In the optical domain, contention can be dealt with in the time, wavelength or space domains. Hybrid contention resolution techniques, which utilize two or all of these domains, can also be used and they outperform single-domain techniques [5]. In the following, we review contention resolution approaches in the three different domains— time, wavelength, and space.

13.2.1 Time Domain - Optical Buffering

As indicated earlier, there is no optical equivalent to electronic RAMs. Therefore, the only method to partially manage the flow of optical packets is by using Fiber Delay Line (FDLs). Programmable delay units can be constructed using a cascade of optical switches and FDLs, as depicted in Figure 13-3. This configuration can delay a signal for various time intervals by sending it through different lengths of FDLs ranging from L to $(1+2+...+2^N)L = (2^{N+1} -1)L$ with L being the minimum step-size. It is therefore possible to delay a signal in steps of nL/c-seconds, where n is the index of refraction of the fiber (~1.45) and c is the speed of light in vacuum (~3 ×10^8 m/s), up to a maximum of $(2^{N+1} -1)nL/c$ seconds.

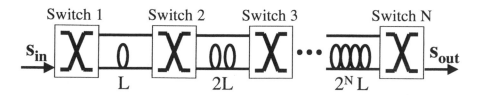

Figure 13-3. Programmable fiber delay line

There are several drawbacks in using FDLs as optical buffers. First, generating even a small delay requires a long length of fiber. For example,

200 m of fiber will generate only 1 μs delay! Second, FDLs are first-in first-out (FIFO) systems with predetermined delays. Once an optical signal is sent into a fiber delay line, it will not be accessible till it comes out of the other end of the line. In addition, unless FDLs are used in conjunction with, say, switches, they tend to be inflexible. If a packet of some path is to be delayed at an input port, subsequent packets of the same path may have to go through the same delay as well. This would degrade the performance of the switching node.

Figure 13-4 shows an architecture in which some of the output ports are connected back to the switch's input through a delay line that acts as a buffer. A packet that cannot be directed to its designated output port due to contention is switched to one of these buffered output ports. The packet returns to the input side of the switch after some specific delay. If the designated output port is available at this time, the packet will be switched to this port. Otherwise, it will keep circulating along the loop until the output port becomes available.

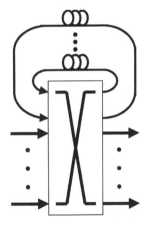

Figure 13-4. Re-circulating fiber delay loop

The advantage of a re-circulating delay line is that the FDLs may be shared by many inputs at the same time. However, this configuration demands additional output ports in the optical switch, which adds to its size, complexity, impairments and cost. There is also a limit on the number of times a packet can be sent to the FDL due to the attenuation encountered by the signal during each circulation.

Contention resolution is not the only application for FDLs in OPS nodes. Packets may also need to be delayed before the optical switch to make sure that the control unit has enough time to process the headers and identify destinations.

13.2.2 Wavelength Domain - Wavelength Conversion

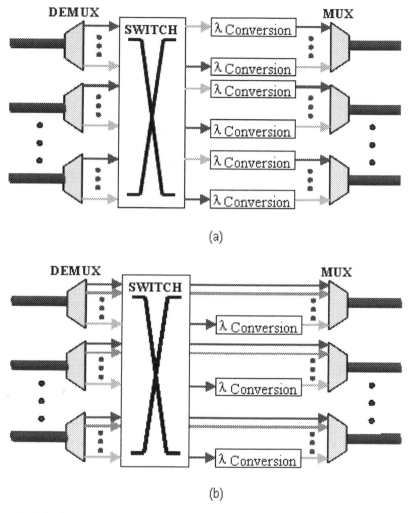

(a)

(b)

Figure 13-5. (a) Use of wavelength conversion for contention resolution in an OPS node, (b)
A form of converter sharing: more efficient use of wavelength converters

Wavelength conversion is perhaps the most efficient contention
resolution technique. It takes place in the optical wavelength domain and
does not delay signals by FDLs or by deflection routing. Figure 13-5(a)
shows a simple architecture that utilizes wavelength conversion to resolve
contention. Note that if contending packets are on different wavelengths,
wavelength conversion is not required. However, if two, or more, packets
have the same wavelength and are heading to the same output port at the
same time, wavelength-converted can be used to resolve contention. Since it

is unlikely that all packets will need to have their wavelength converted simultaneously at any given time, it is possible to share wavelength converters or use fewer numbers of them, as shown in Figure 13-5(b) [6].

An ideal wavelength converter would change the wavelength of an optical signal irrespective of its original wavelength, bit-rate, data format, polarization, or power. However, practical wavelength converters are far from perfect, and depending on the technique being used for the conversion, are sensitive to some or all of the input signal parameters. Other important performance metrics for wavelength converters are low noise figure and high output extinction ratio.

The popular wavelength conversion techniques are based on one of three phenomena [7], namely, gain saturation in semiconductor optical amplifiers (SOAs), interferometric effects, or nonlinear wave-mixing.

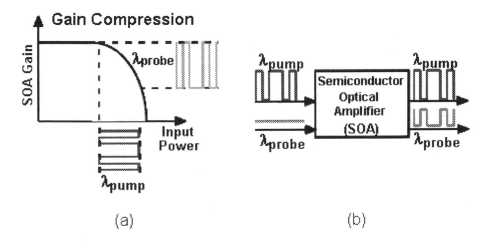

Figure 13-6. (a) Gain saturation in SOA by high power pump, (b) A continuous wave probe signal is inversely modulated by the pump

13.2.2.1 Wavelength Conversion Using Gain Saturation in SOAs

Perhaps the simplest technique for wavelength conversion is to use gain saturation in SOAs. As depicted in Figure 13-6, if the input power of a signal (pump) is large enough, it reduces the gain of the SOA by depleting the carriers. As a result any other signal (probe) propagating through the SOA at the same time sees a significantly reduced gain. This effect is called cross-gain modulation (XGM). As a modulated pump signal drives the SOA into saturation and back at the bit-rate, the inverse data pattern will be imprinted on the continuous wave (CW) probe signal. At the output, a filter is used to obtain the wavelength-converted signal, which is the inverted copy of the original pump signal at the probe wavelength. This technique can only be used for on-off-keying (OOK) data formats since the phase information is lost in the process. Other drawbacks of this technique are signal-to-noise-ratio (SNR) degradation due to Amplified-Spontaneous-Emission (ASE) noise introduced by the SOA, low extinction ratio, and chirp induced on the output signal [8]. Slow gain recovery also limits the bit rate at which XGM in an SOA can be used for wavelength conversion.

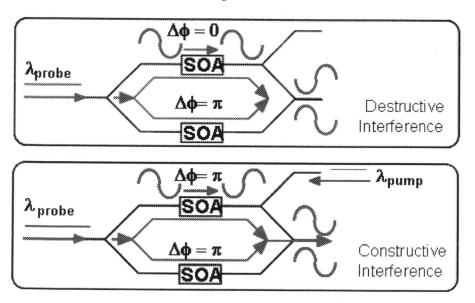

Figure 13-7. Interferometric wavelength conversion. Data-carrying pump disturbs the balance between the two arms

13.2.2.2 Wavelength Conversion by Interferometric Effects

Interferometric wavelength converters exploit the dependency of the refractive index on the carrier density in semiconductors. As the index of refraction changes, the phase of the optical signal propagating through the medium changes as well. This phase modulation can be converted to amplitude modulation by simply causing the original signal to interfere with a replica of itself after it has undergone a specific phase change. Figure 13-7 shows a simple interferometric configuration for wavelength conversion. In the absence of the pump signal, the two arms of the interferometer will have exactly π phase difference (e.g. by adjusting the bias of the SOAs), causing a destructive interference at the output waveguide junction. However, when the pump is present, the carrier concentration in the upper SOA changes and the probe sees a different phase change in the upper arm, resulting in constructive interference. This change in the phase of a probe signal due to the pump signal is called cross-phase modulation (XPM). Due to the sensitivity of the interferometer to carrier density, the output signal can have a sharp rise/fall time as well as a high extinction ratio. However, this technique suffers from stability problems if the interferometer is not an integrated device. It also requires very accurate adjustment of the SOA bias current and is sensitive to temperature changes.

Other configurations of interferometric wavelength converters have also been reported using a nonlinear optical loop mirror (NOLM) as shown in Figure 13-8. In the absence of the pump signal, an undisturbed optical loop simply works as a mirror, reflecting back the probe signal to the input port. This effect is a result of a constructive interference at the input port and destructive interference at the output ports. However, any phase disturbance caused by fiber nonlinearity—Kerr effect, changes this balance. A high-power optical signal can change the index of refraction of the fiber. Thus, whenever there is an optical pump power ("1") coupled to the loop, the output changes to "1". Otherwise, the probe is reflected back to the input port and the output power remains "0". To make a more efficient nonlinear medium, an SOA can be used inside the loop [9]. As a result of this arrangement the signal at λ_{pump} is converted to the wavelength λ_{probe}.

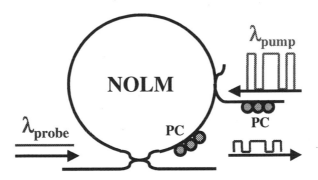

Figure 13-8. Nonlinear optical loop mirror in a wavelength converter configuration

13.2.2.3 Wavelength Conversion by Nonlinear Wave-Mixing

Nonlinear wave-mixing is the only wavelength conversion technique that makes an exact wavelength-shifted copy of a signal (amplitude, frequency, conjugated phase). Wave-mixing can happen as a result of cascaded second-order nonlinearities, in which case it is referred to as difference frequency generation (DFG). LiNbO$_3$ (Lithium Niobate) is a good example of a material with a high second-order nonlinearity, high optical bandwidth, and high dynamic range [10]. Therefore, it can be used in this application. Figure 13.9 shows the DFG mechanism. First, the high power pump signal initiates frequency doubling or second harmonic generation (SHG) inside the material. Then the second harmonic component mixes with the probe signal to generate a wavelength-shifted copy of the probe. It is important to note that the frequency spacing between the pump and the probe dictates the output wavelength. In other words, the output signal's wavelength is a mirror image of the probe wavelength with respect to the pump. Consequently, DFG can be used to wavelength-shift several WDM channels simultaneously.

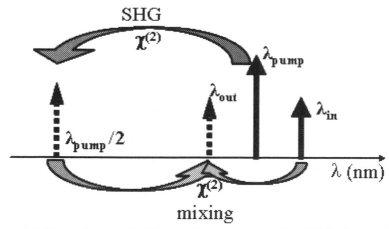

Figure 13-9. The mechanism of Difference Frequency Generation (DFG): first step is SHG and second step is mixing. $\chi^{(2)}$ represents second-order nonlinearity

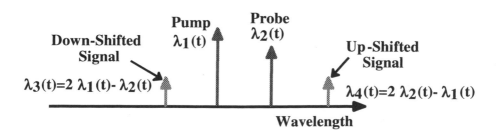

Figure 13-10. Four-Wave Mixing

When third-order nonlinearity is used to generate a new wavelength, the process is called four-wave-mixing (FWM). FWM creates two new frequencies as a result of mixing two strong optical signals, pump and probe as shown in Figure 13-10. These new frequencies are generated at $2f_{pump}$-

f_{probe} and $2f_{probe}-f_{pump}$. An optical filter is then used to select the desired wavelength. FWM can be accomplished inside fiber as well as in SOAs, where the third-order nonlinearity is much higher than in fiber. In general, nonlinear wave-mixing techniques are polarization sensitive, though it is possible to reduce polarization sensitivity by wavelength-shifting each polarization component of the signal separately. Most wave-mixing techniques are capable of wavelength shifting many independent channels simultaneously in a single device, unlike XGM and XPM.

Cross absorption modulation (XAM) in an electro-absorption modulator has also been used for wavelength conversion [11]. The principle is similar to XGM but instead of the gain, the pump signal modulates the absorption seen by the probe signal. As a result, a copy of the data on the pump signal is imprinted on the probe wavelength.

13.2.3 Space Domain – Deflection Routing

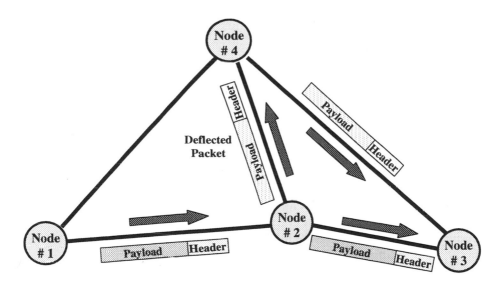

Figure 13-11. A packet which is preferably switched from Node # 2 directly to Node # 3, is now deflected via Node # 4 to Node # 3. Like this, contention is avoided at the expense of traveling a longer route and experiencing some more delay

Deflection routing is a solution to the contention problem at the network level. In this technique, one or more of the contending packets is redirected via other ports to reach its destination over alternative routes, as depicted in the example of Figure 13-11. The deflected packet usually ends up traveling a longer path to its destination. As a result packets may arrive out of order at the destination, and may need to be rearranged. Deflection routing requires sophisticated protocols and control algorithms to make sure packets are not lost inside the network. In effect, it also increases the network load.

13.3 OPTICAL SIGNAL PROCESSING TECHNIQUES

13.3.1 Address Recognition

Header information can be transmitted in several different ways. Baseband header is transmitted as a data field inside a packet either at the same bit-rate as the data, or at a lower rate [12] to relax the speed requirement on the electronics for header detection. It can also be located out-of-band either on a subcarrier on the same wavelength [13], as shown in Figure 13-12, or on a different wavelength, altogether.

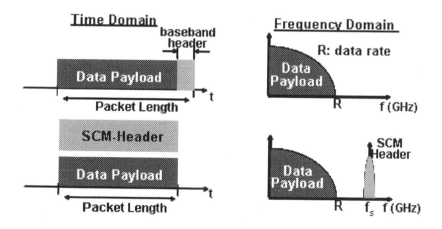

Figure 13-12. Subcarrier header in time and frequency domains

An optical baseband address recognition technique can be built based on optical matched filters (correlators) which determine the correlation between the received header and entries in a routing look-up table. An optical correlator adds weighted versions of the input signal in a tapped-delay-line structure. If the output of the correlator exceeds a threshold level, an address match is detected.

Fiber Bragg Gratings (FBGs) can serve as passive optical correlators. An FBG is nothing but a short section of fiber with a periodic index of refraction along the direction of propagation. Light with a wavelength that is twice the period of the grating (pitch) is reflected and all other wavelengths pass through. Different sections along an FBG array act as partial mirrors which reflect the incoming light with a predetermined delay. The weight of each delayed version of the signal is proportional to the strength of the grating at that point. Fig 13-13 shows an experimental demonstration of a 4-bit correlator configured to match the "11x1" pattern ("x" stands for a "don't care" bit). In order to equalize the optical power reflected off each section of the grating, the reflectivity of each section increases as we go deeper inside the grating [4].

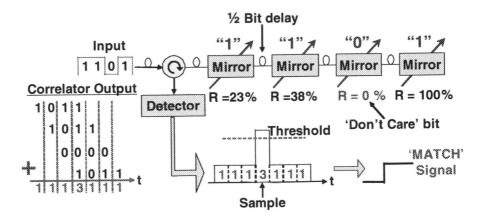

Figure 13-13. Architecture and operating principle of an FBG-array based optical correlator

An address match can also be detected using active time-to-wavelength mapping [14]. In this technique, each address bit is converted to a unique wavelength through a nonlinear wave mixing process (e.g. by using a Periodically-Poled Lithium Niobate waveguide—PPLN), and sent through a wavelength dependent delay as shown in Figure 13-14. The output level will be proportional to the correlation between the incoming and the stored address. Though it is possible to use nonlinearities and/or wavelength conversion in other devices such as semiconductor optical amplifiers, the PPLN waveguide offers the advantage of simultaneous processing of several WDM channels.

The complexity of optical address recognition techniques increases drastically with increasing the length of the address. Conventional internet routers process 32-bit destination addresses and match them to entries in a large routing table. This can be a significant source of latency in the electronic domain as routing tables in the core may contain > 500,000 entries. On the other hand, most of the current optical techniques are capable of processing only a few address bits at a time. However, a recent study [4], [15] shows that by using some software algorithm that determines a small subset of addresses that correspond to the most popular network destinations, a significant amount of traffic can be correctly routed, all-optically, by checking only a few address bits. This "popular" look-up table can be

implemented using a manageable number of optical correlators as shown in
Figure 13-15. Traffic that cannot be processed by optical means will be
processed using conventional electronics.

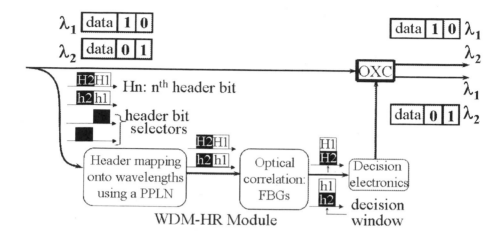

Figure 13-14. Schematic of a WDM header recognition (HR) module. Each address bit is
mapped onto a different wavelength using a PPLN and the address bits are checked for a
match with an address pattern using FBGs. SWT stands for "switch".

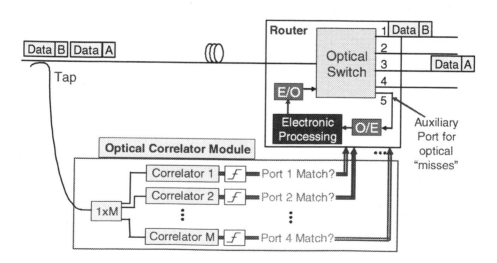

Figure 13-15. Optically-assisted routing uses a limited optical look-up table for popular
entries. The electronic processing module undertakes the routing decision if there is an optical
miss.

The algorithm can be explained as follows. The 500,000-entry table is first reduced to an electronic cache of the 100 most popular entries. As discussed in [15], it is feasible with a routing-table cache containing as few as 100 entries to successfully forward as much as 90% of the traffic entering an IP router. The remaining traffic that cannot be routed optically would be processed with conventional electronics, as indicated in Figure 13-15. Many Internet Protocol (IP) routers typically have only two to four outgoing ports making it fairly simple to group the incoming packets based on their corresponding output port. The algorithm then searches for patterns among these groups for which the output port can be determined by examining only a subset of the address, ideally less than 5 bits. An additional goal is to determine the optimum grouping of these entries so as to minimize the number of optical correlators required, typically <10. For example, the algorithm may find that all incoming packets with the second, fourth and twentieth bits equal to "0", "1" and "1," respectively, should all go to port 1. Thus, the optical correlator needs to be capable of matching only these 3 bits, in their respective positions, instead of all 32.

13.3.2 Label Swapping

Modern electronic routers, such are those based on Multi-Protocol Label Switching (MPLS), rely on the concept of label swapping. Many research efforts have been examining the possibility of adopting the same concept in OPS. Optical labels would carry routing information of the packets and these labels would be swapped with new labels, and new routing information, at each node along the path. This involves erasure of the original label, determining the new label based in part on the information received from the incoming label and writing the new label. The wavelength of the packet, itself, may be regarded as the label and wavelength conversion would be required in this case to change the label. The label may also be processed electronically to determine the output port, wavelength and the bits in the new label, but the payload does not have to undergo any O/E/O conversion. Since the electronic processing modules operate only on the label, it is possible to maintain transparency with respect to the rest of the data packet while using lower bit rate labels pertinent to the processing speeds of electronics.

Various technologies have been explored to implement all-optical label swapping (AOLS) [16]. These usually involve modulation of tunable lasers for selective conversion of the wavelength of packets. SOA based interferometric wavelength converters have been proposed for label swapping at 10 Gbit/s data rate while XPM in fiber was used to demonstrate label swapping at 80 Gbit/s [17]. Since the label need not have the same bit rate or data format as the packet, it is possible to devise various schemes to exploit this flexibility in order to obtain highly efficient and high-speed label swapping techniques.

13.3.3 Synchronization

The switching scheme in OPS can be implemented synchronously (time-slotted) or asynchronously (un-slotted). Most of the research on OPS so far has been performed on fixed-length packets, time-slotted networks which require synchronization. Variable packet size switching (un-slotted) networks, in general, are more susceptible to contention, and have a lower throughput as a result [1].

In general, synchronization is realized between either two incoming packets or an incoming packet and a local clock signal. To synchronize two signals, their bits should be aligned in the time domain. Tunable delay lines can achieve this by accurately delaying one of the signals relative to the other. Tunable optical delay lines can be realized using a set of fibers and switches, as shown in Figure 13-3. To achieve fine synchronization, the resolution of the tunable delay should be much smaller than the bit time. No matter how many sections of fiber are used in a tunable delay line, the tuning of the delay will always remain discrete. It is possible to get a continuous delay line by exploiting fiber dispersion in conjunction with wavelength conversion techniques [18].

In order to adjust a tunable delay, a feedback signal proportional to the phase delay between the two optical bit patterns has to be generated. This process can be accomplished either in the electrical domain or in the optical optical domain. An electrical phase-locked loop (PLL) detects the time delay between the two signals by mixing (or multiplying) them and filtering the output. A D flip-flop that is triggered by the local clock signal generates a delayed replica of the incoming signal. The delay is proportional to the phase difference between the signal and the clock. An XOR operation of the incoming signal and the flip-flop output will provide a feedback of the phase delay.

All-optical synchronization techniques demonstrated so far have been fairly complex and expensive. Most optical synchronization approaches rely on nonlinear interaction between optical signals or the nonlinear response of a semiconductor device in order to generate a timing signal. In one approach, the fast-saturated gain of SOAs is exploited to detect the start of a packet [19], as shown in Figure 13-16. A nonlinear optical loop mirror (NOLM) is then used to enhance the intensity of the synchronized pulse. In another approach, a terahertz optical asymmetric demultiplexer (TOAD) is utilized to extract the clock signal from an incoming packet, which can then be used for synchronization [20]. TOAD has been discussed in chapter 7.

Depending on the architecture of an OPS network, it may be sufficient to only synchronize the time-slots of the optical packets rather than the actual bits. One novel approach for slot synchronization is called Distributed Slot Synchronization [21]. In this technique, synchronization is achieved by adjusting the nodes' transmission time so that packets arrive synchronously at a chosen switching point in the network. Each node has to time its transmission by listening to the master synchronization signal and its own signal reflected from the switching node. This way the synchronization would be independent of transmission lengths. A packet arrival jitter of less than 13 ns has been demonstrated using this technique [21].

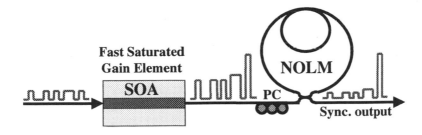

Figure 13-16. Self-synchronization by using SOA as a fast saturating amplifier and unbalanced fiber loop mirror as intensity discriminator

13.3.4 Regeneration

As optical signals travel through fiber, several transmission impairments distort the signal quality. Some of these distortions are due to non-ideal fiber characteristics, including chromatic dispersion, polarization mode dispersion, and fiber nonlinearities. Others are generated by optical components and switches such as polarization dependent loss and crosstalk. Besides these signal-degrading effects, the noise generated by optical amplifiers accumulates as the signal propagates through the network. Conventional electronic routers are immune to these problems as they detect and re-transmit the signal at each node. Since signals go through optical to electrical conversion at each node, all the distortion and noise is removed before it reaches an unacceptable level. OPS networks, however, may require all-optical signal regeneration in order to maintain signal quality on an end-to-end basis.

Regeneration may be realized at different levels. As discussed in chapter 11, an ideal regenerator adjusts the power, the pulse-shape and the timing of an incoming signal, and removes the noise. A 3-R (re-amplification, re-shaping, re-timing) regenerator incorporates all these functionalities except that it may or may not reduce the optical noise. Re-timing process makes clock-recovery an indispensable part of a 3-R regenerator. A 2-R regenerator only amplifies and re-shapes the signal. However, it may not require a complicated clock-recovery module since it does not re-time the signal. This means that timing jitter may accumulate after several 2-R regenerations.

So far regeneration has been performed by O/E/O conversions. All-optical regeneration is a subject of research. Several techniques have been proposed in this regard. Some of these techniques are based on nonlinear optical loop mirrors (NOLMs) [22]. They require relatively long fiber lengths for nonlinear interactions to take effect and result in the desired functionalities. They also suffer from polarization sensitivity. In other techniques, semiconductor based nonlinearity is used to take advantage of its large nonlinear response [23]. The nonlinear coefficient of semiconductors could be several orders of magnitude higher than that of glass and the active area length is generally less than 1 mm. However, the semiconductor fabrication process for optical amplification still has a low yield. In addition, semiconductor optical amplifiers add optical noise and exhibit pattern-dependent gain saturation due to their carrier dynamics [24]. In another approach, an electroabsorption modulator in conjunction with wavelength shifting has been used to build a 3R regenerator [25].

13.4 OPTICAL PACKET SWITCHING TESTBEDS

In order to demonstrate the feasibility of OPS networks, several testbeds have been developed by different groups across the globe. These testbeds were instituted to address implementation issues of OPS and to analyze its advantages over conventional electronic schemes.

The KEOPS (Keys to Optical Packet Switching) project is one of the large collaborative efforts in this regard. The project, which was funded by the European Union, had several European partners from both industry and academia. KEOPS used fixed packet length with a low bit-rate header on the same wavelength. Several key OPS functionalities and technologies were covered and demonstrated by this project, including: SOA-based space switching, Wavelength conversion based on Mach–Zehnder interferometric (MZI) devices and XPM in SOAs at 10 Gbit/s, 2R and 3R optical regeneration based on XGM in a SOAs and XPM in SOA-based MZI structures, tunable wavelength conversion, dispersive fiber for fine synchronization, and switched FDL systems for coarse synchronization [26].

WASPNET (Wavelength Switched Packet Network) is another program executed by a group of British universities and supported by several industrial companies [27]. A number of wavelength-conversion techniques were examined to address the contention problem. Also, several techniques for header transmission, including in-band, out-of-band and a hybrid technique, were investigated.

The All-Optical Network (AON) consortium comprised a number of industrial and academic partners in the USA [28]. The consortium established a testbed to investigate different classes of service for OPS networks. Optical phase-locked loop were also demonstrated at rates higher than 10 GHz.

The CORD (Contention Resolution by Delay Lines) testbed was also a joint effort of academia and industry in the USA [29]. This testbed was made of two nodes operating at 2.488 Gbit/s with subcarrier header.

13.5 OPTICAL BURST SWITCHING

Commercial deployment of OPS faces numerous challenges [3], [30]. One of the biggest challenges is the lack of optical random access memories, which are necessary for contention resolution. FDLs seem to be the only substitute for electronic buffers. However, an FDL works as inflexible first-in first-out (FIFO) memory, and can only buffer an optical packet for a predetermined time interval. Therefore, synchronization is necessary to reduce the packet loss probability[40]. In addition, OPS requires complex processing units and ultra-fast optical switches.

With a vision to reduce the processing required for switching decisions at each node and to avoid FDL and optical signal processing, optical burst switching (OBS) has been introduced [31]-[32].

13.5.1 Operation Principles

We use the description in [30] to explain the operation principles of OBS. In an OBS network, packets are assembled into larger data units, each of which is called a data burst (DB). For every burst, a Burst Header Packet (BHP) is created for control purposes. DB assembly and BHP generation take place in OBS *edge nodes*, i.e. nodes that reside in the edge of the OBS network and interface with non-OBS (e.g. IP) networks. Packets may be grouped together according to common destinations, QoS measures and some assembly criteria. The DB and its BHP are routed on different wavelength channels through OBS *core nodes*. The burst is sent from edge node over a data channel to the following node(s) while its BHP is sent over a control channel to the same node(s). For every group of data channels, one or more control channels are assigned. Depending on the protocol deployed the BHP may be transmitted ahead of the burst introducing an offset time to account for BHP processing in subsequent nodes. While DBs are switched optically and remain in the optical domain, BHPs are converted to the electrical domain at every node for processing. They are converted back to optical domain before transmission to next node(s). A BHP contains all necessary information to configure the optical switching fabrics. At the egress, another OBS edge node disassembles the bursts to their original packets and the latter are forwarded to their designated routers outside the OBS network. Typical edge-node and core-node architectures can be explored in [30].

Like OPS, OBS takes advantage of statistical multiplexing and has therefore the potential to provide better utilization of network resources [33]. In almost all OBS schemes, the BHP is transmitted after the DB is launched by an *offset time*. This offset time varies from one OBS transmission scheme/protocol to another. The BHP is to set the switch fabrics and reserve resources for the incoming DB. The separation between control and data helps to avoid optical buffering, but there is always a chance for burst dropping due to contentions.

OBS is different than OPS in many aspects. Switching in OBS is carried out at the burst granularity, which is an intermediate granularity somewhere between a packet and a circuit (lightpath). OBS is also characterized by a separation between data and control paths and is based on electronic control. OBS control enjoys a reduced processing overhead, compared to OPS, thanks to assembling many packets into a single burst. These differences and others dictate special design and operation considerations in OBS networks. In principle, however, an OPS network may be programmed to work like an OBS network in many ways.

13.5.2 Transmission Protocols

There are a number of DB/BHP transmission schemes in OBS, the variation among which is mainly based on the relative transmission timing between the DB and the BHP. We briefly review these approaches.

In the Tell-and-Wait (TAW) approach, the edge node sends the BHP first to request the resources necessary for DB transmission. If all the nodes, end to end, can accommodate the transmission of this burst, a virtual circuit is set up and the edge node may start transmission. Otherwise a new request has to be submitted again. The drawback of this technique is in the time it takes for transmission, including the delay associated with acknowledgments by all nodes along the path. On the other hand, since transmission does not take place until the network is ready, the burst loss probability can be low. No optical buffering is required in TAW technique.

In the Tell-and-Go (TAG) approach, the burst is sent immediately after sending the BHP. There is no time wasted in this protocol. However if any node along the path cannot accommodate the transmitted burst, the burst has to be dropped or buffered. Some limited buffering would be needed in with this protocol.

The Just-Enough-Time (JET) approach is similar to TAG with the only difference being that the burst is sent with some delay (offset time) after launching the BHP. The offset time should be longer than the total control processing time at the nodes on the route. By the time the burst reaches a node, the switch should be set to direct the burst to the desired output port. Thus, the need for FDLs is considerably reduced. Figure 13-17 shows a conceptual comparison between OBS and OPS in terms of data/control decoupling. The offset time between the BHP and the DB allows the control units to configure switches. A unique feature of JET is the concept of delayed reservation (DR) which enables the reservation of resources from the time the burst arrives to the node, rather than from the time the BHP is processed. By assigning different delays to different classes of traffic, it is possible to support variable quality of service (QoS) in this approach [34]. Relatively higher delays for higher priority class (real time traffic) , with careful design, can give the OBS network more time to configure switches for incoming burst and reduce burst loss probability. This technique, however, adds to the complexity of the scheduler of the switching system [35].

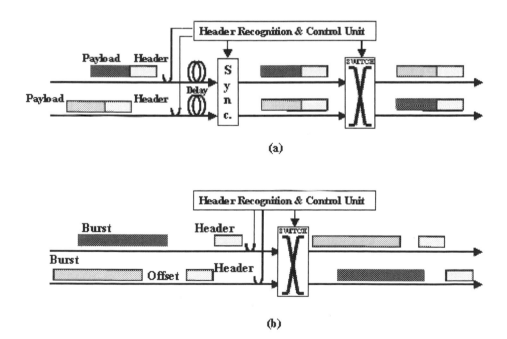

Figure 13-17. Data and control fields: (a) In OPS networks, (b) In OBS networks. Data and control fields are not drawn to scale

JET or TAG, which are sometimes referred to as one-way reservation protocols, are preferable in comparison to TAW, which may be considered a two-way reservation protocol. The latter that requires an acknowledgement to be sent back to the source, thus increasing overall latency. However to support reliable communication at the network layer a two way reservation protocol, providing a negative acknowledgement to the source in case a packet is dropped, maybe required. This enables the source to retransmit if so desired [36].

Recently, a new OBS transmission scheme has been introduced. This scheme is based on group scheduling [37]. This approach is designed to maximize bandwidth utilization as a primary benefit of OBS. Instead of scheduling data bursts on a one-by-one basis, bursts are grouped into periodic transmission windows and each group is scheduled all together in such a way to manage contentions, transmit as many bursts as possible in a given time window and reduce burst loss probability.

Today, OBS is an active and promising area of research. Several efforts have been invested in OBS research and many activities are underway (for example, see [38]-[44]). However, OBS also faces technological and architectural challenges and these challenges must be addressed and overcome before OBS can be deployed [30], [44]. Many issues remain unresolved so far. Despite progress made, these issues include the need for agile tunable wavelength converters and switching technologies that are very fast, reliable and scalable at the same time. Issues of reliability and system cost are important too. In order that OBS be deployed, it must be compatible with existing carrier infrastructures. An entry strategy, which preserves carriers investments, must be introduced, otherwise the startup cost of OBS implementation would not be justifiable.

Acknowledgments

The authors of this chapter would like to acknowledge the valuable assistance of Kashyap K. Merchant during the preparation of this chapter.

REFERENCES

[1] S. Yao, B Mukherjee, and S. Dixit, "Advances in photonic packet switching: an overview," IEEE Commun. Mag., Vol. 38 No. 2, pp. 84-94, Feb. 2000.

[2] M. J. O'Mahony, D. Simeonidou, D. Hunter, and A. Tzanakaki, "The application of optical packet switching in future communication networks," IEEE Commun. Mag., Vol. 39, No. 3, pp. 128-35, March 2001.

[3] T. S. El-Bawab and Jong-Dug Shin, "Optical packet switching in core networks: between vision and reality," IEEE Commun. Mag., Vol. 40, No. 9, pp. 60 –5, 2002.

[4]   A. Willner, D. Gurkan, A. Sahin, J. McGeehan, and M Hauer, "All-optical address recognition for optically-assisted routing in next-generation optical networks," IEEE Commun. Mag., Vol. 41, No. 5, pp. S38-S44, May 2003.

[5]   S. Yao, B. Mukherjee, S J Yoo, and S. Dixit, "All-optical packet-switched networks: a study of contention resolution schemes in an irregular mesh network with variable-size packets," Proc. OptiComm 2000, Dallas, Nov. 2000.

[6]   B. Ramamurthy and B. Mukherjee, "Wavelength conversion in WDM networking," IEEE J. Select. Areas in Commun., Vol. 16, No. 7, pp. 1061 –73, 1998.

[7]   J. Elmirghani and H. Mouftah, "All-optical wavelength conversion: Techniques and applications in DWDM networks," IEEE Commun. Mag., Vol. 38, No. 3, pp. 86-92, March 2000.

[8]   T. Durhuus, B. Mikkelsen, C. Joergensen, S. Lykke Danielsen, and K.E. Stubkjaer, "All-optical wavelength conversion by semiconductor optical amplifiers," IEEE J. Lightwave Technol., Vol. 14, No. 6, pp. 942-54, June 1996.

[9]   D. Nesset, T. Kelly, and D. Marcenac, "All-optical wavelength conversion using SOA nonlincarities," IEEE Commun. Mag., Vol. 36, No. 12, pp. 56-61, Dec. 1998.

[10] I. Brener, M.H. Chou, and M.M. Fejer, "Efficient wideband wavelength conversion using cascaded second-order nonlinearities in LiNbo3 waveguides," Proc. Optical Fiber Communications Conf., Paper FB6, pp.39-41, San Diego, 1999.

[11] A. Hsu and S.L. Chuang, "Wavelength conversion by cross-absorption modulation using an integrated electroabsorption modulator/laser," Summaries of Papers Presented at the Conference on Lasers and Electro-Optics, pp 488, Baltimore, 1999.

[12] R. Fortenberry, A.J. Lowery, W.L. Ha, and R.S. Tucker, "Photonic packet switch using semiconductor optical amplifier gates," Electron. Lett., Vol. 27, No. 14, pp.1305-7, 1991.

[13] M. Vaughn and D. Blumenthal, "All-optical updating of subcarrier encoded packet headers with simultaneous wavelength conversion of baseband payload in semiconductor optical amplifiers," IEEE Photon. Technol. Lett., Vol. 9, No. 6, pp. 827-9, 1997.

[14] D. Gurkan, M. C. Hauer, A. B. Sahin, Z. Pan, S. Lee, A. E. Willner, K. R. Parameswaran, and M. M. Fejer, "Demonstration of multi-wavelength all-optical header recognition using a PPLN and optical correlates," European Conf. Optical Communication, pp. 312-3, Amsterdam, 2001.

[15] J. Bannister, J. Touch, P. Kamath, and A. Patel, "An optical booster for internet routers," Proc. 8th int'l. conf. High Performance Computing, Denver, 2001.

[16] D. J. Blumenthal, J.E. Bowers, L. Rau, Hsu-Feng Chou, S. Rangarajan, Wei Wang, K. N. Poulsen, "Optical signal processing for optical packet switching networks," IEEE Commun. Mag., Vol. 41, No. 2 , pp. S23 - S29, Feb. 2003.

[17] L. Rau, S. Rangarajan, D.J. Blumenthal, H.-F. Chou, Y.-J. Chiu, and J.E. Bowers, "Two-hop all-optical label swapping with variable length 80 Gb/s packets and 10 Gb/s labels using nonlinear fiber wavelength converters, unicast/multicast output and a single EAM for 80- to 10 Gb/s packet demultiplexing," Optical Fiber Communication Conference, pp. FD2-1-FD2-3, Anaheim, 2002.

[18] P. Blixt and J.E. Bowers, "An optical technique for bit and packet synchronization," Lasers and Electro-Optics Society Annual Meeting, LEOS '94, Vol. 2 , pp. 103–4, Boston, 1994.

[19] T. J. Xia, Y.-II. Kao, Y. Liang, J. W. Lou, K. H. Ahn, O. Boyraz, G. A. Nowak, A. A. Said, and M. N. Islam, "Novel self-synchronization scheme for high-speed packet TDM networks," IEEE Photon. Technol. Lett., Vol. 11, No. 2, pp. 269 –71, 1999.

[20] L. J. Wang, H. K. Shi, J. T. Lin, K. J. Guan, and P. Ye, "Clock and frame synchronization recovery based on a terahertz optical asymmetric demultiplexer," IEEE Photon. Technol. Lett., Vol. 10 , No. 6, pp. 881–3, 1998.

[21] R. T. Hofmeister, Chung-Li Lu, Min-Chen Ho, P. Poggiolini, and L. G. Kazovsky, "Distributed slot synchronization (DSS): a network-wide slot synchronization technique for packet-switched optical networks," IEEE J. Lightwave Technol., Vol. 16, No. 12, pp. 2109 –16, 1998.

[22] M. Jinno, "All optical signal regularizing/regeneration using a nonlinear fiber Sagnac interferometer switch with signal-clock walk-off," IEEE J. Lightwave Technol., Vol. 12, No. 9, pp. 1648 –59, 1994.

[23] Y. Ueno, S. Nakamura, and K. Tajima, "Penalty-free error-free all-optical data pulse regeneration at 84 Gb/s by using a symmetric-Mach-Zehnder-type semiconductor regenerator," IEEE Photon. Technol. Lett., Vol. 13, No. 5, pp. 469 –71, 2001.

[24] S. A. Hamilton, B. S. Robinson, T. E. Murphy, S. J. Savage, and E. P. Ippen, "100 Gb/s optical time-division multiplexed networks," IEEE J. Lightwave Technol., Vol. 20, No. 12, pp. 2086 –2100, 2002.

[25] T. Otani, T. Miyazaki, and S. Yamamoto, "40-Gb/s optical 3R regenerator using electroabsorption modulators for optical networks," IEEE J. Lightwave Technol., Vol. 20, No. 2, pp. 195 –200, 2002.

[26] C. Guillemot, M. Renaud, P. Gambini, C. Janz, I. Andonovic, R. Bauknecht, B. Bostica, M. Burzio, F. Callegati, M. Casoni, D. Chiaroni, F. Clerot, S.L. Danielsen, F. Dorgeuille, A. Dupas, A. Franzen, P.B. Hansen, D.K. Hunter, A. Kloch, R. Krähenbühl, B. Lavigne, A. Le Corre, C. Raffaelli, M. Schilling, J.C. Simon, and L. Zucchelli, "Transparent optical packet switching: the European ACTS KEOPS project approach," IEEE J. Lightwave Technol., Vol. 16, No. 12, pp. 2117 –34, 1998.

[27] D. K. Hunter, M. H. M. Nizam, M. C. Chia, I. Andonovic, K. M. Guild, A. Tzanakaki, M. .J. O'Mahony, L. D. Bainbridge, M. F. C. Stephens, R. V. Penty, and I. H. White, "WASPNET: a wavelength switched packet network," IEEE Commun. Mag., Vol. 37, No. 3, pp. 120 –9, 1999.

[28] R. A. Barry, V. W. S. Chan, K. L. Hall, E. S. Kintzer, J. D. Moores, K. A. Rauschenbach, E. A. Swanson, L. E. Adams, C. R. Doerr, S. G. Finn, H. A. Haus, E. P. Ippen, W. S. Wong, and M. Haner, "All-Optical Network Consortium-ultrafast TDM networks," IEEE J. Select. Areas in Commun., Vol. 14, No. 5, pp. 999 -1013, 1996.

[29] I. Chlamtac, A. Fumagalli, L. G. Kazovsky, P. Melman, W. H. Nelson, P. Poggiolini, M. Cerisola, A. N. M. M. Choudhury, T. K. Fong, R. T. Hofmeister, Chung-Li Lu, A. Mekkittikul, D. J. M. Sabido, Chang-Jin Suh, and E. W. M. Wong, "CORD: contention resolution by delay lines," IEEE J. Select. Areas in Commun., Vol. 14, No. 5, pp. 1014 – 29, 1996.

[30] T. S. El-Bawab, A. Agrawal, F. Poppe, L. B. Sofman, D. Papadimitriou, and B. Rousseau, "The evolution to optical-switching based core networks," Optical Networks Magazine, Vol. 4, pp. 7-19, 2003.

[31] J. Turner, "Terabit burst switching," Journal of High Speed Networks, Vol. 8, No. 1, pp. 3-16, 1999.

[32] C. Qiao and M. Yoo, "Optical burst switching (OBS)– a new paradigm for an optical Internet", Journal of High Speed Networks, Vol. 8, No. 1, pp. 69-84, 1999.

[33] L. B. Sofman, A. Agrawal, and T. S. El-Bawab, "Traffic grooming and bandwidth efficiency in packet and burst switched networks," Proceedings of the Applied Telecommunications Symposium (ATS'2003), part of the Advanced Simulation Technologies Conference (ASTC'2003), pp. 3-8, Orlando, Florida, March 30- April 3, 2003.

[34] M. Yoo and C. Qiao, "A new optical burst switching (OBS) protocol for supporting quality of service," All Optical Communication Systems: Architecture, Control and Network Issues, Proc. SPIE, Vol. 3531, pp. 396-405, Nov. 1998.

[35] I. Baldine, G. Rouskas, H. Perres, and D. Stevenson, "JumpStart: A just-in-time signaling architecture for WDM burst-switched networks," IEEE Commun. Mag., Vol. 40, No. 2, pp. 82-9, Feb. 2002.

[36] L. Xu, H.G. Perros, and G. Roukas, "Techniques for optical packet switching and optical burst switching," IEEE Commun. Mag., Vol. 39, No. 1, pp. 136-42, Jan. 2001.

[37] S. Charcranoon, T. S. El-Bawab, H. C. Cankaya, and J.-D. Shin, "Group scheduling for optical burst switched (OBS) networks," The 2003 IEEE Global Telecommunications Conference (GLOBECOM '03), Vol. 5, pp. 2745-9, San Francisco, 1-5 December 2003.

[38] C. Qiao and M. Yoo, "Choices, features and issues in optical burst switching," SPIE Optical Networks Magazine, Vol. 1, No. 2, pp. 36–44, April 2000.

[39] Y. Xiong, M. Vandenhoute and H. C. Cankaya, "Control architecture in optical burst switched WDM Networks," IEEE J. Select. Areas Commun., Vol. 18, No. 10, pp.1838-51, October 2000.

[40] C. Qiao, "Labeled optical burst switching for IP-over-WDM integration," IEEE Commun. Mag., Vol. 38, No. 9, March 2001

[41] M. Düser and P. Bayvel, "Analysis of a dynamically wavelength-routed optical burst switched network architecture," IEEE J. of Lightwave Technol., Vol. 20, No. 4, pp. 574-85, April 2002.

[42] L. B. Sofman, T. S. El-Bawab, and K. Laevens, "Segmentation overhead in optical burst switching," Proceedings of SPIE, Vol. 4874, pp. 101-7, Opticomm 2002, Optical Networking and Communications, 30-31 July 2002, Boston, Massachusetts, USA.

[43] J.-D. Shin, S. Charcranoon, H. C. Cankaya, and T. S. El-Bawab, "Procedures and functions for operation and maintenance in optical burst-switching networks," Proceedings of the 2002 IEEE Workshop on IP Operations and Management (IPOM 2002), pp. 149-53, 29-31 October 2002, Dallas, Texas, USA.

[44] A. Agrawal, T. S. El-Bawab, and L. B. Sofman, "Comparative account of bandwidth efficiency in optical burst switching and optical circuit switching networks," Photonic Network Communications, Vol. 9, No. 3, May 2005.

Chapter 14

CLOSING REMARKS

Tarek S. El-Bawab

We have thus far made quite a journey with optical switching. We discussed the subject, from inception to where it stands today, from physical theories to practical realizations, and from device level to system and network levels. We started in Part I by positioning optical switching in the field of telecommunications and data networks in general, and within the area of switching in particular. In Part II, we covered almost all technologies which have been pursued to build optical switches for communications networks. In Part III, we explored how optical switching fabrics, based on any of these technologies, can be built, controlled, and utilized as the main building blocks of optical network elements. We discussed the role and applications of optical switching in communications networks and examined numerous methods of optical switching.

In retrospective, the perception of optical switching deployment in the telecommunications industry has swung from being imminent to being farfetched. In the early 1990s, the research community developed a considerable interest in optical switching and optical logic and invested a lot of effort in them. Rapid deployment of these technologies was anticipated by many researchers, and was dismissed by others [1]-[2]. However, neither the technology nor the market was ready for optical switching at that time. Indeed, the very need for it in carrier networks at that time was questionable. Considerable progress occurred during the 1990s. By the end of that decade, when the telecommunications industry was undergoing huge growth, the situation appeared to have changed. Optical-switching was actually deployed in field trials where OXCs were operated and tested in several carrier networks. This move was put on hold however by the downturn of the industry in mid 2001. A fundamental question to pose thus is what lies ahead for optical switching?

A key to answering this question is to understand what happened to the telecommunications industry in the period from the late 1990s to mid 2001, the *bubble* or *boom* years, and thereafter. In this chapter we look back at the bubble. We try to analyze briefly what happened, and why it happened. We conclude by trying to look at the future of optical switching.

14.1 WHAT HAPPENED?

Before and during the bubble era, several analyses anticipated traffic rising exponentially for a number of years and predicted sustained growth through the first decade of the 21st century. This was largely based on Internet growth rates and anticipation of new bandwidth hungry applications and services. Some forecasts estimated yearly growth of up to 400%. This resulted in concerns that carriers' networks were not ready for this huge demand, and created fear of a bandwidth drought. Therefore, carriers, especially North American and Pan European, added more optical bandwidth to their networks in the form of dark fiber and WDM transmission systems [3].

While it put to rest concerns about a bandwidth drought, WDM added a layer of optical transmission to a complex stack of network layers and created the challenge of switching and managing the large number of lightpaths it enables. It seemed clear that considerable savings in processes and cost could be made if traffic was managed optically. The industry viewed optical switching as key to counteracting complexity and providing switching relief.

Large investments were poured into research and development of optical technologies, components and networks. Progress in optical switching, in particular, was remarkable. New switching technologies were introduced and old ones reemerged with enhanced potential, thanks to advances in material science and fabrication technologies. Many vendors, established and startups alike, proposed optical cross-connects (OXCs) and/or reconfigurable optical add/drop multiplexers (R-OADMs) based on optical switching fabrics, and a number of field trials were carried out in carriers' networks. Optical packet switching (OPS) and optical burst switching (OBS) attracted considerable interest in the research community. Many researchers pursued them with the genuine belief that they could be deployed in the first decade of the 21st century.

Progress in optical switching was paralleled by progress in numerous optical transmission technologies [4]. Up to 320-wavelength systems are deployed today. Other developments were foreseen, especially in the US long-haul segment. Ultra long haul optical transmission, for 2000-4000 kilometers without 3R regeneration, became visible. Some equipment vendors proposed to open up new transmission windows, i.e. the L-band (1570-1610 nm) and the S-band (1440-1525 nm). Research and development efforts targeted 40 Gbit/sec transmission systems and many argued that these systems would be needed soon.

As a result of these advancements, the vision of the optical layer acquired large mindshare. Major changes in network architecture seemed imminent. New networking principles found their way to standardization bodies, which started to work on optical network control issues and to define frameworks to integrate the IP and optical layers. Many researchers believed that it was time for carriers to let go their SDH/SONET and ATM gears and migrate their network architectures to a two-layer model of the IP and optical layer. The notion of IP over WDM, or IP over optics, gained large popularity among research communities and a new generation of carriers and service providers.

The growth of the telecommunications industry in general, and optical networking in particular, was outstanding and unprecedented. To many, it appeared unbounded too. Telecommunications deregulation in the USA and movements toward privatizing telecommunications in other countries increased competition and demand for bandwidth. Venture capitalists and major industry players continued to steer their investments to optical components and networking technologies. Equipment vendors rushed to increase their manufacturing capacity. Carriers incurred large debts to remain competitive. Large numbers of startup companies emerged as equipment vendors and service providers. Historical Initial Public Offerings (IPOs) and acquisitions took place. Then, to the surprise of many, the industry took a sharp downturn and was largely stalled in a state of uncertainty for more than three years [5]-[8].

14.2 WHY DID IT HAPPEN?

The downturn of telecommunications industry in the late 1990s will probably remain a subject of study and research for some time. There are a number of factors that contribute to telecommunications business conditions. These include market factors, technical factors, operational factors, and regulatory factors, among others. The boom of the telecommunications industry in the late 1990s was particularly driven by predictions of traffic and market growth. Another factor that was instrumental in creating the bubble was the anticipation that a number of emerging technologies would mature in a fairly short period of time. Unfortunately, as we look back now, most of these predictions, be it for traffic, market, or technology, were far from being accurate.

Although the demand for more bandwidth had been admittedly on the rise throughout the 1990s, forecasts had indeed inflated it. Let us take, for example, the domains of Local Exchange and Inter-Exchange Carriers (LECs and IXCs) of the USA. Many new Competitive Local Exchange Carriers (CLECs) emerged and were competing for the same pool of customers. Each of these carriers was predicting growth based on acquiring most of this pool, and therefore growth was being double and triple counted. Each of them was ordering bandwidth from Incumbent LECs (ILECs) and IXCs based on the same predictions, at least in part. The per-customer traffic figures were also inflated. Hence, many established carriers were seeing growth that was not totally real. This created a house of cards, where when one falls all of the others fall as well. As many of the CLECs went out of business, they abandoned the capacity they were buying from ILECs and IXCs. For the latter, much of the capacity added to their networks ended up not being used. The race to prevent a bandwidth drought left their networks with a bandwidth glut. It is not perfectly clear why the industry could not see this coming. Of course, there were some voices that questioned the situation, or tried to rationalize forecasts. However, these were generally dismissed as not accepting the reality of growth in that time of prosperity.

In the past, transmission systems grew by a factor of 4 for each new generation. Optical networks have provided such a huge jump that new WDM transmission systems take longer times to fill. This lowered the demand for new systems, crashed the cost of bandwidth, and eroded the profit margin for carriers as they compete with each other.

Meanwhile, technology promises made were not all kept. For example, a lot had been said about the need for 40 Gbit/sec transmission systems, about developing and deploying them and about the changes they would introduce at the network level. In order for 40 Gbit/s transmission systems to be deployed, several technical and business issues must be addressed. Dispersion (chromatic and polarization mode), optical signal-to-noise ratio, fiber nonlinearities, regeneration span and cost, and packaging issues are among the technical challenges that, despite progress made, are not fully dealt with yet. Also, clear economic benefit must exist for 40 Gbit/s systems to see the light. These systems require closer regenerator spacing and extra per-wavelength correction of transmission impairments, compared to 2.5 and 10 Gbit/s systems. In order to deploy 40 Gbit/s systems, some carriers may have to change their installed fiber base, something they are not usually keen to do. If the cost and complexity of 40 Gbit/s systems outweigh the simplicity of using, say, four 10 Gbit/s systems, deployment becomes questionable. Indeed, it took several years for 10 Gbit/s systems to mature. Yet, 2.5 Gbit/s systems dominate many of today's networks. Thus fast deployment of 40 Gbit/s is not as easy or obvious as many have assumed. A similar argument applies to other innovative transmission systems such as those of the L and S bands. Many researchers anticipated fast introduction of OPS and/or OBS in carriers' networks. These views have neglected technological, architectural, and service challenges and were not based on understanding of how the industry works.

It has been suggested that the complete cycle time for new optical technology, from proof of concept to commercial deployment, is around nine years [6]. It can be argued that this is a conservative estimate. However, attempts by some to reduce this cycle to two or three years have proved to be unwise.

It is of course public knowledge that many unprofessional practices in the telecommunications industry have also contributed to the bubble. These however are not within our scope of discussion and are better covered already by market, economic, and social debates and studies.

The post bubble time witnessed a huge drop in business and investments. As a result of tough economic conditions, several research and development activities were either terminated or put on hold. Very little room, if any, was left for the industry to invest in long-term technologies. Driven by lower profit margins, carriers became forced to seek stronger cases for new investments and to require better return on them. Equipment vendors and service providers had to downsize and many went out of business. A lot of pessimism surfaced, especially in optical networking, which is the segment of the industry that took the hardest hit. Carriers who were most supportive of optical networking had to put this spirit on hold.

14.3 WHERE DO WE STAND TODAY?

The telecommunications industry will undoubtedly recover and resume growth. Today, the downturn seems to have already reached a plateau. A credible indication of full recovery will be when most of the excess bandwidth (the glut) is actually utilized, or when current systems have had their value written off.

Today, there are all kinds of market studies and predictions about the bandwidth glut. The truth however is that it is not perfectly known how big this glut is and how long it will take to get over it. Except for carriers themselves (each on its own), it is difficult to estimate the proportion of installed fiber that is dark versus lit. It is also difficult to know the fill levels of lit fiber (number of traffic-carrying wavelengths and their capacity). The type and level of excess bandwidth varies from carrier to another. This is why it is not possible to predict the recovery pattern accurately. It is also unknown how fast the industry will proceed forward after rebounding and in which direction. There are three factual observations however to keep in mind. These observations are self evident and can help us see the big picture.

First, true drivers for the optical layer, and optical switching, are not based on inaccurate predictions, whether about traffic, market, or technology. Although these predictions were exploited, among other things, by some analysis to push the optical layer toward deployment prematurely, the latter, in its own right, presents a solution to existing network problems. The technical and operational drivers for the vision of the optical layer are genuine. The network architecture needs simplification, agility, flexibility, intelligence, and scalability and the optical layer is a key enabler. The economic recession in the post-bubble time have merely outweighed the problems of the network and lowered their priority for a while.

Second, the progress made in several optical technologies, especially in optical switching, is real and impressive. More progress is of course desirable, but a lot has been accomplished and it is important to appreciate that many of the enablers of the optical layer are possible to realize in practice today.

Third, although its growth is not close to the fourfold rate predicted by some in the past, the Internet does grow and the yearly incremental rise in traffic volume may indeed be increasing. These are not merely claims of new market studies, which do confirm that bandwidth utilization in carrier networks has been rising during downturn, but can be verified by live and factual observations. Computers' processing power continues to increase, leading to newer machines which are generating information bits at higher rates and pumping them into the Internet at elevating speeds. The popularity of the Internet has increased despite the worldwide economic depression and it is now even carrying larger volumes of voice, video, and multimedia traffic. Applications like Voice over IP (VoIP) and IPTV are already deployed. Every day, more people are using Digital Subscriber Lines (DSLs) and broadband access. New mobile/cellular applications continue to unfold at a fast pace [9]. The result of all these is more traffic poured into carriers' networks, and this trend will continue. These are credible signs that there is traffic growth and that it is not insignificant. The problem for carriers is that this growth is not matched by revenue growth. This is a business model problem.

Hence, it should be a matter of time only before the excess bandwidth will be utilized and demand for more core bandwidth and flexible optical network elements resumes. Unlike core networks, access and metro networks are having bottlenecks and need more bandwidth. While many developed regions in the world complain of a bandwidth surplus, other regions are underprovided with bandwidth [6]. Thus, the fundamentals for continued growth do exist and optical networks are essential for this growth.

Although the time frame of its appearance was not rationally estimated by many, the optical layer is not an illusion.

14.4 FUTURE PERSPECTIVE

Thus, there are reasons to believe that the bandwidth glut will eventually disappear. What telecommunications is all about is the need of people to communicate with each other in numerous forms. This need has been increasing and will neither reverse course, nor slowdown. We firmly believe that the telecommunications industry will recover, sooner or later, and resume growth. This recovery may be imminent or may take a little more time to be noticeable. Nevertheless, the industry will not get back to the chaos of the bubble years, at least not for some long time.

Upon recovery, the telecommunications theatre will reveal new players' setup and different market dynamics. Industry players will adopt new business models and will utilize new technologies. The downturn has changed a lot. As it has reached a plateau, a new state of equilibrium will result. Unfortunately, economic difficulties have forced many companies, equipment vendors and carriers alike, to either shrink or go out of business. The truth however, regrettably, is that the industry was opened up during bubble years for adventures, uncalculated risks, and greedy practices. This led to an unhealthy environment and resulted in too many players for a given market size. The fact that many of the players were actually unfit made the situation worse. Therefore, harsh corrections became inevitable. Today, surviving carriers have seriously cut back their capital expenditures. There may not be room for many more cuts to be made and they could be close to resuming investments in new technologies.

Some researchers in the optical networking community had argued that the role of ATM and SDH/SONET in today's architectures will be taken over by the IP and optical layers and that the network will soon be largely based on a simple two-layer architecture. Clearly, this is not how events have unfolded so far. The chances are high that SDH/SONET, in particular, may actually remain in carrier networks for a relatively long time. This is due to their success, wide-scale deployment, and to the fact that they constitute cornerstone in carriers' transport infrastructures. Also, these veteran transport technologies continue to demonstrate flexibility and adaptation ability over the years. Next Generation (NG) SDH/SONET, with generalized mapping of variable-length multi-protocol packets into SDH/SONET frames, can enable standard elastic data transport over existing telecommunications infrastructures and make them more data friendly. Hence, it offers carriers new rewards for their existing investments in SDH/SONET, while improving the performance of the networks based on them. Therefore, NG SDH/SONET has the potential to revive SDH/SONET and keep them going for more time.

It is unclear so far if a similar argument can be made for ATM. There is concern that ATM is not going to be scalable beyond OC-48 port rates. While this rate can keep ATM desirable for a number of years, it is likely that it will be a limiting factor in the future when higher data rates are required. There was a perception among the networking community during the bubble years that OC-192 ports would dominate carrier networks and that OC-768 (40 Gbit/s) ports were to follow. This generated doubts about the future of ATM. Since ATM has been instrumental in securing traffic engineering tools for carriers' data networks, router vendors started to equip their IP routers with traffic engineering capabilities. From a networking perspective, ATM has not been used to deliver B-ISDN or to merge voice and data traffic as originally anticipated. Today, other approaches are pursued for somewhat similar purposes. For example, voice, and even video, is carried over IP. Nevertheless, ATM continues to enjoy popularity among most carriers worldwide and OC-48 is still one of the dominant transport rates, especially in the metro domain. This is the case today, but may change in future. At any rate, it is important to note that DSL will keep ATM going for some more time[1].

Most experts would agree today that while the two-layer model of the IP and optical layers may be realized in parts of the network in the near future, SDH/SONET and ATM will continue to be active layers for some time. This view however is challenged by the increasing interest in Ethernet as a wide area network (WAN) technology.

As we try to look into the future, we must realize that the strong drivers to have an optical layer still exist and that the evolution of transport networks to embrace such a layer is ultimately inevitable. The network will become simpler, more efficient, and will perform better by incorporating optical-switching based architectures. Optical networking is instrumental to enable the transport infrastructure to support widespread deployment of broadband access, mobile/cellular applications, multimedia, virtual private networks (VPN), and wavelength-based services. However, the vision of the optical layer makes more sense by embracing gradual evolution scenarios based on a combination of realistic market demands and actual technological progress. Development of optical networking elements should also focus now on cost effectiveness and meeting imminent carrier needs, more than on bandwidth efficiency.

While they continue to represent important areas of research, OPS and OBS are not ready for deployment today and it is premature to claim

---

[1] DSALMs, Digital Subscriber Line Access Multiplexers, are used at service providers' switching centers to link customers DSLs to ATM.

otherwise or to predict when they will be ready. OCS is mature and is starting to be deployed as discussed in chapters 11 and 12. It had been suggested during the boom years that very large scale OXCs, with thousands of input/output ports, were needed. Today, it has become clear that the demand for small and medium sized OXCs and optical switching systems will probably pick up first. Larger OXCs will follow thereafter when all enabling technologies and market conditions are ready.

Recent progress in numerous optical switching technologies represents a wealth of advancements which should be capitalized upon while paving the way for a realistic role for optical switching in future. One of the challenges ahead for optical switching is to continue the effort to build even faster and more scalable switches. Optical switching is the cornerstone in the vision of the optical layer and will play a major role in the future. It is an essential ingredient of the solution to many network problems. This is equally valid for the short term with a blend of current and future architectures, and in the long term when more parts of the network may be based on new paradigms, such as IP over optics. The inherent merits of optical switching constitute, without doubt, strong motivations to utilize it in the foreseeable future. Not so long ago, the industry was riding the optical networking bandwagon and was getting ready for the changeover to the optical layer. We will eventually get there.

Optical switching has been an active subject of research for decades. In practice, however, it was envisioned as being just around the corner in good times, but viewed as farfetched in times of caution and economic difficulty. Therefore, a fresh look at optical switching is needed today. This book is a genuine attempt to do this service to the telecommunications, data, and optical networking communities. As the industry overcomes the challenge of getting back on track, and gather the lessons of past and present, it will find it necessary to re-examine the case for optical switching. The primary contribution of this book is to help us take on this case.

REFERENCES

[1] H. S. Hinton and J. E. Midwinter (Eds), "Photonic Switching," IEEE Press, 1990
[2] E. Nussbaum, "Communication network needs and technologies- a place for photonic Switching?," IEEE J. Select. Areas Commun., Vo. 6, No. 7, pp. 1036-43, August 1988.
[3] A D. Cavendish, "Evolution of Optical Transport Technologies: From SONET/SDH to WDM," IEEE Commun. Mag., Vol. 38, No. 6, pp. 164-72, June 2000.
[4] T. El-Bawab, A. Agrawal, F. Poppe, L. Sofman, D. Papadimitriou and B. Rousseau, "The evolution to optical-switching-based core networks," Optical Network Magazine, SPIE/Kluwer, Vol. 4, No. 2, pp. 7-19, March/April 2003.
[5] P. Bernstein, "What's wrong with Telecom," IEEE Spectrum, Vol. 40, No. 1, pp. 26-9, January 2003.

[6] A. Houghton, "Supporting the rollout of broadband in Europe: optical network research in the IST program," IEEE Commun. Mag., Vol. 41, No. 9, pp. 58-64, September 2003.

[7] T. El-Bawab, "On the potential of optical switching in communication networks, Proceedings of SPIE, Vol. 5247, pp. 111-4, Part of ITCom 2003 (Information Technologies and Communications), 7-11 September 2003, Orlando, Florida, USA.

[8] "A survey of telecoms", IEEE Engineering Management Review, Vol. 31, No. 4, 4th Quarter, 2003

[9] S. M. Cherry, "What's right with telecom," IEEE Spectrum, Vol. 40, No. 1, pp. 30-4, January 2003.

Index

Acousto-optic effect, 29, 39, 83–88, 91
Acousto-optic polarization converter
(AOPC), 89–90, 92–93, 96, 98–104
Acousto-optic switch(es), 88–90, 93-97, 99,
102, 104-105, 343, 387
Acousto-optic switching, 29, 33, 340
ACTS. *See* Advanced Communications
Technologies and Services
Actuation, 132, 170–172, 174–176, 178,
191, 246, 248, 253, 255, 259
electromagnetic, 246, 248, 255
electrostatic, 178, 246
thermal, 132, 246, 248, 259
Actuator(s), 171–173, 175–178, 259–261
Advanced Communications Technologies
and Services (ACTS), 343–344
Algorithm(s), 307–331
control, 250, 279, 307–310, 312–313, 331
looping, 327–330
matching and graph coloring, 322
path searching, 309–310
rearrangement, 309, 315-319
All-optical label swapping (AOLS), 419
All-optical network(s), 26, 383, 400, 423
All-optical switch(es), 24, 26, 32, 230,
234–240, 262–263, 405
All-optical technologies/processes, 26, 234,
417, 420-422
All-optical wavelength conversion, 220,
227, 240
Amplified Spontaneous Emission (ASE),
222, 352
Anchoring forces, 158
Anisotropic, 83, 88, 90, 141–142, 152
AOLS. *See* all-optical label swapping
AOPC. *See* Acousto-optic polarization
converter
ARPANET, 16–18
Arrayed waveguide device, 49
Arrayed Waveguide Grating (AWG), 113,
132, 165, 207, 226–227, 352
ASE. *See* Amplified Spontaneous Emission
ASON. *See* Automatically Switched Optical
Network

ASTN. *See* Automatic Switched Transport
Network
Asynchronous Transfer Mode (ATM),
18–22, 340, 346, 348, 349
ATM switch(es), 22, 346, 348, 357
ATM. *See* Asynchronous Transfer Mode
Augmented model, 366–367
Automatic Switched Transport Network
(ASTN), 365
Automatically Switched Optical Network
(ASON), 364
AWG. *See* Arrayed Waveguide Grating

Backplane(s), 347, 349
Balanced bridge, 51, 60
Bandwidth glut, 434–435, 437
Banyan network, 71
Bar state, 50, 56–59, 61, 64–65, 67, 72, 88,
104, 117, 119, 129, 132, 135, 150–151,
252, 264, 277, 310, 312
Beam splitting, 150
Beam steering, 49, 176, 178, 184, 188–191,
193–196, 206, 252
Beneš algorithm, 313–314
Beneš architecture(s), 294, 300, 357
Beneš network(s), 290–296, 298, 300,
327-328
dilated, 294–296
Beneš switch, 328–330
dilated, 328–330
Bi-directional rings, 391
Bipartite graph(s), 323–324
Bipartite graph coloring, 323–327
Bipartite multigraph(s), 323–326
Birefringence, 88, 90, 98–100, 112, 122,
128, 130, 141–143, 146–149, 152, 238
electrically controlled birefringence
(ECB) cell(s), 149
B-ISDN. *See* Broadband Integrated Service
Digital Networks
Blocking, 7, 13, 70–71, 202, 206, 276, 286,
308–309, 311–314, 318, 338, 356, 359,
395
nonblocking (non-blocking), 7, 48,
70–73, 133, 169, 227, 250, 279, 286,